I0053788

An Introduction to Geometric Algebra and Geometric Calculus

M. D. Taylor
Professor Emeritus
Department of Mathematics
University of Central Florida
Orlando, FL 32816
E-mail: taylorseries77@gmail.com

ii

Copyright ©2021
Michael D. Taylor

Contents

Preface

An important tool for would-be mathematicians (and some other disciplines) is a knowledge of multivariable analysis. This has two major aspects, the analytical $\varepsilon - \delta$ side and the algebraic-geometric. There is a strong feeling in the mathematical community that the "right" way to master multivariable analysis ought to revolve about differential forms on manifolds. (This is on the algebraic-geometric side.)

Unfortunately, despite many excellent texts, such as [4], [13], and [25], the theory of differential forms has not succeeded in embedding itself in the consciousness of mathematicians and users of mathematics in the way that, say, linear algebra or differential equations has, nor in a way that would seem merited by its excellent properties and sophistication. (Exceptions to this can be found in subcommunities such as differential geometry or theoretical mathematical physics.)

This brings us to a second topic: geometric algebra. This concept grows from the work of H. Grassmann and W. K. Clifford (see [16, 5]) and is perhaps more properly thought of as being Clifford algebra in a particular setting. There is a vigorous, yet not widely visible, movement in the mathematical physics community to adopt geometric algebra and its attendant analytical machinery as the standard tool kit for analysis on manifolds. The great exponent of the movement over many years has been David Hestenes, and the book *Clifford Algebra to Geometric Calculus*, [22], by Hestenes and his colleague Garret Sobczyk, can appropriately be described as the "Bible" of the movement.

Hestenes is quite clear in his belief that geometric algebra (or, more precisely, geometric calculus or geometric function theory) is at least as powerful as differential form theory. Actually he says more than that; in [21], for example, we find, "I invite you, instead, to join me in proclaiming that Geometric Algebra is no less than a universal mathematical language for precisely expressing and reasoning with geometric concepts." The breadth and depth of his claims is impressive. A good summary of them can be found in [18, 20], while the evidence to support them is on display in [22].

Nor is his a lone voice in claiming a tremendous breadth, power, and potential importance for the machinery of geometric algebra and geometric calculus. Dorst et al., in *Geometric Algebra for Computer Science*, [9], say, "Geometric algebra is a powerful and practical framework for the representation and solution of geometrical problems. We believe it to be eminently suitable to those subfields of computer science in which such issues occur: computer graphics, robotics, and computer vision." From Doran and Lasenby, *Geometric Algebra for Physicists*, [8], with regard to mathematical tools for physics, we hear that "In this book we describe what we believe to be the most powerful available mathematical system developed to date." From Abłamowicz and Sobczyk's *Lectures on Clifford (Geometric) Algebras*, [1], we hear that, "Clifford (geometric) algebra offers a unified algebraic framework for the direct expression of the geometric ideas underlying the great mathematical theories of linear and multilinear algebra, projective and affine geometries, and differential geometry." And John Snygg, in *A New Approach to Differential Geometry Using Geometric Algebra*, [41], proclaims that, "The fact that Clifford algebra (otherwise known as "geometric algebra") is not deeply embedded in our current curriculum is an accident of history."

However writings on geometric algebra and geometric calculus at an introductory level are still in a process of being developed. Alan Macdonald has a particularly useful introduction and survey [33]. Macdonald has also brought out two books, [31, 32], that are suitable for undergraduates at the freshman, sophomore level. There is a text by Doran and Lasneby, [8], that is aimed more at the physics student than the mathematician or the general potential user of geometric algebra. The book [9] is targeted at computer scientists and provides a panoramic view of how to apply geometric algebra to various geometries. Garret Sobczyk's [42] approaches geometric algebra as an extension of the number concept. John Snygg, in [41], offers a serious attempt to use geometric algebra and geometric calculus as a tool in developing differential geometry. Two other offerings on the internet are [46] and the extensive notes [3].

There have been efforts to develop software for computation with geometric algebra and many examples can be turned up by simply running the phrase *geometric algebra software* on a search engine. Or a good list can be found at the website Geometric Algebra Software - Geometric Algebra Explorer with URL https://ga-explorer.netlify.com/index.php/ga-software/.

The purpose of these notes is to provide a limited exposure to the ideas of geometric algebra/calculus and to simultaneously use them as a vehicle to introduce the reader to some of the important concepts of multivariable analysis. (The emphasis here is on the geometric-algebraic side, and we do very little with $\varepsilon - \delta$ arguments.) Our target audience is mathematics students with the hope that they will also be accessible to a larger group, for example, physics, engineering, computer science students, etc. The required preparation is linear algebra, multivariable calculus from a decent introductory calculus course, and, of course, a modicum of mathematical sophistication.

Here are two interesting aspects of this enterprise:

This approach to multivariable analysis can be carried out in an unusually geometrical and physical way. That is, the machinery lends itself well to visualization and intuition, even in a higher-dimensional setting. In connection with that, an important genesis of these notes was the article [27] in *The American Mathematical Monthly* in which the idea of equivalence classes of oriented parallelepipeds is used to motivate and construct the space of k-vectors $\Lambda^k \mathbb{R}^n$ and the wedge product.

The second aspect is that the level of abstraction seems to be less high than is usually required for studying similar concepts. That may be partially because we have chosen to restrict ourselves to $\mathbb{R}^n$ equipped with the standard inner product of Euclidean space rather than dealing with more general Riemannian or pseudo-Riemannian manifolds. Let the reader judge for himself or herself the truth of these impressions.

Finally, I wish to acknowledge, on the one hand, the great personal encouragement and help I have received from Alan Macdonald in becoming acquainted with geometric algebra and geometric calculus and in preparing these notes and, on the other hand, the towering role of David Hestenes in making this material known to the world. I must also note the encouragement I have received from Stephen Kennedy as well as comments from certain anonymous reviewers which have led to improvements in the manuscript. If there are errors or shortcomings in what I have done, then the credit for those is wholly and solely mine.

Chapter 1

A little orientation

1.1 Things it would be well to know beforehand

Here is a somewhat more detailed list of the preparation one should have to read these notes.

Vector space axioms (over $\mathbb{R}$).
Linear independence, linear dependence, basis, dimension.
Linear transformation; representation as matrix.
Determinant; connection with independence.
$\mathbb{R}^n$ as a vector space.
Distance formula; magnitude of a vector.
Dot product in $\mathbb{R}^n$.
$$a \cdot b = |a|\,|b|\,\cos\theta = \sum_{i=1}^{n} a_i b_i.$$
Orthogonality.
Orthonormal basis.
Gram-Schmidt orthogonalization process.
Schwarz inequality.
Orthogonal transformation; rotation.
Lines, planes, $(n-1)$-dimensional analogues in $\mathbb{R}^n$; geometric significance of the equation

$$a \cdot (x - p) = 0 \quad \text{where } a, x, p \in \mathbb{R}^n.$$

Chain rule for partial derivatives of multivariable functions.
The chain rule as matrix multiplication.
Tangent vectors to curves.
The directional derivative.
The gradient and its significance in terms of rate of increase.
How to set up multivariate integrals as iterated integrals over simple regions.

1.2   Notations and conventions

We shall try to follow these conventions:

$\mathbb{R}$ is the real numbers.

$\mathbb{R}^n$ is the set of ordered n-tuples of real numbers.

$\mathbb{R}^n = \mathbb{R} \times \cdots \times \mathbb{R}$ n times.

$\mathcal{I}$ is the unit interval $[0,1]$.

$\mathcal{I}^n$ is the n-dimensional unit cube $\mathcal{I} \times \cdots \times \mathcal{I}$.

Real numbers (scalars): Greek — α, β, etc.

Vectors, multivectors: Lower case Roman — a, b, x, etc.

Points, matrices: Upper case Roman — A, B, X, etc.

We shall take the *standard basis of* $\mathbb{R}^n$ to be $\{e_1, \ldots, e_n\}$ where

$$e_1 = (1,0,0,\ldots,0),$$
$$e_2 = (0,1,0,\ldots,0),$$
$$\ldots$$
$$e_n = (0,0,\ldots,0,1).$$

In $\mathbb{R}^3$ one often sees the standard basis vectors e_1, e_2, e_3 denoted $\mathbf{i}, \mathbf{j}, \mathbf{k}$ respectively.

1.3   A word about presentation of proofs

Although we try to give a careful discussion of concepts and the logic behind them, there is a tendency in the first six chapters to move longer proofs to Appendix A. This is partly to keep things moving, partly because we wish to stress conceptual understanding over an examination of details.

Exceptions to this strategy occur when we spell out the construction of the geometric product and when we present the proof of the fundamental theorem of geometric calculus. Also, once we have gotten into Chapter 7, we tend to present the details of the proofs in all their messy glory.

1.4   Matrices and determinants

In all that follows, we restrict ourselves to vectors in $\mathbb{R}^n$ and make essential use of the dot product: Recall that if $a,b \in \mathbb{R}^n$ with $a = (\alpha_1, \ldots, \alpha_n)$ and $b = (\beta_1, \ldots, \beta_n)$, then $a \cdot b = \alpha_1 \beta_1 + \cdots + \alpha_n \beta_n$.

It will frequently be useful to talk about an $n \times k$ matrix $A = (a_1, \ldots, a_k)$ where each $a_i \in \mathbb{R}^n$. The matrix entries are not really specified without first giving a basis for the space. Thus, if we have in mind the basis $\{f_1, \ldots, f_n\}$ for $\mathbb{R}^n$ and we can write $a_i = \alpha_{1i} f_1 + \cdots + \alpha_{ni} f_n$ for each i, then we mean the matrix

$$A = (a_1, \ldots, a_k) = \begin{pmatrix} \alpha_{11} & \alpha_{12} & \ldots & \alpha_{1k} \\ \alpha_{21} & \alpha_{22} & \ldots & \alpha_{2k} \\ \vdots & \vdots & & \vdots \\ \alpha_{n1} & \alpha_{n2} & \ldots & \alpha_{nk} \end{pmatrix}. \tag{1.1}$$

In (1.1) we have written A as $(a_1, \ldots, a_k)$ and we can think of each a_i as a column vector or column matrix,

$$a_i = \begin{pmatrix} \alpha_{1i} \\ \vdots \\ \alpha_{ni} \end{pmatrix}$$

where the particular basis we are using, namely $\{f_i\}_{i=1}^n$, is in the back of our mind. Sometimes we want to write matrices in terms of row vectors. If, for example, we wrote the transpose of A, we might well put down

$$A^T = \begin{pmatrix} a_1 \\ \vdots \\ a_k \end{pmatrix} = \begin{pmatrix} \alpha_{11} & \alpha_{21} & \cdots & \alpha_{n1} \\ \alpha_{12} & \alpha_{22} & \cdots & \alpha_{n2} \\ \vdots & \vdots & & \vdots \\ \alpha_{1k} & \alpha_{2k} & \cdots & \alpha_{nk} \end{pmatrix}.$$

It is often convenient to use a basis $\{f_1, \ldots, f_n\}$ for $\mathbb{R}^n$ having the property that $\{f_1, \ldots, f_k\}$ is a basis for $\text{span}\{a_1, \ldots, a_k\}$, in which case

$$A = (a_1, \ldots, a_k) = \begin{pmatrix} \alpha_{11} & \cdots & \alpha_{1k} \\ \vdots & & \vdots \\ \alpha_{k1} & \cdots & \alpha_{kk} \\ 0 & \cdots & 0 \\ \vdots & & \vdots \\ 0 & \cdots & 0 \end{pmatrix}.$$

We may then talk about the matrix

$$A' = \begin{pmatrix} \alpha_{11} & \cdots & \alpha_{1k} \\ \vdots & & \vdots \\ \alpha_{k1} & \cdots & \alpha_{kk} \end{pmatrix}$$

associated with the vector subspace $\text{span}\{a_1, \ldots, a_k\}$, and, by an abuse of notation, we write $A' = (a_1, \ldots, a_k)$ even though A and A' are different size matrices.

We will usually write our matrices with respect to orthonormal bases because of an important connection between matrix multiplication and the dot product: If $A = (a_1, \ldots, a_k)$ and $B = (b_1, \ldots, b_m)$, where each a_i and b_j belongs to $\mathbb{R}^n$, and the matrices are specified with respect to an orthonormal basis for $\mathbb{R}^n$, then

$$B^T A = \begin{pmatrix} b_1 \\ \vdots \\ b_m \end{pmatrix} (a_1 \ldots a_k) = \begin{pmatrix} b_1 \cdot a_1 & \cdots & b_1 \cdot a_k \\ \vdots & & \vdots \\ b_m \cdot a_1 & \cdots & b_m \cdot a_k \end{pmatrix} \tag{1.2}$$

where B^T is, of course, the transpose of the $m \times k$ matrix B. A nice feature of this last matrix is that it is independent of the choice of basis. We also feel free to write it as $B^T A = (b_i \cdot a_j)$ or $(b_i \cdot a_j)_{m \times k}$.

We give exercises that establish (1.2) in detail.

The reader is also asked to establish in the exercises the useful fact that if either $\{a_1, \ldots, a_k\}$ or $\{b_1, \ldots, b_k\}$ is linearly dependent, then $\det(a_i \cdot b_j) = 0$. Because of this, later on in these notes, it sometimes makes sense to only deal with the case where $\{a_1, \ldots, a_k\}$ and $\{b_1, \ldots, b_k\}$ are sets of linearly independent vectors.

We usually treat the determinant in $\mathbb{R}^n$ as function of n vectors. Thus we write $\det(a_1, \ldots, a_n)$ where we may imagine we are dealing with the $n \times n$ matrix $A = (a_1 \ldots a_n)$ in which each a_i is a (column) vector. We recall from linear algebra the useful fact that the following properties *completely characterize* the determinant:

1. $\det(e_1, \ldots, e_n) = 1$.

2. $\det(a_1, \ldots, a_n)$ is linear in each variable a_i.

3. If we switch any two of the vectors a_i, then the determinant changes sign. For example,
$$\det(a_1, a_2, a_3, \ldots, a_n) = -\det(a_2, a_1, a_3, \ldots, a_n).$$

4. If $i \neq j$ and λ is a scalar, then
$$\det(a_1, \ldots, a_{i-1}, a_i + \lambda a_j, a_{i+1}, \ldots, a_n) = \det(a_1, \ldots, a_{i-1}, a_i, a_{i+1}, \ldots, a_n).$$

(One can easily check these properties on 2×2 matrices.) It will also be convenient to keep in mind that $\det(A) = \det(A^T)$ for any square matrix A.

Exercises 1.4.

1. We let $a_1, a_2 \in \mathbb{R}^3$ be the vectors
$$a_1 = 5e_1 + e_2 - 6e_3$$
$$a_2 = 3e_1 - 5e_3,$$
and we construct a new basis for $\mathbb{R}^3$,
$$f_1 = e_1 + e_2 - 2e_3$$
$$f_2 = e_1 - e_2$$
$$f_3 = e_3.$$
Set
$$A = (a_1 \, a_2) = \begin{pmatrix} 5 & 3 \\ 1 & -5 \\ -6 & 0 \end{pmatrix}$$
in terms of the standard basis. Now find $A' = (a_1 \, a_2)$ in terms of f_1, f_2, f_3.

2. Suppose that $\{u_1, \ldots, u_n\}$ is an orthonormal basis for $\mathbb{R}^n$.

(a) Show that

$$\sum_{j=1}^{n} (e_i \cdot u_j)(e_k \cdot u_j) = \delta_{ik} \text{ (Kronecker's delta).}$$

(Hint: Expand e_i in terms of u_j and vice versa.)

(b) Suppose that $a, b \in \mathbb{R}^n$ and that we have expanded them in terms of both $\{e_i\}_{i=1}^{n}$ and $\{u_j\}_{j=1}^{n}$:

$$a = \sum_{i=1}^{n} \alpha_i e_i = \sum_{i=1}^{n} \gamma_i u_i,$$

$$b = \sum_{i=1}^{n} \beta_i e_i = \sum_{i=1}^{n} \delta_i u_i.$$

We know that $a \cdot b \overset{\text{def.}}{=} \sum_{i=1}^{n} \alpha_i \beta_i$. Show that

$$\sum_{i=1}^{n} \alpha_i \beta_i = \sum_{i=1}^{n} \gamma_i \delta_i.$$

(c) We choose two sets of vectors from $\mathbb{R}^n$: $a_1, \ldots, a_k$ and $b_1, \ldots, b_m$. Let us form the $n \times k$ and $n \times m$ matrices $A = (a_1 \cdots a_k)$ and $B = (b_1 \cdots b_m)$ where the entries for a_i and b_j come from their expansions in terms of e_k. Next let us form the matrices $A' = (a_1 \cdots a_k)$ and $B' = (b_1 \cdots b_m)$ where the entries now come from the expansions of the vectors in terms of u_i. Show that $B^T A = B'^T A'$.

3. Show that if $\{a_1, \ldots, a_k\}$ or $\{b_1, \ldots, b_k\}$ is linearly dependent, it follows that $\det(a_i \cdot b_j) = 0$.

Chapter 2

Maps and manifolds

One of our long-range goals is to carry out "calculus on curved spaces" where the curved spaces can have any finite dimension. In this chapter we shall review and extend our knowledge of calculus of several variables in the very "flat" space $\mathbb{R}^n$. We shall also discuss simple examples of the sort of objects that we want to use as "curved spaces" or "manifolds."

In some respects, our approach will be shamelessly simple-minded: We shall treat limits and continuity in an intuitive manner and avoid rigorous $\varepsilon - \delta$ arguments. We shall imagine, for the most part, that we are dealing with elementary functions such as sin, cos, arctan, polynomials, etc. where it is obvious how to evaluate limits and derivatives. However we may occasionally have to make a more delicate argument involving an ε or two.

Our first step must be to review or learn a minuscule amount of point-set topology; that is, we must have some feeling for what are open or closed sets in $\mathbb{R}^n$. These concepts play a basic if sometimes quiet and unnoticed role in calculus on $\mathbb{R}^n$.

2.1 Open sets, closed sets

We introduce the notion of open set because when we take limits and let $x \to a$ where x and a are points in $\mathbb{R}^n$, we want a language that helps us to see that x can approach a from "all possible directions."

Recall that if $a \in \mathbb{R}^n$ and $a = (\alpha_1, \ldots, \alpha_n) = \alpha_1 e_1 + \cdots + \alpha_n e_n$, then

$$|a| = \sqrt{\alpha_1^2 + \cdots + \alpha_n^2}.$$

We think of $|a|$ as the *magnitude* or *norm* of the vector a; it is an extension of the notion of absolute value of a number in $\mathbb{R}$. If $a, b \in \mathbb{R}^n$, then we interpret $|a - b|$ as the distance from a and b.

If $a \in \mathbb{R}^n$ and ε is a positive real number (usually being thought of as being "small"), then by $B(a, \varepsilon)$, the (open) *ball or disk centered at a with radius ε*, we mean the set of points $x \in \mathbb{R}^n$ whose distance from a is less than ε. Symbolically,

$$B(a,\varepsilon) = \{x \in \mathbb{R}^n : |x-a| < \varepsilon\}.$$

(See Figure 2.1.)

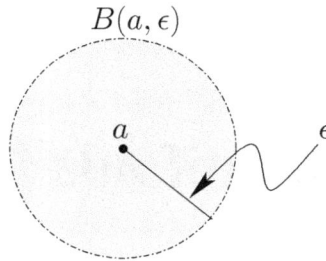

Figure 2.1: Disk centered at a with radius ε

Definition 2.1. We say that a subset A of $\mathbb{R}^n$ is *open in* $\mathbb{R}^n$ if and only if it has the property that for every $a \in A$ there exists an $\varepsilon > 0$ such that $B(a,\varepsilon) \subseteq A$.

Example 2.1. Consider the infinite rectangle $R = (0,\infty) \times (0,\infty)$ in $\mathbb{R}^2$. If we choose any $a = (\alpha_1, \alpha_2) \in R$, then $\alpha_1, \alpha_2 > 0$. Set $\varepsilon = \min(\alpha_1, \alpha_2)$. This is a positive number. Then it is easily seen that $B(a,\varepsilon) \subseteq R$. (See Figure 2.2)

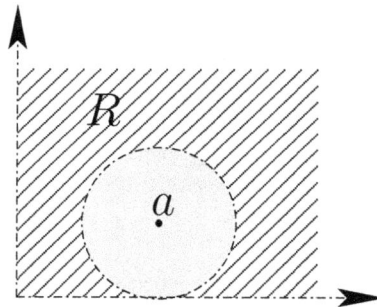

Figure 2.2: R is an open set

We can, if we wish, extend this example to $\mathbb{R}^n$, set $R = (0,\infty) \times \cdots \times (0,\infty) = (0,\infty)^n$ and repeat the proof. For $a = (\alpha_1, \ldots, \alpha_n) \in R$, we can take ε to be $\min\{\alpha_1, \ldots, \alpha_n\}$ as before. However we can no longer draw pictures and appeal to visual intuition for justification, so the proof (which we do not exhibit) that $x \in B(a,\varepsilon)$ implies $x \in (0,\infty)^n$ would need to depend on inequalities and algebra.

Definition 2.2. We say that a subset A of $\mathbb{R}^n$ is *closed in* $\mathbb{R}^n$ if and only if its complement, the set $\mathbb{R}^n - A$, is open in $\mathbb{R}^n$.

Example 2.2. The interval $\mathbb{J} = [0,1]$ is closed in $\mathbb{R}$. To see this, first notice that $\mathbb{R} - \mathbb{J} = (-\infty, 0) \cup (1, \infty)$. Any $x \in \mathbb{R} - \mathbb{J}$ is a member of either $(-\infty, 0)$ or $(1, \infty)$. If $x \in (-\infty, 0)$, then set $\varepsilon = -x$, which must be a positive number, and note that

$$B(x, \varepsilon) = (2x, 0) \subseteq (-\infty, 0) \subseteq \mathbb{R} - \mathbb{J}.$$

There is a similar proof if $x \in (1, \infty)$.

Sometimes it will be convenient to talk about a set being open or closed not in $\mathbb{R}^n$ but in some manifold $\mathcal{M}$ or, more generally, in some subset A of $\mathbb{R}^n$.

Definition 2.3. Suppose that $A \subseteq \mathbb{R}^n$. We say that a subset B of A is *open (closed) in A* if and only if there is a subset U of $\mathbb{R}^n$ such that U is open (closed) in $\mathbb{R}^n$ and $B = A \cap U$.

Example 2.3. Let A be the plane $\chi_1 + \chi_2 + \chi_3 = 0$ in $\mathbb{R}^3$. If B is the set of $x = \chi_1 e_1 + \chi_2 e_2 + \chi_3 e_3$ such that $x \in A$ and $\chi_1 > 0$, then B is open in A. This is because $B = A \cap U$ where U is the set of all x such that $\chi_1 > 0$ and U is open in $\mathbb{R}^3$.

The following term is sometimes useful:

Definition 2.4. If $x \in \mathbb{R}^n$ and A is a subset of $\mathbb{R}^n$, we say that A is a *neighborhood* of x provided there is an $\varepsilon > 0$ such that $B(x, \varepsilon) \subseteq A$. If A is also an open set, we can call A an *open neighborhood of x*.

Example 2.4. Suppose A is the closed disk $\chi_1^2 + \chi_2^2 \leq 1$ in $\mathbb{R}^2$. This is a neighborhood of the point $(\chi_1, \chi_2) = (1/2, 0)$ but not of the point $(1, 0)$. Since it is not an open set, it is not an open neighborhood of $(1/2, 0)$.

Exercises 2.1.

1. Show that $B(a, \varepsilon)$ is an open set.

2. Let A be the set of $x = (\chi_1, \chi_2)$ in $\mathbb{R}^2$ such that $\chi_1 = 0$. Show that A is closed in $\mathbb{R}^2$. (It may be helpful to draw a picture.)

3. Show that the set of $x = \chi_1 e_1 + \chi_2 e_2 + \chi_3 e_3$ such that $\chi_1 > 0$ is open in $\mathbb{R}^3$.

4. Suppose that U and V are open subsets of $\mathbb{R}^n$. Show that $U \cap V$ is an open subset of $\mathbb{R}^n$.

5. Suppose that A is a subset of $\mathbb{R}^n$ and U and V are open subsets of A. Show that $U \cap V$ is an open subset of A.

6. Is the empty set, $\emptyset$, an open or closed subset of $\mathbb{R}^n$? Or neither?

7. Give a proof that the set $R = (0, \infty)^n$ in Example 2.1 is an open set in $\mathbb{R}^n$.

2.2   Differentiability and the chain rule

Let us start with a little review. We assume familiarity with standard introductory calculus up through functions of several variables, their derivatives and integrals.

To say that f is continuous at $x_0 \in \operatorname{dom} f$ means that

$$\lim_{x \to x_0} f(x) = f(x_0).$$

If f is continuous at every point x of its domain, we say that it is a $\mathcal{C}^0$ function.

If $\operatorname{ran} f$ (the *range of* f) lies in $\mathbb{R}^n$, then we can write f in the unique form

$$f(x) \;=\; \big(\phi_1(x), \ldots, \phi_n(x)\big) \;=\; \phi_1(x)\, e_1 + \cdots + \phi_n(x)\, e_n$$

where each ϕ_i is a real-valued function and $x \in \operatorname{dom} f$. Let $(\chi_1, \ldots, \chi_k)$ be the coordinates we use to designate a point x in $\mathbb{R}^k$. If the domain of f is a subset of $\mathbb{R}^k$, then by the partial derivative of f at x with respect to χ_i we mean

$$\frac{\partial f}{\partial \chi_i}(x) \;=\; \lim_{\tau \to 0} \frac{1}{\tau}\big(f(x + \tau e_i) - f(x)\big) \;=\; \sum_{j=1}^{n} \frac{\partial \phi_j}{\partial \chi_i}(x)\, e_j \qquad (2.1)$$

assuming the limit exists.

Definition 2.5. Suppose that $f : U \to \mathbb{R}^n$ where U is an open set in $\mathbb{R}^m$ and $r = 1, 2, \ldots$. If

$$\frac{\partial^r f}{\partial \chi_{i_1} \cdots \partial \chi_{i_r}}(x)$$

exists and is continuous for for all $x \in U$ and for all $i_1, \ldots, i_r$ (where $i_1, \ldots, i_r$ are chosen from $\{1, \ldots, m\}$ and where repetitions are permissible in the list $i_1, \ldots, i_r$), then we say that f is a $\mathcal{C}^r$ function or, equivalently, that it is *r-times continuously differentiable*. If f is $\mathcal{C}^r$ for all positive r, then we say it is $\mathcal{C}^\infty$.

More generally, if $A \subseteq \mathbb{R}^m$ and we say that $f : A \to \mathbb{R}^n$ is $\mathcal{C}^r$ on A, then we shall mean that for every $a \in A$ there exist an open subset U of $\mathbb{R}^m$ and a function $F : U \to \mathbb{R}^n$ such that

1. $a \in U$,

2. $F(x) = f(x)$ for all $x \in A$,

3. F is $\mathcal{C}^r$ on U.

(The function F is not necessarily unique.) We may refer to this situation by saying that f has a $\mathcal{C}^r$ extension to an open neigborhood of every point of A.

The following result shows how the continuity and continuous differentiability of f is linked to those properties for its components and that these properties are independent of our choice of basis.

Proposition 2.1. *Suppose that* $f : U \to \mathbb{R}^n$, *where* U *is an open subset of* $\mathbb{R}^m$, *and* $\{u_i\}_{i=1}^n$ *is a fixed basis for* $\mathbb{R}^n$. *Let us expand* f *thus:* $f = \sum_{i=1}^n \psi_i u_i$ *where each* ψ_i *is a real-valued function on* $\mathrm{dom}(f)$. *Then* f *is* $\mathcal{C}^r$ *for* $r = 0, 1, \dots$ *if and only if* ψ_i *is* $\mathcal{C}^r$ *for each* i.

We leave the proof of the case $r = 1$ as an exercise.

Example 2.5. If $x = \chi_1 e_1 + \chi_2 e_2 \in \mathbb{R}^2$ and $f : \mathbb{R}^2 \to \mathbb{R}^3$ is defined by

$$f(x) = \cos(\chi_1^2 + \chi_2^2)\, e_1 + \exp(\chi_1^2 - \chi_2^2)\, e_2 + (\chi_2 - \chi_1)\, e_3,$$

then f is $\mathcal{C}^\infty$.

We are interested in a somewhat different property than a function being $\mathcal{C}^r$. The following definition is taken from [35]:

Definition 2.6. If U is an open subset of $\mathbb{R}^m$, we say that $f : U \to \mathbb{R}^n$ is *differentiable at* $x \in U$ if there is a linear transformation $f'(x) : \mathbb{R}^m \to \mathbb{R}^n$ and a function $g : V \to \mathbb{R}^n$ where V is a neighborhood of 0 in $\mathbb{R}^m$ such that

$$f(x+v) - f(x) = f'(x)(v) + g(v) \quad \text{for } v \in V$$

and

$$\lim_{v \to 0} \frac{g(v)}{|v|} = 0.$$

f is *differentiable on* U if it is differentiable at every $x \in U$. Given a point x in the domain of f, we call the map $v \mapsto f'(x)v$ the *differential (or first differential) of* f.

The linear transformation $f'(x)$ is unique. When we use symbolism such as $f'(x)v$ or $f'(x)(v)$ or $[f'(x)]v$, we mean that the linear transformation $f'(x)$ is acting on the vector v. Notice that it is not f' that is the linear transformation but $f'(x)$; we get a different linear transformation at every point x.

Proof of the following useful result can be found in references such as [24] or [35].

Proposition 2.2. *If* $f : U \to \mathbb{R}^n$, *where* U *is an open subset of* $\mathbb{R}^m$, *is* $\mathcal{C}^1$, *then* f *is differentiable on* U.

Recall that the *directional derivative* of a real-valued function ϕ at the point $x \in \mathbb{R}^n$ in the direction of the vector v is

$$\lim_{\lambda \to 0} \frac{1}{\lambda} \left(\phi(x + \lambda v) - \phi(x) \right) \tag{2.2}$$

where $\lambda \in \mathbb{R}$. (If ϕ is $\mathcal{C}^1$, this is known to reduce to the gradient of ϕ dotted with v, that is, $\mathrm{grad}\,\phi(x) \cdot v$.) Clearly we can generalize this idea to $f : U \to \mathbb{R}^n$ where U is an open subset of $\mathbb{R}^m$:

$$\partial_v f(x) \stackrel{\text{def.}}{=} \lim_{\lambda \to 0} \frac{1}{\lambda} \left(f(x + \lambda v) - f(x) \right) \tag{2.3}$$

where $\lambda \in \mathbb{R}$. In particular, if $x = (\chi_1, \ldots, \chi_m)$, a point in $\mathbb{R}^m$, then

$$\frac{\partial f}{\partial \chi_i}(x) \; = \; \partial_{e_i} f(x). \tag{2.4}$$

Here is a very useful connection between the directional derivative and $f'(x)$:

Proposition 2.3. *If $f \colon U \to \mathbb{R}^n$, where U is an open subset of $\mathbb{R}^m$, is differentiable at x, then $\partial_v f(x) = f'(x)v$, where $v \in \mathbb{R}^m$, assuming the limits exist.*

Proof. Because f is differentiable we can write

$$\frac{1}{\lambda}\left(f(x+\lambda v) - f(x)\right) \; = \; \frac{1}{\lambda}\left(f'(x)(\lambda v) + g(\lambda v)\right).$$

Since $f'(x)$ is a linear transformation, we have $(1/\lambda)\,f'(x)(\lambda v) = f'(x)\,v$; and by the role of g in the definition of differentiability, $g(\lambda v)/\lambda \to 0$ as $\lambda \to 0$. It follows that

$$\lim_{\lambda \to 0} \frac{1}{\lambda}\left(f(x+\lambda v) - f(x)\right) \; = \; f'(x)v. \qquad \square$$

The following is an immediate and useful result of Proposition 2.3:

Corollary 2.1. *The directional derivative $\partial_v f(x)$ is linear in v. That is, $\partial_{\lambda v} f(x) = \lambda \, \partial_v f(x)$ and $\partial_{v_1 + v_2} f(x) = \partial_{v_1} f(x) + \partial_{v_2} f(x)$ where λ is a scalar and v, v_1, v_2 are vectors.*

In particular, if we recall that $\{e_i\}_{i=1}^n$ is the standard basis for $\mathbb{R}^n$, if we represent an arbitrary point in $\mathbb{R}^n$ by $x = (\chi_1, \ldots, \chi_n)$, and if we choose a vector $v = (\xi_1, \ldots, \xi_n) = \xi_1 e_1 + \cdots + \xi_n e_n$, then via Equation 2.4, we have

$$\partial_v f \; = \; \sum_{i=1}^n \xi_i \, \partial_{e_i} f \; = \; \sum_{i=1}^n \xi_i \, \frac{\partial f}{\partial \chi_i}.$$

Example 2.6. Consider the function $f : \mathbb{R}^2 \to \mathbb{R}^2$ given by

$$f(\rho \, e_1 + \theta \, e_2) \; = \; \rho \cos(\theta) \, e_1 + \rho \sin(\theta) \, e_2.$$

This is the function that turns the polar coordinates of a point in the plane into its corresponding cartesian coordinates. Let

$$p \; = \; (\rho, \theta) \; = \; \rho \, e_1 + \theta \, e_2.$$

We think of p as a typical point in the domain of f and compute

$$\left(f'(p)\right)(e_1) \; = \; \partial_{e_1} f(p) \; = \; \frac{\partial f}{\partial \rho}(p) \; = \; \cos(\theta) \, e_1 + \sin(\theta) \, e_2,$$

$$\left(f'(p)\right)(e_2) \; = \; \partial_{e_2} f(p) \; = \; \frac{\partial f}{\partial \theta}(p) \; = \; -\rho \sin(\theta) \, e_1 + \rho \cos(\theta) \, e_2.$$

Since $f'(p)$ is a linear transformation and we know how it operates on the basis vectors e_1, e_2, we can compute $\left(f'(p)\right)(v)$ on any vector $v = \xi_1 e_1 + \xi_2 e_2$.

Example 2.7. Recall that if ϕ is a real-valued function on $\mathbb{R}^n$, then the gradient of ϕ is given by

$$\text{grad}\phi(x) = \sum_{i=1}^{n} \frac{\partial \phi}{\partial \chi_i}(x)\, e_i.$$

Our function is a map $\phi : U \to \mathbb{R}$ where U is an open subset of $\mathbb{R}^n$, and we assume it is $\mathcal{C}^1$ so that the linear transformation $\phi'(x)$ will exist. Let $v = \sum_{i=1}^{n} \xi_i e_i$, a vector in $\mathbb{R}^n$. Then

$$\left[\phi'(x)\right]v = \partial_v \phi(x) = \sum_{i=1}^{n} \xi_i \frac{\partial \phi}{\partial \chi_i}(x) = \left(\text{grad}\phi(x)\right) \cdot v.$$

We give another useful result but ask the reader to prove this one. It can be thought of as a generalization to higher dimensions of the simple fact in introductory calculus that if $\phi : \mathbb{R} \to \mathbb{R}$ and $\phi(\tau) = \lambda \tau$, then $\phi'(\tau) = \lambda$.

Proposition 2.4. *If f is a linear transformation, then for every $x \in dom(f)$, we have $f'(x) = f$.*

One can describe linear transformations by showing how they act on vectors, particularly basis vectors, or by using matrices. We shall, for the most part, avoid the use of matrices.

There are two reasons for this. First, to the extent that you avoid matrices, you avoid making a choice of a particular coordinate system and you deal with your vectors and mappings in way that is *intrinsic*. (That is, it is independent of the representation of your vectors or mappings in some particular coordinate system. This seems especially appropriate when dealing with, say, physics; there are no coordinate systems painted on space-time.) The second reason is that as we get into geometric algebra, it is actually often easier to compute how $f'(x)$ acts on a multivector. Matrices simply do not fit comfortably into this expanded setting.

Nevertheless, at this point, for the sake of completeness and because it is so familiar to many readers in mathematics, we explore the representation of $f'(x)$ by a matrix.

Proposition 2.5. *Suppose that U is an open subset of $\mathbb{R}^m$, that $f : U \to \mathbb{R}^n$ is $\mathcal{C}^1$, and that f has the expansion $f = \sum_{i=1}^{n} \phi_i e_i$ in terms of the standard basis. Each ϕ_i is, of course, a real-valued function. If $v = \sum_{i=1}^{m} v_i e_i$ and $y = \sum_{j=1}^{n} \xi_j e_j$ where v_i and ξ_j are scalars, then the equation $f'(x)v = y$ is equivalent to the matrix calculation*

$$\begin{pmatrix} \frac{\partial \phi_1}{\partial \chi_1}(x) & \cdots & \frac{\partial \phi_1}{\partial \chi_m}(x) \\ \cdots & & \\ \frac{\partial \phi_n}{\partial \chi_1}(x) & \cdots & \frac{\partial \phi_n}{\partial \chi_m}(x) \end{pmatrix} \begin{pmatrix} v_1 \\ \vdots \\ v_m \end{pmatrix} = \begin{pmatrix} \xi_1 \\ \vdots \\ \xi_n \end{pmatrix}.$$

We leave the proof of this as an exercise.

Definition 2.7. The $n \times m$ matrix $\left(\partial \phi_i / \partial \chi_j\right)$ in Proposition 2.5 is called the *Jacobian matrix of f* with respect to the standard basis.

Remark 2.1. Notice that we may think of the $n \times m$ Jacobian matrix of f at $x = (\chi_1, \dots, \chi_m)$ as

$$\left(\frac{\partial f}{\partial \chi_1}(x) \cdots \frac{\partial f}{\partial \chi_m}(x) \right)$$

where $\partial f / \partial \chi_i$ is treated as a column vector.

Example 2.8. Consider again the function $f : \mathbb{R}^2 \to \mathbb{R}^2$ of Example 2.6:

$$f(\rho\, e_1 + \theta\, e_2) = \rho \cos(\theta)\, e_1 + \rho \sin(\theta)\, e_2.$$

The Jacobian matrix for f is

$$\begin{pmatrix} \cos(\theta) & -\rho \sin(\theta) \\ \sin(\theta) & \rho \cos(\theta) \end{pmatrix}.$$

The chain rule of multivariable calculus can be thought of as a special case of matrix multiplication or as a composition of linear transformations.

Proposition 2.6 (The chain rule). *Suppose that we are given* $\mathcal{C}^1$ *maps f and g thus:*

$$\mathbb{R}^p \xrightarrow{g} \mathbb{R}^q \xrightarrow{f} \mathbb{R}^r$$

where it is understood that the domains of f and g are open subsets of $\mathbb{R}^q$ and $\mathbb{R}^p$ respectively. Let us expand f and g in terms of the standard bases thus,

$$f = \sum_{i=1}^{r} \phi_i\, e_i \quad and \quad g = \sum_{j=1}^{q} \psi_j\, e_j,$$

and let us suppose that $x \in \mathrm{dom}\, g$ and $y \in \mathrm{dom}\, f$ and $y = g(x)$. We also understand that $x = (\chi_1, \dots, \chi_p) \in \mathbb{R}^p$ and $y = (v_1, \dots, v_q) = g(x) \in \mathbb{R}^q$. Then the following statements hold and are equivalent:

1. $(f \circ g)'(x) = f'(y)\, g'(x)$ *where the right-hand side of this equation is a composition of two linear transformations.*

2.

$$\begin{pmatrix} \dfrac{\partial (\phi_1 \circ g)}{\partial \chi_1}(x) & \cdots & \dfrac{\partial (\phi_1 \circ g)}{\partial \chi_p}(x) \\ & \cdots & \\ \dfrac{\partial (\phi_r \circ g)}{\partial \chi_1}(x) & \cdots & \dfrac{\partial (\phi_r \circ g)}{\partial \chi_p}(x) \end{pmatrix}$$
$$= \begin{pmatrix} \dfrac{\partial \phi_1}{\partial v_1}(y) & \cdots & \dfrac{\partial \phi_1}{\partial v_q}(y) \\ & \cdots & \\ \dfrac{\partial \phi_r}{\partial v_1}(y) & \cdots & \dfrac{\partial \phi_r}{\partial v_q}(y) \end{pmatrix} \begin{pmatrix} \dfrac{\partial \psi_1}{\partial \chi_1}(x) & \cdots & \dfrac{\partial \psi_1}{\partial \chi_p}(x) \\ & \cdots & \\ \dfrac{\partial \psi_q}{\partial \chi_1}(x) & \cdots & \dfrac{\partial \psi_q}{\partial \chi_p}(x) \end{pmatrix}.$$

A proof of Proposition 2.6 can be found in [35].

Exercises 2.2.

1. Prove Proposition 2.1 for $r = 1$.

2. Set

$$\phi(\tau) = \begin{cases} \tau^2/2 & \text{for } \tau \geq 0 \\ -\tau^2/2 & \text{for } \tau < 0. \end{cases}$$

   (a) Show that ϕ is $\mathcal{C}^1$ but not $\mathcal{C}^2$. What is $\phi'(\tau)$?

   (b) What is the maximal r for which $\phi(\tau)^2$ is a $\mathcal{C}^r$ function?

3. If ϕ is the function of Exercise 2, then find the maximal r for which each of the following is a $\mathcal{C}^r$ function:

   (a) $\cos(\tau) e_1 + \sin(\tau) e_2$.

   (b) $\tau^3 e_1 - (|\tau|^2 + 1) e_2$.

   (c) $\exp(\tau_1 \tau_2) e_1 - \tau_1^2 e_2 + \phi(\tau_2) e_3$.

4. Prove Proposition 2.5. Since $f'(x)$ is a linear transformation, it is sufficient to prove this for arbitrary e_k. This can be done by replacing v in the equation in Definition 2.6 by τe_k, where τ is a nonzero scalar, dividing both sides of the equation by τ, and letting τ go to zero.

5. Assume all the functions below are at least $\mathcal{C}^1$ and find the linear transformation $f'(x)$ in two ways: First compute $\partial_{e_i} f(x)$ for all relevant e_i. This is the same as finding $(f'(x)) e_i$, and a linear transformation is determined by its action on basis vectors. Second, give the Jacobian matrix for f at x.

   (a) $f(\theta) = \cos(\theta) e_1 + \sin(\theta) e_2$ where $\theta \in \mathbb{R}$.

   (b) $f(\tau) = \phi_1(\tau) e_1 + \cdots + \phi_n(\tau) e_n$ where $\tau \in \mathbb{R}$ and each ϕ_i is real-valued.

   (c) $f(\chi_1 e_1 + \chi_2 e_2) = \chi_1 e_1 + \chi_2 e_2 + \phi(\chi_1, \chi_2) e_3$ where ϕ is real-valued.

   (d) $f(\chi_1, \chi_2) = (\chi_1^2 - \chi_2^2) e_1 + 2\chi_1 \chi_2 e_2$.

   (e) $f(\chi_1 e_1 + \cdots \chi_m e_m) = \chi_i$.

   (f) $f(\chi_1, \chi_2) = \phi(\chi_1, \chi_2) e_1 + \phi(\chi_2, \chi_1) e_2$ where ϕ is real-valued.

6. Prove that if f is a linear transformation, then for every $x \in \text{dom}(f)$, we have $f'(x) = f$.

2.3 Manifolds

We want to carry out calculus not merely on $\mathbb{R}^n$ but on more general "curved spaces." The usual setting for such activities is a *manifold*. An n-dimensional manifold is often informally described as being "locally like $\mathbb{R}^n$."

However manifolds, if described properly, are certain kinds of topological spaces and do not necessarily lie in any $\mathbb{R}^n$ or euclidean space. This requires more mathematical machinery than we are willing to assume or develop at this time. Also, some of the spaces on which we wish to operate, such as "cells"—which we describe later—are not quite proper manifolds but are better thought of as "manifolds-with-corners."

A little later we will say more about the standard, precise definition of a manifold. However, what we shall work with is not that definition but a *description* of manifolds embedded in $\mathbb{R}^n$, a description which will be sufficiently precise to permit us to prove theorems and will cover not only manifolds but manifolds with boundaries and manifolds-with-corners (at least if they are embedded in $\mathbb{R}^n$).

To gain intuition, we begin by giving simple examples of things we want to call manifolds.

2.3.1 Hyperplanes

One of the simplest examples of a manifold is a plane and its generalizations to higher dimensions. If we refer to a plane in three-dimensional space, we think of this flat, two-dimensional object which can be specified by an equation of the form $\alpha_1 \chi_1 + \alpha_2 \chi_2 + \alpha_3 \chi_3 = \beta$. If it passes through the origin, then it is a 2-dimensional vector subspace of $\mathbb{R}^3$ and $\beta = 0$. See Figure 2.3.

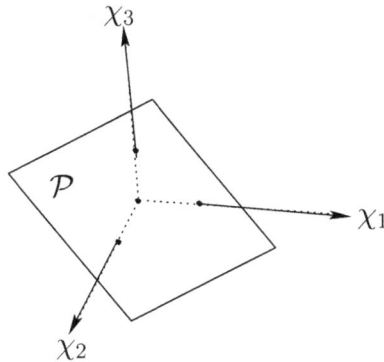

Figure 2.3: A 2-dimensional plane in $\mathbb{R}^3$

More generally, we will think of a *k-dimensional hyperplane* (or, more briefly, a *k-plane*) $\mathcal{P}$ in $\mathbb{R}^n$ as being a translate of a k-dimensional vector subspace of $\mathbb{R}^n$.

Definition 2.8. We say that $\mathcal{P}$ in $\mathbb{R}^n$ is a k-dimensional *hyperplane* provided there is a point p and linearly independent vectors $a_1, \ldots, a_k$ such that $x \in \mathcal{P}$ if and only if x is of the form

$$x = p + \tau_1 a_1 + \cdots + \tau_k a_k$$

where $\tau_1, \ldots, \tau_k \in \mathbb{R}$. We call $\mathcal{P}$ the *translate by p* of the k-dimensional vector subspace spanned by $a_1, \ldots, a_k$.

Of course, if $k = 1$, we might prefer to refer to $\mathcal{P}$ as a *line* rather than a 1-dimensional hyperplane; and if $k = 2$, the preferred term for it would simply be a *plane*.

Example 2.9. Consider the plane $\mathcal{P}$ described by the equation $\chi_1 + \chi_2 + \chi_3 = 1$ in $\mathbb{R}^3$. This is obtained by translating the 2-dimensional vector space V which satisfies the equation $\chi_1 + \chi_2 + \chi_3 = 0$. Choosing at random, two vectors that span V are $a_1 = e_1 - e_2$ and $a_2 = e_2 - e_3$. Next we choose an arbitrary point p in $\mathcal{P}$, say, $p = (1, 0, 0)$. The translation of V to $\mathcal{P}$ is then given by $x \mapsto x + p = x + e_1$. If $x = (\xi_1, \xi_2, \xi_3) \in V$ and $x + p = (\chi_1, \chi_2, \chi_3)$, then we check that we must have

$$\begin{aligned}
\xi_1 + \xi_2 + \xi_3 &= 0, \\
\chi_1 &= \xi_1 + 1, \ \chi_2 = \xi_2, \ \chi_3 = \xi_3, \\
\chi_1 + \chi_2 + \chi_3 &= 1.
\end{aligned}$$

Thus $x + p$ is a typical point of $\mathcal{P}$. In terms of our definition of hyperplane, $x + p$ has the form

$$x + p = p + \tau_1 a_1 + \tau_2 a_2 = e_1 + \xi_1 (e_1 - e_2) + (\xi_1 + \xi_2)(e_2 - e_3).$$

An important characteristic of a k-dimensional hyperplane $\mathcal{P}$ is that it possesses a *parametrization* $g : \mathbb{R}^k \to \mathcal{P}$. If $\mathcal{P}$ passes through the point p and is the translate of the vector space spanned by the k linearly independent vectors $a_1, \ldots, a_k$, then one parametrization is

$$g(\tau_1, \ldots, \tau_k) = p + \tau_1 a_1 + \cdots \tau_k a_k.$$

Of course $\mathcal{P}$ has an infinite number of such parametrizations since we have an infinite number of possible choices for the point p and the spanning set $\{a_1, \ldots, a_k\}$. We think of $(\tau_1, \ldots, \tau_k)$ as being *coordinates* assigned to the point $x = g(\tau_1, \ldots, \tau_k)$ in the hyperplane $\mathcal{P}$. Note that we can also write $x = (\chi_1, \ldots, \chi_n)$ since x is a point in $\mathbb{R}^n$ and that we also refer to $\chi_1, \ldots, \chi_n$ as coordinates of x. The τ_i's are part of a *local coordinate system* that exists only in the hyperplane, while the χ_i's can be thought of as something more global, inherited from the space $\mathbb{R}^n$ that contains $\mathcal{P}$.

(We do not give a definition of *parametrization*, but it will be clear later what properties we want it to have.)

2.3.2 The half-space $\mathbb{H}^n$

The simplest example of a manifold with a boundary is the half-space $\mathbb{H}^n$. We define this to be

$$\mathbb{H}^n \stackrel{\text{def.}}{=} \{(\chi_1, \ldots, \chi_n) \in \mathbb{R}^n : \chi_1 \geq 0\}.$$

Of course, conceptually, it does not matter whether we define this by specifying $\chi_i \geq 0$ or $\chi_i \leq 0$ for any given value of i; each of these would give us simple variations on the same underlying idea.

Given a point $a \in \mathbb{H}^n$, we say that a is an *interior point* of $\mathbb{H}^n$ provided there is an open ball $B(a, \varepsilon)$ centered at a such that $B(a, \varepsilon) \subseteq \mathbb{H}^n$. If there is no such ball, that is, if for every $\varepsilon > 0$, $B(a, \varepsilon)$ contains points both in $\mathbb{H}^n$ and outside $\mathbb{H}^n$, then we call a a *boundary point* of $\mathbb{H}^n$. In Figure 2.4, we see that a is an interior point of $\mathbb{H}^2$ since we

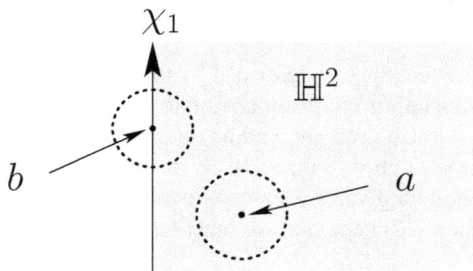

Figure 2.4: Interior and boundary point in $\mathbb{H}^2$

can find an $\varepsilon > 0$ such that $B(a, \varepsilon) \subseteq \mathbb{H}^2$, but b is a boundary point since every $B(b, \varepsilon)$ must hang a little bit outside $\mathbb{H}^2$.

2.3.3 Parallelepipeds and simplexes

Some other simple examples of bounded manifolds (with corners) are parallelepipeds and simplexes.

Definition 2.9. A *k-dimensional parallelepiped* $\mathscr{A}$ in $\mathbb{R}^n$ with base point $P \in \mathbb{R}^n$ and "edges" $a_1, a_2, \ldots, a_k$, linearly independent vectors in $\mathbb{R}^n$, consists of all points of the form
$$P + \tau_1 a_1 + \tau_2 a_2 + \cdots + \tau_k a_k \quad \text{where} \quad 0 \leq \tau_1, \tau_2, \ldots, \tau_k \leq 1.$$

We show a 2- and a 3-dimensional parallelepiped in Figure 2.5. An important thing

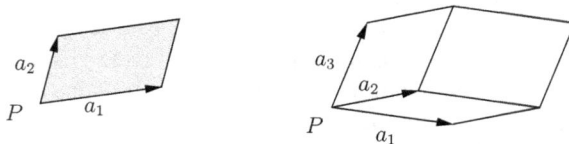

Figure 2.5: 2- and 3-dimensional parallelepipeds

to notice here is that the map
$$(\tau_1, \ldots, \tau_k) \overset{g}{\mapsto} P + \tau_1 a_1 + \tau_2 a_2 + \cdots + \tau_k a_k$$

is a "parametrization" of the parallelepiped.

Example 2.10. Consider the 2-dimensional parallelepiped (parallelogram) $\mathcal{P}$ in $\mathbb{R}^3$ with basepoint $P = (1,1,1)$ and edges

$$a_1 = -e_3 \quad \text{and} \quad a_2 = -e_1 - e_2.$$

See Figure 2.6. We construct a parametrization $g : \mathcal{I}^2 \to \mathcal{P}$ by

$$\begin{aligned} x = g(\tau_1, \tau_2) &= P + \tau_1 a_1 + \tau_2 a_2 \\ &= (1 - \tau_2) e_1 + (1 - \tau_2) e_2 + (1 - \tau_1) e_3. \end{aligned}$$

The vertices of the parallelepiped are found by evaluating g at the vertices of $\mathcal{I}^2$:

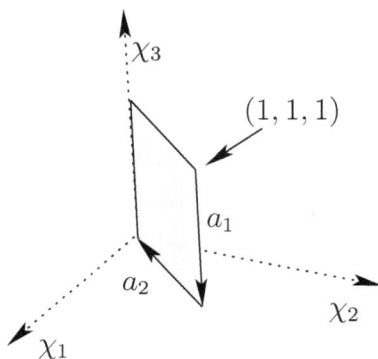

Figure 2.6: The 2-parallelepiped of Example 2.10

$$\begin{aligned} g(0,0) &= (1,1,1) = P, \\ g(1,0) &= (1,1,0), \\ g(0,1) &= (0,0,1), \\ g(1,1) &= 0 \text{ (origin)}. \end{aligned}$$

We see that the center of the unit square maps to the center of the parallelogram:

$$g\left(\tfrac{1}{2}, \tfrac{1}{2}\right) = \left(\tfrac{1}{2}, \tfrac{1}{2}, \tfrac{1}{2}\right).$$

Now sometimes we want to work backward from the $\mathbb{R}^3$ coordinates of $x = (\chi_1, \chi_2, \chi_3)$, a point in $\mathcal{P}$, to the τ_i-coordinates that are assigned to x by g. To see how we would do this, note that we must have $x = g(\tau_1, \tau_2)$. This amounts to

$$\chi_1 e_1 + \chi_2 e_2 + \chi_3 e_3 = (1 - \tau_2) e_1 + (1 - \tau_2) e_2 + (1 - \tau_1) e_3.$$

We can then solve for the τ_i's in terms of the χ_j's:

$$\tau_1 = 1 - \chi_3 \quad \text{and} \quad \tau_2 = 1 - \chi_1 = 1 - \chi_2.$$

Now we turn our attention to simplexes.

Before defining *simplex*, we remark that when we say a subset U of $\mathbb{R}^n$ is *convex*, we mean it has the property that whenever A and B are points in U, then U must contain the line segment that runs from A to B. See Figure 2.7.

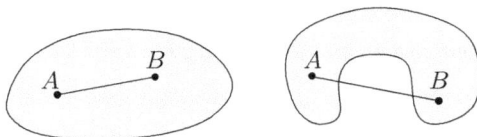

Figure 2.7: Sets convex and not convex

Definition 2.10. The simplex S determined by $A_0, A_1, \ldots, A_k$, points in $\mathbb{R}^n$, is

$$S = \left\{ \tau_0 A_0 + \cdots + \tau_k A_k \, : \, \tau_i \in \mathbb{R}, \, 0 \leq \tau_i, \, \sum_{i=0}^{k} \tau_i = 1 \right\}.$$

S is the smallest convex set containing $A_0, \ldots, A_k$. We say that $A_0, \ldots, A_k$ are the *vertices* of S. We denote this simplex as $A_0 A_1 \cdots A_k$.

The simplest kind of simplex is a point, A_0, a 0-simplex. Next is a 1-simplex, $A_0 A_1$, the line segment from A_0 to A_1. A 2-simplex is a triangle while a 3-simplex is a tetrahedron. See Figure 2.8.

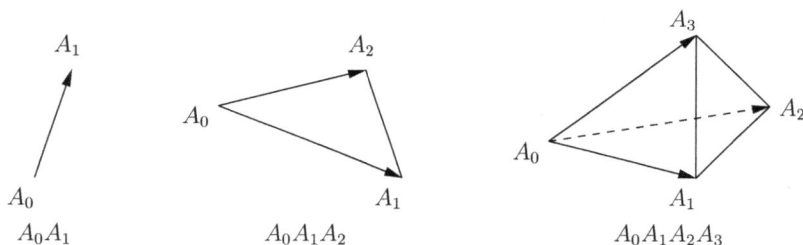

Figure 2.8: k-simplexes

Just as in the cases of hyperplanes and parallelepipeds, we can set up "nice" parametrizations of simplexes. To do this, it is convenient to work with a "standard" or "canonical" n-simplex: We declare this to be the simplex S_n determined by the $n+1$ points $0, e_1, e_2, \ldots, e_n$. Equivalently, this is the set bounded by the coordinate hyperplanes $\chi_i = 0$ for $i = 1, \ldots, n$ and the hyperplane $\chi_1 + \cdots + \chi_n = 1$. See the picture of S_2 in Figure 2.9.

It can be shown that any S_k is the set of $(\tau_1, \ldots, \tau_k) \in \mathbb{R}^k$ such that $\tau_1 + \cdots \tau_k \leq 1$ and $\tau_1, \ldots, \tau_k \geq 0$. Now let $A_0, \ldots, A_k$ be points chosen from $\mathbb{R}^n$. In general we want the points chosen in such a way that the vectors $A_0 A_1$, $A_0 A_2$, $\ldots$, $A_0 A_k$ are linearly

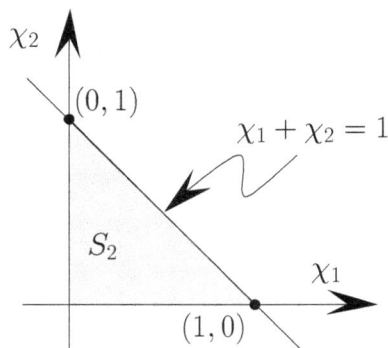

Figure 2.9: The standard 2-simplex

independent. Otherwise the resultant k-simplex will be "degenerate." We then set up a parametrization $g : S_k \to A_0 \cdots A_k$ thus:

$$x = g(\tau_1, \ldots, \tau_k) = \tau_0 A_0 + \tau_1 A_1 + \cdots \tau_k A_k$$

where

$$\tau_0 \overset{\text{def.}}{=} 1 - \tau_1 - \cdots - \tau_k.$$

Notice that we must have $\sum_{i=0}^{k} \tau_k = 1$ and $\tau_0 \geq 0$. We may think of $(\tau_1, \ldots, \tau_k)$ as being the *coordinates* assigned the point x in $A_0 \cdots A_k$ by the parametrization. (The $(k+1)$-tuple $(\tau_0, \tau_1, \ldots, \tau_k)$ is referred to as the *barycentric coordinates* of x.)

2.3.4 Cells

By a *p-cell* we shall mean, roughly speaking, a one-to-one continuously differentiable image of $\mathfrak{I}^p$. We shall give a more precise definition in a moment.

Thus we expect a p-cell $\mathcal{M}$ in $\mathbb{R}^n$ to be something that has a *parametrization* $x : \mathfrak{I}^p \to \mathcal{M}$ where x is, at the least, a $\mathcal{C}^1$ map and is one-to-one. Visually, we expect a 2-cell to look a bit like a piece of rectangular cloth flapping in the wind and a 3-cell to look something like a deformed rubber cube. See Figure 2.10.

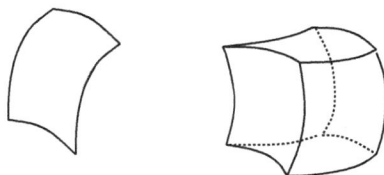

Figure 2.10: 2-cell and 3-cell

Example 2.11. The set of points $x = (\chi_1, \chi_2, \chi_3)$ satisfying $\chi_3 = \chi_1^2 + 1$ is a 2-dimensional surface in $\mathbb{R}^3$. We construct a 2-cell $\mathcal{M}$ in this surface by projecting the unit square $\mathcal{I}^2$ in the $\chi_1\chi_2$-plane straight up into the curved surface. See Figure 2.11. We take as our parametrization the map $x : \mathcal{I}^2 \to \mathcal{M}$ given by

$$x(\tau_1, \tau_2) = \tau_1 e_1 + \tau_2 e_2 + (\tau_1^2 + 1) e_3 = (\chi_1, \chi_2, \chi_3).$$

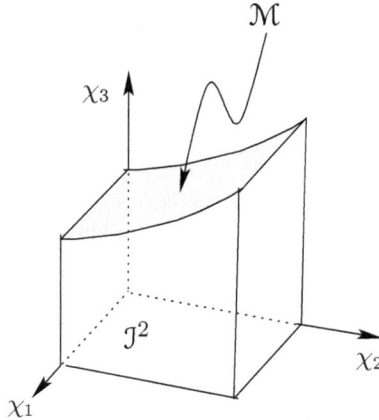

Figure 2.11: 2-cell formed from a projection

We now give a careful definition of what we mean by a cell:

Definition 2.11. If we say that $\mathcal{M}$, a subset of $\mathbb{R}^n$, is a $\mathcal{C}^r$ p-cell (where $p, r \geq 1$), then we mean that there is associated with $\mathcal{M}$ a family of maps $x, y, \ldots : \mathcal{I}^p \to \mathcal{M}$, *parametrizations* of $\mathcal{M}$, such that the following hold for all x and y:

1. x is one-to-one and onto.

2. x and x^{-1} are $\mathcal{C}^r$.

3. For all $t \in \mathcal{I}^p$, the set of vectors $\{x'(t)e_1, \ldots, x'(t)e_p\}$ is a linearly independent set.

4. $x^{-1} \circ y$ and $y^{-1} \circ x$ are $\mathcal{C}^r$.

Example 2.12. Let $\mathcal{M}$ be the $\mathcal{C}^\infty$ 1-cell (arc) lying in $\mathbb{R}^2$ along the parabola $\chi_2 = \chi_1^2$ between the points $(-1, 1)$ and $(1, 1)$. (See Figure 2.12.) Let $x : \mathcal{I} \to \mathcal{M}$ be the parametrization

$$x(\tau) = (2\tau - 1) e_1 + (2\tau - 1)^2 e_2 = x.$$

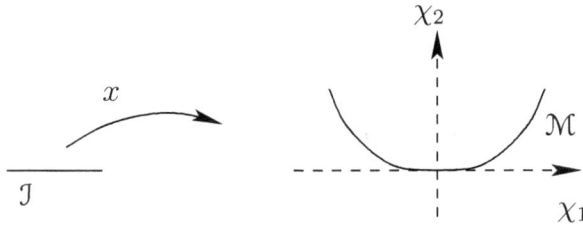

Figure 2.12: A 1-cell in $\mathbb{R}^2$

(Notice that we are playing fast and loose with the symbolism here, that x has two different meanings. On the one hand it is a map, on the other, a point in $\mathbb{R}^2$.) This is a $\mathbb{C}^\infty$ map, and

$$x^{-1}(x) = g^{-1}(\chi_1 e_1 + \chi_2 e_2) = \frac{\chi_1 + 1}{2} = \tau$$

which can clearly be thought of as a $\mathbb{C}^\infty$ map on the open set $\mathbb{R}^2$. The linear transformation $x'(\tau) : \mathbb{R} \to \mathbb{R}^2$ for a fixed τ is given by

$$\left[x'(\tau) \right] \xi = (2 e_1 + 4 \tau e_2) \xi$$

which is one-to-one.

Remark 2.2. Sometimes it is convenient to parameterize a p-cell using a one-to-one map $x \colon \mathcal{R} \to \mathcal{M}$ where $\mathcal{R}$ is a p-dimensional rectangle other than $\mathcal{J}^p$. If, for example, $\mathcal{M}$ is a 2-cell, we can parameterize it using $\mathcal{R} = [\alpha_1, \beta_1] \times [\alpha_2, \beta_2]$ where it is understood that $\alpha_i < \beta_i$ and otherwise $x \colon \mathcal{R} \to \mathcal{M}$ satisfies all the other requirements of Definition 2.11. We could replace this particular parametrization with a new one $y \colon \mathcal{J}^2 \to \mathcal{M}$ by setting $y = x \circ f$ where

$$f(\tau_1, \tau_2) = \left((1 - \tau_1)\alpha_1 + \tau_1 \alpha_2, (1 - \tau_2)\beta_1 + \tau_2 \beta_2 \right).$$

2.3.5 Manifolds, charts, and tangent vectors

We want to briefly consider the way *manifold* is usually defined. We do not talk about the most general definition since we consider only manifolds embedded in some $\mathbb{R}^n$. We also restrict ourselves to those that are $\mathbb{C}^k$ manifolds with $k \geq 1$; those are the ones on which we can do calculus.

We shall not use the following definition, but we think the reader should be exposed to it. We shall instead shortly introduce and use a description of a slightly broader yet clearly closely related concept which includes such objects as manifolds with boundaries and corners.

We say that $\mathcal{M}$, a subset of $\mathbb{R}^n$, is a p-dimensional $\mathbb{C}^k$ manifold provided the following are true:

First, we have a given collection of maps $x \colon U \to \mathbb{R}^p$ such that the following hold:

1. U is an open subset of $\mathcal{M}$ and $x(U)$ is an open subset of $\mathbb{R}^p$.

2. x is one-to-one.

3. x and x^{-1} are both $\mathcal{C}^k$ maps.

4. If $\{x_\alpha : \alpha \in A\}$ is the collection of these special maps x, where A is an indexing set for these maps, and for each index α, we have $x_\alpha : U_\alpha \to \mathbb{R}^p$, then

$$\mathcal{M} = \bigcup_{\alpha \in A} U_\alpha.$$

Second, if x_α and x_β are are any two of these given maps, then $x_\alpha \circ x_\beta^{-1}$ is a $\mathcal{C}^k$ map between open sets in $\mathbb{R}^p$ (provided, of course, that the domains of x_α and x_β have a nonempty intersection).

We call each of these given maps x a *chart* on $\mathcal{M}$ and x^{-1} a *parametrization*. (In a short while, when we get to our slightly different version of a manifold, we shall use the word *chart* in a different but closely related way.)

An example of a 2-manifold in $\mathbb{R}^3$ would be the unit sphere $\mathcal{M}$ described by the equation $|x| = 1$ or, equivalently, $\chi_1^2 + \chi_2^2 + \chi_3^2 = 1$. See Figure 2.13.

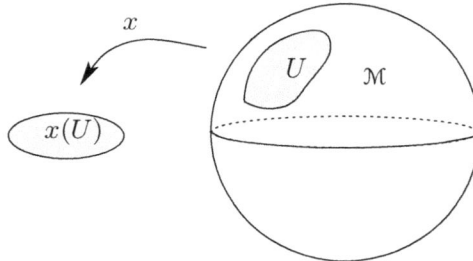

Figure 2.13: Sphere with chart x

In the spirit of the description given above, one can also talk about a *manifold with boundary* $\mathcal{M}$ by simply replacing $\mathbb{R}^p$ with $\mathbb{H}^p$. In this case, a chart is a map $x : U \to \mathcal{M}$ where U is an open subset of $\mathbb{H}^p$. If $a \in \mathcal{M}$ has the property that $x(b) = a$ where b is a boundary point of $\mathbb{H}^p$, we then call a a *boundary point* of $\mathcal{M}$.

We now begin to consider a conception of manifold which is somewhat different from the standard one, the one we just considered. It will, however, include the standard manifolds (at least the ones on which the concept of a dot product or *inner product* makes sense, the *Riemannian manifolds*), and it will also include what we have referred to as *manifolds-with-corners*.

In our new description, we think of manifolds as being a kind of landscape with the property that at every point we can find a family of "tangent vectors" that tell us whether we are going "uphill" or "downhill" and whether we are at a "mountain peak," a "valley," a "pass," or some more exotic structure. A little more specifically, this description will include the following:

1. For every point of the manifold, a coordinate system that can be "pasted" over a neighborhood of that point.

2. A space of tangent vectors at every point of the manifold.

3. A way of calculating directional derivatives for functions defined on the manifold, particularly directional derivatives with respect to tangent vectors.

Let us begin by supposing that $\mathcal{M}$ is a subset of $\mathbb{R}^n$, nothing more, and define the concept of a tangent vector to $\mathcal{M}$.

By a *curve* in $\mathbb{R}^n$ we mean a continuous map $c : [\alpha, \beta] \to \mathbb{R}^n$ where $[\alpha, \beta]$ is an interval in $\mathbb{R}$. By a change of parameter, we can replace $[\alpha, \beta]$ by $[0, 1]$, $[-1, 1]$, or whatever other interval may be convenient. Of course, this discussion still makes sense if we replace the closed interval $[\alpha, \beta]$ by an open interval (α, β).

We say that c is a *curve in* $\mathcal{M}$ provided $c(\tau) \in \mathcal{M}$ for every $\tau \in [\alpha, \beta]$. The curve is $\mathcal{C}^k$ provided the map c is $\mathcal{C}^k$. If c is one-to-one and both c and c^{-1} are continuous, we say it is an *arc*.

Definition 2.12. [A restricted definition of tangent vector] We say that $a \in \mathbb{R}^n$ is a *tangent vector to* $\mathcal{M}$ *at the point* $x_0 \in \mathcal{M}$ provided there is as a $\mathcal{C}^1$ curve in $\mathcal{M}$, $c : [0, 1] \to \mathcal{M}$, such that $x_0 = c(0)$ and

$$a = \lim_{\tau \to 0} \frac{c(\tau) - c(0)}{\tau} = c'(0).$$

It is convenient to expand this definition. The set of tangent vectors to $\mathcal{M}$ at a point $x_0 \in \mathcal{M}$ in the sense of Definition 2.12 can be thought of as generating a vector space which we call a *tangent space*.

Definition 2.13. [Expanded definition of tangent vector] If x_0 is a point in $\mathcal{M}$, then by $T_{x_0}\mathcal{M}$, *the tangent space to* $\mathcal{M}$ *at* x_0, then we mean the set of vectors $a \in \mathbb{R}^n$ which can be written in the form

$$a = \lambda_1 a_1 + \cdots \lambda_r a_r$$

where each λ_i is a scalar and each a_i is a tangent vector to $\mathcal{M}$ at x_0. That is, $T_{x_0}\mathcal{M}$ is the set of finite linear combinations of tangent vectors to $\mathcal{M}$ at x_0 which are generated by curves in $\mathcal{M}$. We now refer to *all* the elements of $T_{x_0}\mathcal{M}$ as *tangent vectors to* $\mathcal{M}$ *at* x_0.

It should be understood that there are elements of $T_{x_0}\mathcal{M}$ which are *not* tangent vectors to $\mathcal{M}$ at x_0 in the sense of Definition 2.12. For example, take $\mathcal{M}$ to be the rectangle in Figure 2.14 and suppose it lies in $\mathbb{R}^2$. It is visually clear that $-e_1$ and $-e_2$ must be tangent vectors to $\mathcal{M}$ at the point x_0 and are generated by curves drawn in $\mathcal{M}$. However a cannot be realized as a tangent vector by a curve drawn within $\mathcal{M}$, though it can be written as a linear combination of the such vectors, $a = \lambda_1(-e_1) + \lambda_2(-e_2)$.

We now give a different version of the idea of a chart, one that is not associated (at first) with any $\mathcal{M}$ embedded in $\mathbb{R}^n$:

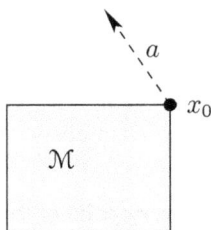

Figure 2.14: A tangent vector that does not correspond to a curve in $\mathcal{M}$

Definition 2.14. We say that x is a $\mathcal{C}^k$ *p-dimensional chart (or p-chart) in* $\mathbb{R}^n$ (where $k, p \geq 1$ and $p \leq n$) provided the following hold:

1. $x\colon U \to \mathbb{R}^n$ is a one-to-one $\mathcal{C}^k$ map where U is an open set in $\mathbb{R}^p$.

2. x^{-1} is a $\mathcal{C}^k$ map on $x(U)$.

3. For all $t_0 \in U$, if we consider the linear transformation $x'(t_0)$ and if $\{u_i\}_{i=1}^{p}$ is a basis for $\mathbb{R}^p$, then $\{[x'(t_0)]u_i\}_{i=1}^{p}$ is a set of linearly independent vectors in $\mathbb{R}^n$.

We also refer to x as a *coordinate patch* or just as a *patch*. In addition, it is sometimes convenient to refer to the set $x(U)$ as well as the map x as a *chart*, *patch*, or *coordinate patch*. In practice there should be no confusion between the set and the map.

We have to make some comments about this definition.

First, each $x(U)$ is a piece of $\mathbb{R}^p$ hanging in $\mathbb{R}^n$. (See Figure 2.15.) If t is a point in

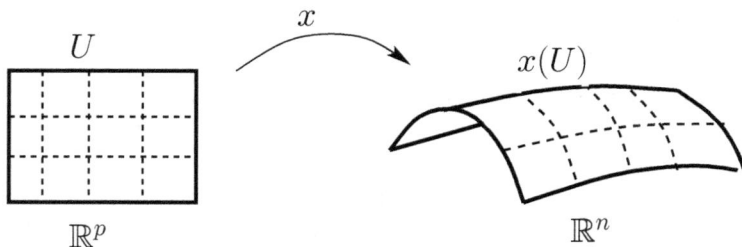

Figure 2.15: A p-chart in $\mathbb{R}^n$

$\mathbb{R}^p$ and $t = (\tau_1, \ldots, \tau_p)$, then we think of x as "assigning the coordinates $(\tau_1, \ldots, \tau_p)$" to the point $x(t)$ lying in $x(U)$. The fact that x is one-to-one means that each point in $x(U)$ has a unique set of coordinates (at least with respect to this particular chart).

Second, although the map $x\colon U \to x(U)$ is one-to-one and $x^{-1}\colon x(U) \to U$ is a perfectly well-defined map, if we talk about x^{-1} being $\mathcal{C}^k$, this means that at every point $x_0 = x(t_0) \in x(U)$, we must be able to able to construct an extension of x^{-1} to a $\mathcal{C}^k$ map y on some open neighborhood of x_0. (Recall that an open neighborhood of x_0 is an open set V in $\mathbb{R}^n$ such that $x_0 \in V$.) In general, the extended map y is neither one-to-one nor unique.

Suppose, for example that $U = \mathbb{R}$ and the chart $x \colon U \to \mathbb{R}^2$ is $x(\tau) = (\tau, \tau)$ so that $x(U)$ is the 45° line through the origin in the plane. Then $x^{-1}(\tau, \tau) = \tau$, and two possible extensions of $x^{-1} \colon x(U) \to \mathbb{R}$ are $y(\tau_1, \tau_2) = \tau_1$ and $y(\tau_1, \tau_2) = \tau_2$.

However as we shall see, the fact that this extension y is neither one-to-one nor unique is not usually a problem since we are, for the most part, ultimately concerned only with what happens right on $x(U)$, not off of it in the larger $\mathbb{R}^n$. We shall, in general, make no reference to the (invisible) extension y and cheerfully refer only to x^{-1}.

Our third remark is that although it is not obvious, conditions 2 and 3 of Definition 2.14 are equivalent. This can be shown by using the inverse function theorem, a result usually derived in an advanced calculus or introductory real analysis course. (See, for example, [24, 35].) We have not shown the equivalence here because it requires more knowledge than we wish to assume, and we are not concerned to be as concise or logically elegant as we could be. On the other hand, we would like to have both of these conditions available for use later on, so we simply state them both as part of the defintion.

Fourth, condition 3 is sometimes described as saying that $x'(t_0)$ has rank p for all $t_0 \in U$. There is also a geometric meaning attached to this condition: It implies that "small" sets $K \subseteq U$ containing t_0 that have positive p-dimensional volume will be mapped to sets $x(K)$ in $\mathbb{R}^n$ that also have positive p-dimensional volume. This is an insight which would be more striking except for the fact that we have yet to define p-dimensional volume.

If $p = 1$, then U can be taken to be an interval and $x(U)$ a curve (possibly quite squiggly and twisty) in $\mathbb{R}^n$. We can think of 1-dimensional volume as length, and condition 3 says that an interval of positive length must map to a curve of positive length (though not necessarily the same length). If $p = 2$, then a U of positive area lying in $\mathbb{R}^2$ will map to an $x(U)$ of positive area in $\mathbb{R}^n$. For $p = 3$, p-dimensional volume is simply volume in the usual sense. We will explore the idea of p-dimensional volume for arbitrarily high p later.

We now introduce tangent vectors to a chart.

Definition 2.15. If $x \colon U \to \mathbb{R}^n$ is a $\mathcal{C}^k$ p-dimensional chart ($k \geq 1$) and $x_0 = x(t_0)$ is a point in $x(U)$, then we say that $a \in \mathbb{R}^n$ is a *tangent vector to the chart at the point x_0* provided there is a vector $b \in \mathbb{R}^p$ such that

$$a = [x'(t_0)]b = \partial_b x(t_0).$$

As part of this definition, we also introduce the notation $T_{x_0} x$ for the set of all tangent vectors to the chart x at the point x_0. This a temporary notation; we shall soon be rid of it. However we note the interesting fact that $T_{x_0} x$ is a vector space. It is, indeed, $[x'(t_0)](\mathbb{R}^p)$, the image of $\mathbb{R}^p$ under the linear transformation $x'(t_0)$.

It is not hard to see intuitively why we call a in the definition a *tangent* vector. Notice that

$$a = \partial_b x(t_0) = \lim_{\lambda \to 0} \frac{1}{\lambda} \left(x(t_0 + \lambda b) - x(t_0) \right). \tag{2.5}$$

We see that $v = x(t_0 + \lambda b) - x(t_0)$ is a vector running from the point $x_0 = x(t_0)$ to $x(t_0 + \lambda b)$, another point in $x(U)$. (Figure 2.16.) As $\lambda \to 0$, the point $x(t_0 + \lambda b)$

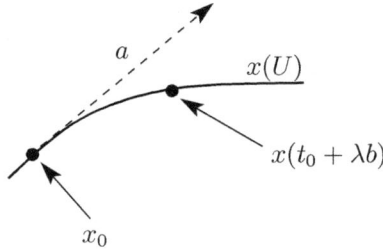

Figure 2.16: Tangent vector to chart at x_0

approaches x_0 and v becomes ever more nearly tangent to $x(U)$ at x_0. Unfortunately v approaches 0; however since we divide it by λ which also approaches 0, the limit can be both nonzero and a tangent vector to $x(U)$.

Notice how Definition 2.15 differs from Definition 2.12 which defined tangent vectors to a subset $\mathcal{M}$ of $\mathbb{R}^n$. Yet the two are really close together. If we define a curve c in $x(U)$ by $c(\lambda) = x(t_0 + \lambda b)$, then Equation (2.5) becomes

$$a = \lim_{\lambda \to 0} \frac{c(\lambda) - c(0)}{\lambda}$$

so that we see a is a tangent vector to the set $x(U)$ in the sense of Definition 2.12.

Before giving an example of tangent vectors, we note that if we have a chart $x \colon U \to \mathbb{R}^n$ and x induces coordinates $(\chi_1, \ldots \chi_p)$ on $x(U)$, then on the basis of Equation (2.4), we can use the notation

$$\frac{\partial x}{\partial \chi_i} = \partial_{e_i} x \tag{2.6}$$

for the tangent vector associated with a curve in $x(U)$ along which χ_i is increasing and all the other coordinates are constant. Now on the basis of (2.4), if x_0 is a point in $x(U)$ and $x_0 = x(t_0)$, then we should consider $\partial x / \partial \chi_i$ as a function of $t_0 \in U$. However it is convenient (as we see later) to treat $\partial x / \partial \chi_i$ as though it is defined on points of $x(U)$. We therefore abuse notation and write $(\partial x / \partial \chi_i)(x_0)$ when we really mean $(\partial x / \partial \chi_i)(t_0)$. In practice, this should not lead to confusion.

Example 2.13. We know that $\chi_1^2 + \chi_2^2 + \chi_3^2 = \rho^2$ (where ρ is a positive constant) is a sphere of radius ρ in $\mathbb{R}^3$. Let U be the set of points (χ_1, χ_2) in $\mathbb{R}^2$ such that $\chi_1^2 + \chi_2^2 < \rho^2$; this is an open set in $\mathbb{R}^2$, an open disk. Let us choose a point $(\chi_1, \chi_2) \in U$ and project the point $(\chi_1, \chi_2, 0)$ in the $\chi_1 \chi_2$-plane upward till it hits the point $x = (\chi_1, \chi_2, \chi_3)$ in the sphere. (See Figure 2.17.) This defines a chart $x \colon U \to \mathbb{R}^3$ where

$$x(\chi_1, \chi_2) = (\chi_1, \chi_2, \chi_3) \quad \text{and} \quad \chi_3 = \sqrt{\rho^2 - \chi_1^2 - \chi_2^2}.$$

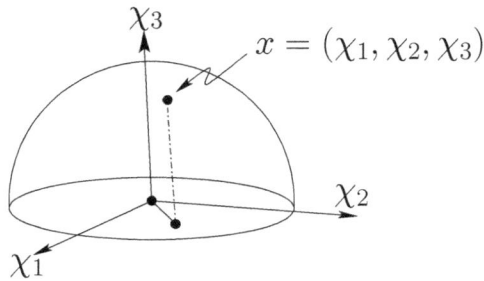

Figure 2.17: An upper half-sphere

We see that $x(U)$ is the (open) upper half-sphere. Then we note that for a point $x = (\chi_1, \chi_2, \chi_3)$ on $x(U)$, we have

$$\frac{\partial \chi_3}{\partial \chi_i} = -\frac{\chi_i}{\chi_3} \quad \text{for } i = 1, 2,$$

and we make use of the fact that we can write $x = \chi_1 e_1 + \chi_2 e_2 + \chi_3 e_3$ to compute

$$\frac{\partial x}{\partial \chi_i} = e_i - \frac{\chi_i}{\chi_3} e_3 \quad \text{for } i = 1, 2.$$

This gives us the two tangent vectors associated with the induced coordinates χ_1 and χ_2 on $x(U)$.

We now describe properties we want a manifold to have; this is the promised alternate description of a manifold which we shall use in these notes. These properties should be regarded as requirements we place on a manifold rather than a definition.

If we say that $\mathcal{M}$ is $\mathcal{C}^k$ p-dimensional manifold in $\mathbb{R}^n$, we assume the following:

1. $\mathcal{M}$ is a subset of $\mathbb{R}^n$.

2. Associated with each point $x_0 \in \mathcal{M}$, there is a collection of $\mathcal{C}^k$ p-charts $x \colon U \to \mathbb{R}^n$ such that the following further conditions hold for each such chart:

3. $x_0 \in x(U)$.

4. $x(U) \cap \mathcal{M}$ is an open subset of $\mathcal{M}$.

5. $T_{x_0} x = T_{x_0} \mathcal{M}$.

6. We say of such a chart x that it *covers* x_0 and that x is a *chart on or coordinate patch on or parametrization of* $\mathcal{M}$.

The intuition here is that a p-manifold is a thing that can be "covered" by "small pieces" of $\mathbb{R}^p$, patches or charts, and that the tangent vectors to the patches can all be generated by curves in $\mathcal{M}$.

Because of the condition $T_{x_0}x = T_{x_0}\mathcal{M}$, we shall, from this point on, use only the notation $T_{x_0}\mathcal{M}$ to talk about tangent spaces.

The condition that $x(U) \cap \mathcal{M}$ is an open subset of $\mathcal{M}$ means that we can do the following:

Suppose U_0 is an open subset of U such that we still have $x_0 \in x(U_0)$. Then if we look at the restriction of x to U_0, it follows that $x|_{U_0} : U_0 \to \mathbb{R}^n$ will still be a chart on $\mathcal{M}$ and it will be one that still covers x_0. This has the sometimes useful implication that one can always take a chart that covers a point x_0 on a manifold and "shrink" it to a chart that is as small as one wishes.

There are two different kinds of points in a manifold:

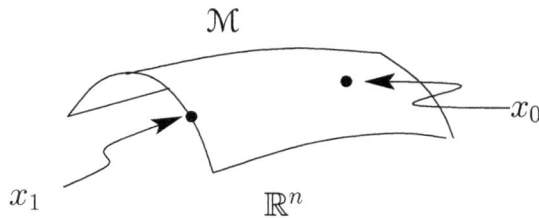

Figure 2.18: Example of interior and boundary points

We say that x_0 is an *interior point of* $\mathcal{M}$ if there is a chart $x : U \to \mathbb{R}^n$ that covers x_0 and $x(U) \subseteq \mathcal{M}$. If there is no such chart—that is, if every chart x that covers x_0 has the property that $x(U) \not\subseteq \mathcal{M}$—then we say x_0 is a *boundary point of* $\mathcal{M}$. We denote the set of boundary points of a manifold as $\partial\mathcal{M}$.

Example 2.14. From Definition 2.11, we know that we can think of a 2-cell $\mathcal{M}$ in $\mathbb{R}^n$ as the image of $x : \mathcal{I}^2 \to \mathbb{R}^n$. All of the points $(\tau_1, \tau_2) \in \mathcal{I}^2$ such that $\tau_i = 0$ or 1 for some i will map under x to a boundary point of $\mathcal{M}$. All points such that $0 < \tau_i < 1$ for both values of i will map to an interior point of the cell. In Figure 2.18, for example, we see that x_0 is an interior point of the 2-cell $\mathcal{M}$ while x_1 is a boundary point.

In conclusion, we wish to say something about directional derivatives and manifolds.

Suppose that ϕ is a real-valued function which is at least $\mathcal{C}^1$ and is defined on the manifold $\mathcal{M}$. The definition of the directional derivative, $\partial_v \phi(x_0)$ is given in Equation (2.2). However if v is a tangent vector to $\mathcal{M}$ at x_0, there is another way to define the directional derivative that is attractive because it is carried out entirely within $\mathcal{M}$. It goes like this:

Suppose we can find a curve or arc in $\mathcal{M}$, $c : J \to \mathcal{M}$, where J is an interval in the reals, $x_0 = c(\tau_0)$, and $c'(\tau_0) = v$. Then it seems reasonable to expect the directional derivative of ϕ at x_0 in the direction v to be

$$\lim_{\tau \to \tau_0} \frac{\phi(c(\tau)) - \phi(x_0)}{\tau - \tau_0}. \tag{2.7}$$

However Equation (2.2), when recalled, is

$$\partial_v\phi(x_0) = \lim_{\lambda \to 0} \frac{\phi(x_0 + \lambda v) - \phi(x_0)}{\lambda}, \tag{2.8}$$

and this has the property that points of the form $x_0 + \lambda v$ may not be in $\mathcal{M}$ at all. There is no problem about evaluating ϕ at such points, because ϕ is at least $\mathcal{C}^1$; thus the domain of ϕ can be extended to some "small" open neighborhood of x_0 in $\mathbb{R}^n$ and $x_0 + \lambda v$ will lie in such an neighborhood for sufficiently small values of λ.

It is a useful fact—whose proof we give in Section A.1 of Appendix A—that these two different concepts of the directional derivative yield the same result:

Proposition 2.7. *Suppose ϕ is a real-valued $\mathcal{C}^1$ function defined on an open subset of $\mathbb{R}^n$. Let x_0 be a point in the domain of ϕ and v be a vector in $\mathbb{R}^n$. Let $c\colon J \to \mathbb{R}^n$ be a $\mathcal{C}^1$ curve in $\mathbb{R}^n$ such that $c(\tau_0) = x_0$ and $c'(\tau_0) = v$. Then (2.7) and (2.8) both give the same result.*

Exercises 2.3.

1. The manifold $\mathcal{M}$ is a helix in $\mathbb{R}^3$ with a chart

   $$x(\tau) = \cos(\tau)e_1 + \sin(\tau)e_2 + \tau e_3.$$

   (a) Sketch a picture of $\mathcal{M}$.
   (b) Compute the tangent vector $\partial x / \partial \tau$ (see Equation (2.6)) as a function of τ.
   (c) Give a basis for $T_{x_0}\mathcal{M}$ at the point $x_0 = (1/\sqrt{2}, 1/\sqrt{2}, \pi/4) \in \mathcal{M}$.

2. Let $\mathcal{M}$ be the 2-manifold in $\mathbb{R}^3$ consisting of the points (χ_1, χ_2, χ_3) satisfying $\chi_3 = \chi_1^2 + \chi_2^2$.

   (a) Find a $\mathcal{C}^\infty$ chart $x\colon \mathbb{R}^2 \to \mathcal{M}$ for $\mathcal{M}$.
   (b) Find a basis for $T_{x_0}\mathcal{M}$ at the point $x_0 = (1/2, 1/2, 1/2) \in \mathcal{M}$.

3. If $\phi(\chi_1, \dots, \chi_n)$ is a $\mathcal{C}^1$ real-valued function on $\mathbb{R}^n$, then the set of points in $\mathbb{R}^{n+1}$, $(\chi_1, \dots, \chi_n, \chi_{n+1})$, which satisfy $\chi_{n+1} = \phi(\chi_1, \dots, \chi_n)$, is a $\mathcal{C}^1$ n-manifold $\mathcal{M}$. Show that given a point $x_0 = (\chi_{01}, \dots, \chi_{0,n+1})$ on $\mathcal{M}$, then a basis for $T_{x_0}\mathcal{M}$ is

   $$\left\{ e_i + \left(\frac{\partial \phi}{\partial \chi_i}(\chi_{01}, \dots, \chi_{0n}) \right) e_{n+1} \ : \ i = 1, \dots, n \right\}.$$

4. Let $\mathcal{M}$ be the 1-manifold in $\mathbb{R}^2$ defined by the equation

   $$\frac{\chi_1^2}{\alpha_1^2} + \frac{\chi_2^2}{\alpha_2^2} = 1$$

   where α_1, α_2 are positive real constants. Find a collection of $\mathcal{C}^\infty$ charts $x\colon U \to \mathcal{M}$ which cover $\mathcal{M}$; that is, find charts such that given any $x_0 \in \mathcal{M}$ there will be at least one chart x with the property that $x_0 \in x(U)$.

Chapter 3

Simple k-vectors

This and the succeeding chapter introduce the *wedge product* or *exterior product*. This is usually done in an abstract algebraic setting, but we choose a geometric one because we think a pictorial understanding can be an aid to intuition and to application when using this mathematical machinery.

A device that is sometimes used to introduce vectors is to describe them as directed line segments, or, more precisely, as equivalence classes of directed line segments. Two directed line segments a and b are said to represent the same vector provided they have the same direction and same length. Thus in Figure 3.1, a and b represent the same vector, but a, c, d, and e all represent different vectors.

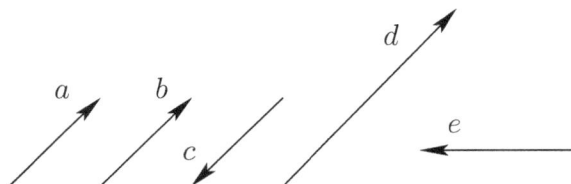

Figure 3.1: Vectors visualized as directed line segments

It turns out that not only directed line segments represent vectors. We can also think of equivalence classes of k-dimensional parallelepipeds as representing new kinds of "vectors." We will say that two parallelepipeds represent the same "vector" provided they have the same *volume* and *orientation*. Since we may not know what these terms mean for arbitrary dimensions, our first concern will be to define them.

These oriented parallelepipeds will ultimately be seen to be examples of something that we will call *k-vectors*. We will be able to add them, multiply by scalars, take "dot" products, and multiply them using an operation called the *wedge product* or *exterior product* that is a kind of generalization of the cross product of vectors in $\mathbb{R}^3$. This construction is marvelously useful in working with geometry and calculus in spaces of arbitrarily high dimension.

Our presentation in this chapter will often be more descriptive than deductive. That is, we will often state results without proofs, though proofs will be supplied in the appendices. Details of the construction of k-vectors and the wedge product can be found in many places; for example, [4], [25], [27], [31], [34], [35], and [43].

3.1   Volume

What should we mean by the *volume* of a p-dimensional parallelepiped? Certain answers are obvious: If $a = (\alpha_1, \ldots, \alpha_n)$ is a vector in $\mathbb{R}^n$, then we take its "volume" to be its length, namely, $|a| = \sqrt{\sum_{i=1}^{n} \alpha_i^2}$. More generally, if $a_1, \ldots, a_p$ are *orthogonal* vectors, then they determine a rectangular "box" $\mathcal{A}$, (Figure 3.2) and we would expect its p-dimensional volume to be $|a_1| \cdots |a_p|$, that is, the product of the lengths of the sides.

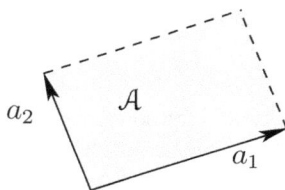

Figure 3.2: Two-dimensional "volume" of a rectangular box is $|a_1| \, |a_2|$

How do we generalize the notion of "volume" to a parallelepiped of arbitrarily high dimension including the case where the edges might not be orthogonal? We claim that the following is a good way to do this:

Definition 3.1. If $\mathcal{A}$ is a p-dimensional parallelepiped with edges $a_1, \ldots, a_p$, we set its *p-dimensional volume* to

$$\mathrm{vol}(\mathcal{A}) \stackrel{\text{def.}}{=} \sqrt{\det \begin{pmatrix} a_1 \cdot a_1 & \cdots & a_1 \cdot a_p \\ & \cdots & \\ a_p \cdot a_1 & \cdots & a_p \cdot a_p \end{pmatrix}}.$$

We also feel free to write $\mathrm{vol}(a_1, \ldots, a_p) = \mathrm{vol}(\mathcal{A})$.

An obvious first question is, why would we expect this definition to give us volume? A second question is this: Can we be sure that $\det(a_i \cdot a_j) \geq 0$? Does it always make sense to take the square root?

We do not answer the first question, but we give some examples which will hopefully assure the reader we are on the right track:

Example 3.1. Let a be a vector in $\mathbb{R}^n$. Then we form the 1×1 matrix $(a \cdot a)$ and take the square root of its determinant:

$$\sqrt{\det(a \cdot a)} = \sqrt{\det(|a|^2)} = \sqrt{|a|^2} = |a|.$$

This is the length of a.

Example 3.2. Let $\mathcal{A}$ be a parallelepiped with orthogonal edges $a_1, \ldots, a_p$. Since the vectors are orthogonal, we have

$$a_i \cdot a_j = \begin{cases} |a_i|^2 & \text{if } i = j, \\ 0 & \text{if } i \neq j. \end{cases}$$

It follows that

$$\det(a_i \cdot a_j) = \det \begin{pmatrix} |a_1|^2 & 0 & \cdots & 0 \\ 0 & |a_2|^2 & \cdots & 0 \\ \cdots & & & \\ 0 & \cdots & 0 & |a_p|^2 \end{pmatrix} = |a_1|^2 \cdots |a_p|^2.$$

So $\sqrt{\det(a_i \cdot a_j)} = |a_1| \cdots |a_p|$, the product of the lengths of the sides of $\mathcal{A}$.

Example 3.3. Let $\mathcal{A}$ be a parallelogram in $\mathbb{R}^n$ with edges a_1 and a_2 and angle θ between a_1 and a_2 (where $0 \leq \theta \leq \pi$).

Now a_1 and a_2 lie in some 2-dimensional subspace U of $\mathbb{R}^n$. Let us choose an orthonormal basis $\{u_1, u_2\}$ of U with the property that u_1 points in the same direction as a_1. That is, we may assume $a_1 = |a_1| u_1$. We may also suppose that $a_2 = |a_2| \big((\cos \theta) u_1 + (\sin \theta) u_2 \big)$. We readily see from Figure 3.3 and our knowledge of elementary geometry and trigonometry that the area of $\mathcal{A}$, that is, its 2-dimensional volume, should be $|a_1| |a_2| \sin \theta$.

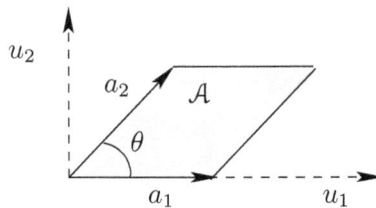

Figure 3.3: Parallelogram $\mathcal{A}$ in $\mathbb{R}^n$

We want to check that our new definition of p-dimensional volume gives us the same answer. We calculate

$$\det \begin{pmatrix} a_1 \cdot a_1 & a_1 \cdot a_2 \\ a_2 \cdot a_1 & a_2 \cdot a_2 \end{pmatrix} = \det \begin{pmatrix} |a_1|^2 & |a_1| |a_2| \cos \theta \\ |a_1| |a_2| \cos \theta & |a_2|^2 \end{pmatrix}$$
$$= |a_1|^2 |a_2|^2 - |a_1|^2 |a_2|^2 \cos^2 \theta$$

$$= |a_1|^2 |a_2|^2 \sin^2 \theta.$$

That is, $\sqrt{\det(a_i \cdot a_j)} = |a_1| |a_2| \sin \theta$, precisely as desired.

The reader who is interested in seeing a greater justification of our definition of volume should consult [40] (look under Gram determinant) or [35].

We now settle the question of the sign of $\det(a_i \cdot a_j)$.

Proposition 3.1. *If $a_1, \ldots, a_p$ are vectors in $\mathbb{R}^n$ and A is the $p \times p$ matrix $(a_i \cdot a_j)$, then $\det A \geq 0$.*

Proof. The vectors $a_1, \ldots, a_p$ must lie in some p-dimensional subspace U of $\mathbb{R}^n$. Let $\{u_i\}_{i=1}^P$ be an orthonormal basis for U. Let us write each a_i in terms of this basis, thus: $a_i = \sum_{j=1}^P \alpha_{ij} u_j$. Now form the $p \times p$ matrix

$$B = (a_1 \ldots a_p) = \begin{pmatrix} \alpha_{11} & \cdots & \alpha_{p1} \\ \cdots & & \\ \alpha_{1p} & \cdots & \alpha_{pp} \end{pmatrix}$$

where $a_1, \ldots, a_p$ are written as column vectors in terms of the basis $\{u_i\}_{i=1}^P$. It follows that

$$B^T B = \begin{pmatrix} a_1 \cdot a_1 & \cdots & a_1 \cdot a_p \\ \cdots & & \\ a_p \cdot a_1 & \cdots & a_p \cdot a_p \end{pmatrix} = A$$

where B^T is the transpose of B. Since we are dealing with square matrices, we have

$$\det A = \det(B^T B) = \left(\det B \right)^2 \geq 0. \qquad \square$$

Exercises 3.1.

1. The volume of an n-dimensional parallelepiped in $\mathbb{R}^n$ can be found by a simpler formula than that of Definition 3.1. If $a_1, \ldots, a_n \in \mathbb{R}^n$, show that $\mathrm{vol}(a_1, \ldots, a_n) = |\det(a_1, \ldots, a_n)|$.

2. In $\mathbb{R}^3$, if $a_1 = e_1$ and $a_2 = e_2 + e_3$, compute the 2-volume (area) of the parallelogram (a_1, a_2).

3. If $a_i = e_1 + \cdots + e_i$, then what is $\mathrm{vol}(a_1, \ldots, a_k)$?

4. If $a(\theta) = \cos(\theta) e_1 + \sin(\theta) e_2$, what is $\mathrm{vol}(e_1, a(\theta))$?

5. A parallelogram $\mathcal{A}$ in $\mathbb{R}^4$ has edges

$$\begin{aligned} a_1 &= \beta e_1 + \beta e_2 + \beta e_3 + \beta e_4, \\ a_2 &= \beta e_1 - \beta e_2 + \beta e_3 - \beta e_4. \end{aligned}$$

Find the 2-volume (area) of $\mathcal{A}$ in terms of β.

6. Recall or look up the definition of vector product $a_1 \times a_2$ in $\mathbb{R}^3$. Show that if $a_1, a_2 \in \mathbb{R}^3$, then $|a_1 \times a_2| = \text{vol}(a_1, a_2)$.

7. Let $a_1, \ldots, a_k \in \mathbb{R}^n$ where $1 \leq k \leq n$. Form the $n \times k$ matrix $A = (a_1, \ldots, a_k)$ with respect to some fixed basis of $\mathbb{R}^n$. Show that

$$\text{vol}(a_1, \ldots, a_k) = \sqrt{\det(A^T A)}.$$

(Hint: You cannot take the determinant of A because it is not a square matrix.)

3.2 Orientation

We now need the idea of *orientation* for a parallelepiped. Actually, we shall not define what we mean by orientation; rather we shall develop a precise idea of having the *same* and *opposite* orientation.

Consider the directed line segments in Figure 3.1. Intuitively, we see that a, b, and d have the same orientation, that c has the opposite orientation to a, b, and d. Because it is not parallel to them, the orientation of e is not comparable to that of a, b, c, and d.

If we translate the directed line segments to the origin and think of them as vectors, we can say that a, b, and d are comparable because they all lie in a common 1-dimensional vector subspace V of $\mathbb{R}^n$ while e is non-comparable because it is not in V.

Also, a, b, and d have the same orientation because $a \cdot b > 0$, $a \cdot d > 0$, and $b \cdot d > 0$; while c has the opposite orientation because each of the dot products $a \cdot c$, $b \cdot c$, and $d \cdot c$ is negative.

We generalize this idea to ordered k-tuples of vectors:

Definition 3.2. Let $(a_1, \ldots, a_k)$ and $(b_1, \ldots, b_k)$ be two ordered k-tuples of vectors from $\mathbb{R}^n$.

1. If $\{a_1, \ldots, a_k\}$ and $\{b_1, \ldots, b_k\}$ are both linearly dependent sets, then we say that $(a_1, \ldots, a_k)$ and $(b_1, \ldots, b_k)$ are considered to have the same orientation, the 0-*orientation*. If $\{a_1, \ldots, a_k\}$ is linearly independent, then we say that $(a_1, \ldots, a_k)$ has *nonzero* orientation.

2. Suppose $\{a_1, \ldots, a_k\}$ and $\{b_1, \ldots, b_k\}$ are both linearly independent sets and lie in the same k-dimensional vector subspace V of $\mathbb{R}^n$. Then $(a_1, \ldots, a_k)$ and $(b_1, \ldots, b_k)$ have the *same orientation* provided $\det(a_i \cdot b_j)_{k \times k} > 0$. If, on the other hand, $\det(a_i \cdot b_j)_{k \times k} < 0$, they have *opposite orientations*. (Note: It is easily seen that we cannot have $\det(a_i \cdot b_j) = 0$.)

3. In all other circumstances, the orientations of the two k-tuples are considered *non-comparable*.

We can think of orientation in two and three dimensions as being the quality of *right-handedness* or *left-handedness*. Let us draw some pictures of this in two dimensions.

Let a_1 and a_2 be two vectors lying in a plane in some $\mathbb{R}^n$. The vectors have an angle of θ between them, and to make life very simple, we can imagine that there is a coordinate system in the plane and that in that system, a_1 is the same thing as e_1. If a_2 is also e_2, that is, $\theta = \pi/2$, then there is a certain "handedness" to the ordered pair of vectors (a_1, a_2).

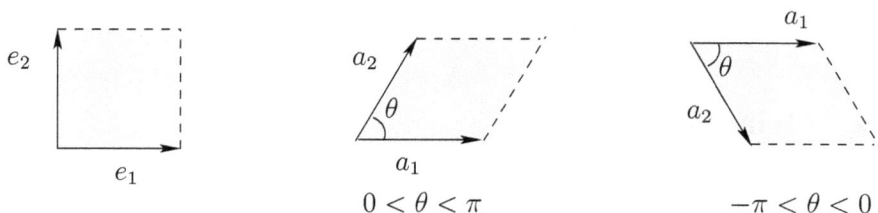

$$0 < \theta < \pi \qquad\qquad -\pi < \theta < 0$$

Figure 3.4: Having the same and opposite "handedness" as (e_1, e_2)

If we look at Figure 3.4, we see that when $0 < \theta < \pi$, the ordered pair of vectors (a_1, a_2) has the same "handedness" as (e_1, e_2). Referring to Definition 3.2, we calculate that

$$\det \begin{pmatrix} a_1 \cdot e_1 & a_1 \cdot e_2 \\ a_2 \cdot e_1 & a_2 \cdot e_2 \end{pmatrix} = a_2 \cdot e_2 > 0$$

so that in accord with our definition, (a_1, a_2) and (e_1, e_2) have the same orientation. That is, our (intuitive) notion of having the same "handedness" and (mathematical) definition of having the same orientation are in agreement.

On the other hand, $a_2 \cdot e_2 < 0$ when $-\pi < \theta < 0$, so it is easily seen that (a_1, a_2) and (e_1, e_2) have opposite orientations in this case.

We recall that an *equivalence relation* on a set X is a relation $\sim$ satisfying the following for all $x, y, z \in X$:

1. $x \sim x$.

2. If $x \sim y$, then $y \sim x$.

3. If $x \sim y$ and $y \sim z$, then $x \sim z$.

Not surprisingly, as the following proposition attests, having the same orientation is an equivalence relation:

Proposition 3.2. *Suppose that $\{a_1, \ldots, a_k\}$, $\{b_1, \ldots, b_k\}$, and $\{c_1, \ldots, c_k\}$ are sets of vectors in $\mathbb{R}^n$. Then the following hold:*

1. $\det(a_i \cdot a_j)_{k \times k} \geq 0$.

2. If $(a_1, \ldots, a_k)$ and $(b_1, \ldots, b_k)$ have the same orientation, then $(b_1, \ldots, b_k)$ and $(a_1, \ldots, a_k)$ have the same orientation.

3. *If* $(a_1,\ldots,a_k)$ *and* $(b_1,\ldots,b_k)$ *have the same orientation and if* $(b_1,\ldots,b_k)$ *and* $(c_1,\ldots,c_k)$ *have the same orientation, then* $(a_1,\ldots,a_k)$ *and* $(c_1,\ldots,c_k)$ *have the same orientation.*

We leave the proof as an exercise. (Hint: Consider Exercise 1.)

Exercises 3.2.

1. Show that if $a_1,\ldots,a_k$ and $b_1,\ldots,b_k$ are vectors in a k-dimensional subspace V of $\mathbb{R}^n$ and we write the $k \times k$ matrices $A = (a_1,\ldots,a_k)$ and $B = (b_1,\ldots,b_k)$ in terms of an orthonormal basis of V, then $\det(a_i \cdot b_j)_{k \times k} = \det(A^T B)$.

2. Suppose $\{a_1,\ldots,a_k\}$ and $\{b_1,\ldots,b_k\}$ are both linearly independent sets and lie in the same k-dimensional vector subspace V of $\mathbb{R}^n$. Show that $\det(a_i \cdot b_j)_{k \times k} \neq 0$.

3. Do (e_1, e_2) and $(e_1 + e_2, e_1 - e_2)$ have the same orientation?

4. If $a_1, a_2 \in \mathbb{R}^n$ and

$$
\begin{aligned}
b_1 &= a_1 \cos(\theta) + a_2 \sin(\theta), \\
b_2 &= -a_1 \sin(\theta) + a_2 \cos(\theta),
\end{aligned}
$$

how do the orientations of (a_1, a_2) and (b_1, b_2) compare?

5. Suppose that $a_k = \sum_{i=1}^k e_i$. Compare the orientations of the ordered m-tuples $(e_1,\ldots,e_m)$ and $(a_1,\ldots,a_m)$.

6. Suppose a_1, a_2 are linearly independent vectors in $\mathbb{R}^2$. Do (a_1, a_2) and $(a_1, a_1 + a_2)$ have the same orientation?

7. Do the ordered pairs $(e_1 + e_3, e_2 - e_3)$ and $(-e_1 + e_2 + e_3, e_1 + e_2)$ have the same orientation?

8. Suppose that a_1, a_2 are vectors in $\mathbb{R}^n$ and λ is a nonzero scalar. How do the orientations of (a_1, a_2) and $(\lambda a_1, a_2)$ compare?

9. Prove Proposition 3.2. (Hint: Recall that for square matrices we have $\det(A^T) = \det(A)$ and $\det(AB) = \det(A)\det(B)$.)

3.3   Definition of a simple k-vector

We take the intuitive notion of vectors as equivalence classes of directed line segments and extend it to 2-, 3-, and higher-dimensional objects. Since we identify the parallelepiped with its edges written as a sequence, we give the definition for ordered sequences of vectors.

Definition 3.3. By the *simple k-vector* $a_1 \wedge \cdots \wedge a_k$, where $a_1, \ldots, a_k \in \mathbb{R}^n$ and $k \geq 1$, we mean the set of all ordered k-tuples $(b_1, \ldots, b_k)$ such that $(a_1, \ldots, a_k)$ and $(b_1, \ldots, b_k)$ have the same orientation and volume. If $(a_1, \ldots, a_k)$ has the 0-orientation, then we write $a_1 \wedge \cdots \wedge a_k = 0$. We take $\mathbb{R}$ to be the set of simple 0-vectors.

It follows from Proposition 3.3 (to appear in a moment) that $a_1 \wedge \cdots \wedge a_k = 0$ is equivalent to $\mathrm{vol}(a_1, \ldots, a_k) = 0$.

Example 3.4. Consider various oriented parallelograms lying in a common plane as in Figure 3.5. The orientations are represented by curved arrows.

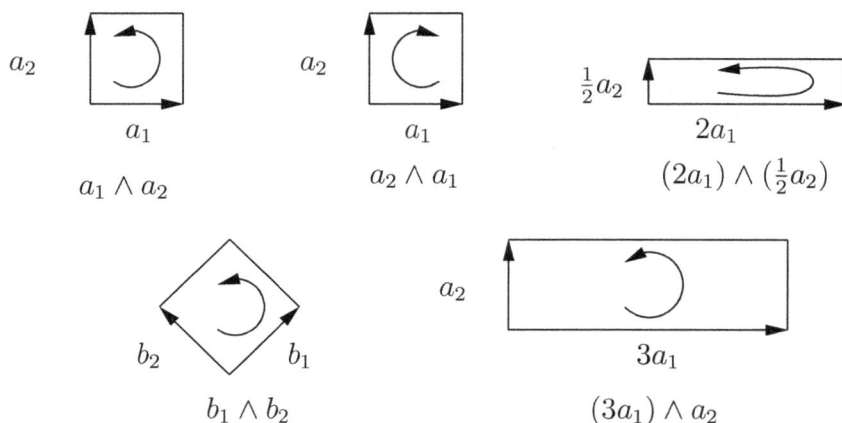

Figure 3.5: Oriented parallelograms, some equivalent, some not

Some of the oriented parallelgrams are equivalent, and represent the same simple 2-vector. Some are not. We have $a_1 \wedge a_2 = (2a_1) \wedge (\frac{1}{2}a_2) = b_1 \wedge b_2$. We may think of $b_1 \wedge b_2$ as obtained from $a_1 \wedge a_2$ by a rotation in the plane; neither orientation nor area is changed. On the other hand, $a_1 \wedge a_2 \neq a_2 \wedge a_1$ or $(3a_1) \wedge a_2$— in the first case because of different orientations and in the second because of different areas.

Example 3.5. In $\mathbb{R}^3$ we see that $e_1 \wedge e_2 \neq e_1 \wedge e_3$ since these simple 2-vectors do not lie in a common 2-dimensional vector subspace.

Remark 3.1. Here is an important point: We use the word *vector* in the phrase *simple k-vector*, yet simple k-vectors do *not* belong to a vector space. (At least not yet.) If one is dealing with objects that live in a vector space, then one has to be able to multiply them by scalars and add them. We have no way to do either of those things with simple k-vectors. In the next chapter, we shall remedy this deficiency and show how to "embed" simple k-vectors in a vector space.

Here is an useful connection between the volume of a simple k-vector and linear independence:

Proposition 3.3. *Suppose $a_1, \ldots, a_k \in \mathbb{R}^n$. Then the following are equivalent:*

1. *$vol(a_1, \ldots, a_k) > 0$.*

2. *$a_1 \wedge \cdots \wedge a_k \neq 0$.*

3. *$a_1, \ldots, a_k$ are linearly independent.*

The proof is in Appendix A.2.

Another way to think of the equivalence of such parallelepipeds as $(a_1, \ldots, a_k)$ and $(b_1, \ldots, b_k)$ is that there is a linear transformation f that carries each edge a_i to the edge b_i and this transformation satisfies $\det(f) = 1$. That is $\det(f) = 1$ amounts to f being volume and orientation-preserving.

For example, if we take $(a_1, a_2) = (e_1, e_2)$ and $(b_1, b_2) = (2e_1, (1/2)e_2)$, then the map given by $f(\chi_1, \chi_2) = (2\chi_1, (1/2)\chi_2)$ takes each a_i to b_i. (Figure 3.6.) It is clear

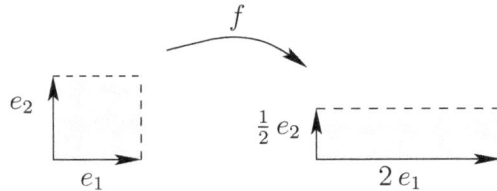

Figure 3.6: Transformation between representatives of 2-vector

in this case that $a_1 \wedge a_2 = b_1 \wedge b_2$. We see that the matrix of f in terms of the standard basis is

$$[f] = \begin{pmatrix} 2 & 0 \\ 0 & \frac{1}{2} \end{pmatrix}$$

and that $\det(f) = 1$.

Here is a general and formal statement of this property:

Proposition 3.4. *Let $\{a_1, \ldots, a_k\}$ and $\{b_1, \ldots, b_k\}$ be sets of linearly independent vectors in $\mathbb{R}^n$ that span the same vector subspace V. Let A and B be the $k \times k$ matrices $A = (a_1, \ldots, a_k)$ and $B = (b_1, \ldots, b_k)$ in terms of some orthonormal basis of V. Let $f : V \to V$ be the unique linear transformation satisfying $f(a_i) = b_i$ for all i. Then the following are equivalent:*

1. *$a_1 \wedge \cdots \wedge a_k = b_1 \wedge \cdots \wedge b_k$.*

2. *$\det A = \det B$.*

3. *$\det f = 1$.*

The proof is given in Appendix A.2.

Exercises 3.3.

1. Show that if $a_1 \wedge \cdots \wedge a_k \neq 0$ and $b_1 \wedge \cdots \wedge b_k \neq 0$ and at the same time $a_1 \wedge \cdots \wedge a_k = b_1 \wedge \cdots \wedge b_k \neq 0$, then $\{a_i\}_{i=1}^k$ and $\{b_i\}_{i=1}^k$ both span the same k-dimensional subspace.

2. If a_1, a_2 are linearly independent vectors in $\mathbb{R}^n$, do we have $a_1 \wedge a_2 = (a_1 + a_2) \wedge a_2$? What if a_1 and a_2 are dependent?

3. Consider the rotation of $\mathbb{R}^2$ by an angle of θ. If we apply this rotation to a vector $a = \alpha_1 e_1 + \alpha_2 e_2$ to obtain a vector $b = \beta_1 e_1 + \beta_2 e_2$, then in terms of matrices and the standard basis, this is

$$\begin{pmatrix} \beta_1 \\ \beta_2 \end{pmatrix} = \begin{pmatrix} \cos(\theta) & -\sin(\theta) \\ \sin(\theta) & \cos(\theta) \end{pmatrix} \begin{pmatrix} \alpha_1 \\ \alpha_2 \end{pmatrix}.$$

Show that if b_i is obtained from a_i by such a rotation, then $b_1 \wedge b_2 = a_1 \wedge a_2$.

3.4 Operations with simple k-vectors

We present several useful operations that can be carried out on simple k-vectors. These are defined on representations of the simple k-vectors, not directly on the equivalence class which is what we take to be the formal definition of the simple k-vector, and because of this, one must always worry about whether or not the operation is *well-defined*. We do not check on such questions here in this section but instead show in Appendix A.2 that these operations are well-defined.

3.4.1 Scalar multiplication

Suppose we replace a parallelogram by one in which one of the edges has been "stretched" by a factor of λ. In Figure 3.7, we construct from $a_1 \wedge a_2$ the simple 2-vectors $(\lambda a_1) \wedge a_2$ and $a_1 \wedge (\lambda a_2)$. It seems clear pictorially that these should be equivalent parallelograms, having the same area and orientation.

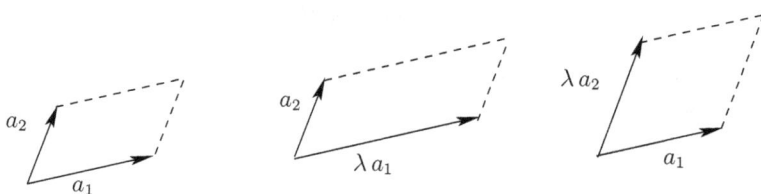

Figure 3.7: Multiplying edges of a parallelogram by λ

The idea of replacing one of the vectors a_i in a simple k-vector $a_1 \wedge \cdots \wedge a_k$ by λa_i where λ is a scalar is both natural and useful. The following is proved in Appendix A.2:

Proposition 3.5. *Let $a_1, \ldots, a_k$ be vectors in $\mathbb{R}^n$ and λ be a scalar. Set*

$$a = a_1 \wedge \cdots \wedge a_i \wedge \cdots \wedge a_k,$$
$$b = a_1 \wedge \cdots \wedge \lambda a_i \wedge \cdots \wedge a_k.$$

Then the following hold:

1. *If $\lambda = 0$, then $b = 0$.*

2. *$vol(b) = |\lambda| vol(a)$.*

3. *If $\lambda \neq 0$, then $a \neq 0$ if and only if $b \neq 0$.*

4. *Suppose $a, b \neq 0$. Then a and b have the same orientation if and only if $\lambda > 0$. If $\lambda < 0$, then they have opposite orientations.*

The fact that we have this result suggests the following definition:

Definition 3.4. If $a_1, \ldots, a_k$ are vectors in $\mathbb{R}^n$ and λ is a scalar, then we define multiplication of the simple k-vector $a_1 \wedge \cdots \wedge a_k$ by λ by

$$\lambda (a_1 \wedge \cdots \wedge a_k) \stackrel{\text{def.}}{=} a_1 \wedge \cdots \wedge (\lambda a_i) \wedge \cdots \wedge a_k$$

where i can assume any of the values $1, \ldots, k$.

Of course we have to show this operation is well-defined.

When we raise this question, we have in mind the following thought: Suppose we know that $a_1 \wedge a_2 = b_1 \wedge b_2$; that is, the oriented parallelograms (a_1, a_2) and (b_1, b_2) both represent the same simple 2-vector. How can we be sure that, for example, $(\lambda a_1) \wedge a_2 = b_1 \wedge (\lambda b_2)$? Proposition 3.5 makes this easy to prove:

Proposition 3.6. *Let $(a_1, \ldots, a_k)$ and $(b_1, \ldots, b_k)$ be two sequences of vectors in $\mathbb{R}^n$ that represent the same simple k-vector. That is, $a_1 \wedge \cdots \wedge a_k = b_1 \wedge \cdots \wedge b_k$. Let λ be a real number. If we replace a_i by λa_i and b_j by λb_j in their respective simple k-vectors, then we have*

$$a_1 \wedge \cdots \wedge \lambda a_i \wedge \cdots \wedge a_k = b_1 \wedge \cdots \wedge \lambda b_j \wedge \cdots \wedge b_k.$$

Proof. This is a matter of simply checking the various cases. For example, suppose $a_1 \wedge \cdots \wedge a_k = b_1 \wedge \cdots \wedge b_k \neq 0$ and $\lambda > 0$. Set

$$a = a_1 \wedge \cdots \wedge a_k,$$
$$b = b_1 \wedge \cdots \wedge b_k,$$
$$a' = a_1 \wedge \cdots \wedge \lambda a_i \wedge \cdots \wedge a_k,$$
$$b' = b_1 \wedge \cdots \wedge \lambda b_j \wedge \cdots \wedge b_k.$$

We know that $a_1 \wedge \cdots \wedge a_k$ and $b_1 \wedge \cdots \wedge b_k$ have the same volume and orientation. By Proposition 3.5,

$$\text{vol}(a') = |\lambda| \text{vol}(a) = |\lambda| \text{vol}(b) = \text{vol}(b').$$

Similarly, a', a, b, and b' all have the same orientation. Thus $a' = b'$. The other cases work the same way. $\square$

It turns out that switching any two factors of $a = a_1 \wedge \cdots \wedge a_k$ introduces a scalar factor of -1 and thus a change of orientation. (We prove this in Proposition A.1 of Appendix A.2.) So for a simple 2-vector,

$$a_1 \wedge a_2 = (-1) a_2 \wedge a_1.$$

Example 3.6. Recall that $\{e_i\}_{i=1}^n$ is the standard basis of $\mathbb{R}^n$ and consider the 2-vectors $e_2 \wedge e_1$ and $(-1)e_1 \wedge e_2$. They lie in the same plane, namely $\mathrm{span}(e_1, e_2)$, so we need merely compare their volumes and orientations. By Proposition 3.6, we may take $(-1)e_1 \wedge e_2$ to be $(-e_1) \wedge e_2$. It is easily checked from the definition of volume that

$$\mathrm{vol}(e_2, e_1) = \mathrm{vol}(-e_1, e_2) = 1.$$

Comparing orientations, we see that

$$\det \begin{pmatrix} e_2 \cdot (-e_1) & e_1 \cdot (-e_1) \\ e_2 \cdot e_2 & e_1 \cdot e_2 \end{pmatrix} = 1,$$

so the orientations are the same. Therefore $e_2 \wedge e_1 = -e_1 \wedge e_2$.

Notation: We will, from this point on, adopt the sensible policy of writing $-a$ for $(-1)a$ whenever a is a simple k-vector. We see that if $a \neq 0$, then a and $-a$ have the same volume and opposite orientations.

For a general statement of what happens when one changes the order of factors of a simple k-vector, it is useful to have the idea of the sign of a permutation:

Suppose $\sigma = (i_1, \ldots, i_k)$ is a permutation or rearrangement of $(1, 2, \ldots, k)$. Clearly every such permutation or rearrangement can be carried out by a finite number of interchanges of indices, two at a time. Thus if we want to go from $(1, 2, 3, 4)$ to $(4, 3, 2, 1)$, we can carry out the sequence of interchanges

$$(1,2,3,4) \mapsto (1,2,4,3) \mapsto (1,4,2,3) \mapsto$$
$$(4,1,2,3) \mapsto (4,2,1,3) \mapsto (4,2,3,1)$$

or the sequence

$$(1,2,3,4) \mapsto (4,2,3,1) \mapsto (4,3,2,1)$$

or several others.

It is a result of abstract algebra that the number of interchanges to get from one arrangement of indices to another is either always even or always odd. If the number of interchanges from $(1, 2, \ldots, k)$ to $(i_1, \ldots, i_k)$ is even, then we call $\sigma = (i_1, \ldots, i_k)$ an *even permutation*. If the number of interchanges is odd, then σ is an *odd permutation*. By *the sign of* σ, we mean

$$\mathrm{sgn}(\sigma) \overset{\text{def.}}{=} \begin{cases} +1 & \text{if } \sigma \text{ is an even permutation,} \\ -1 & \text{if } \sigma \text{ is an odd permutation.} \end{cases}$$

Equivalently, $\mathrm{sgn}(\sigma) = (-1)^r$ where r is the number of interchanges of indices, two at a time, needed to convert $(1, \ldots, k)$ to $(i_1, \ldots, i_k)$. (More precisely, r is only determined up to some integral multiple of 2.)

Knowing that interchanging any two vectors in a simple k-vector introduces a factor of -1, the following result is then obvious:

Proposition 3.7. *Let $a_1,\ldots,a_k \in \mathbb{R}^n$. If σ is the permutation $(i_1,\ldots,i_k)$ of the k-tuple $(1,\ldots,k)$, then*

$$a_{i_1} \wedge \cdots \wedge a_{i_k} = sgn(\sigma)\, a_1 \wedge \cdots \wedge a_k.$$

Example 3.7. We have

$$\begin{aligned}
a_1 \wedge a_2 \wedge a_3 &= -a_1 \wedge a_3 \wedge a_2, \\
b_1 \wedge b_2 \wedge b_3 &= -b_3 \wedge b_2 \wedge b_1, \\
c_1 \wedge c_2 \wedge c_3 \wedge c_4 &= c_4 \wedge c_3 \wedge c_2 \wedge c_1.
\end{aligned}$$

Next is a proposition that exhibits two important facts:

First, it shows an important connection between determinants and the wedge product.

Second, the set of simple k-vectors $a_1 \wedge \cdots \wedge a_k$ where all the a_i lie in some k-dimensional vector space U behaves as though it were, in a sense, a one-dimensional vector space. That is, if $b_1 \wedge \cdots \wedge b_k$ is a k-vector all of whose factors b_i lie in U and $b_1 \wedge \cdots \wedge b_k \neq 0$, then every such $a_1 \wedge \cdots \wedge a_k$ must be a scalar multiple of $b_1 \wedge \cdots \wedge b_k$; that is, $a_1 \wedge \cdots \wedge a_k = \lambda\,(b_1 \wedge \cdots \wedge b_k)$ for some scalar λ. To put this another way, the ray $\lambda\,(b_1 \wedge \cdots \wedge b_k)$ where λ ranges through all real numbers, gives us every possible simple k-vector lying in U.

We shall find, as we go on, that this is a very useful fact. It will, for example, be very helpful in defining what we mean by the orientation of a k-dimensional manifold and in understanding the change-of-variables formula for multiple integrals.

Proposition 3.8. *Let $a_1,\ldots,a_k$ and $b_1,\ldots,b_k$ be vectors in $\mathbb{R}^n$ such that $a_i = \sum_{j=1}^{k} \gamma_{ij}\, b_j$ for all i. Set $C = (\gamma_{ij})_{k \times k}$, a $k \times k$ matrix. Then*

$$a_1 \wedge \cdots \wedge a_k = \det(C)\,(b_1 \wedge \cdots \wedge b_k).$$

Intuitively, this result is saying that if we look at two k-dimensional parallelepipeds lying in a common k-dimensional vector space, then we can deform one into the other—this deformation may involve both rotating and stretching vectors—and $|\det(C)|$ is the ratio of their volumes while the sign of $\det(C)$ tells us whether or not they have the same orientation.

We give a proof of Proposition 3.8 in Appendix A.2. Here we present an example:

Example 3.8. Let u_1 and u_2 be two orthonormal vectors in $\mathbb{R}^n$ and suppose we have a_1 and a_2 such that

$$a_i = \gamma_{i1}\, u_1 + \gamma_{i2}\, u_2.$$

Set $a = a_1 \wedge a_2$ and $u = u_1 \wedge u_2$ and

$$C = (a_1, a_2) = \begin{pmatrix} \gamma_{11} & \gamma_{21} \\ \gamma_{12} & \gamma_{22} \end{pmatrix}.$$

We would like to see that, as claimed by Proposition 3.8, we have $a = \det(C)\, u$.

First we check the volume condition: Since $\{u_1, u_2\}$ is an orthonormal set, we see that

$$\text{vol}(u) = \sqrt{\det(u_i \cdot u_j)} = 1.$$

By Proposition 3.5, multiplying a k-vector by a scalar changes its volume by the magnitude of that scalar, so we have

$$\text{vol}\big(\det(C)u\big) = |\det(C)|.$$

On the other hand,

$$\text{vol}(a)^2 = \det(a_i \cdot a_j) = \det(CC^T) = \det(C)^2.$$

So we have equality of volumes. To compare orientations, let us attach the scalar $\det(C)$ to the first factor of $\det(C)\, u = (\det(C)u_1) \wedge u_2$. Then we compute

$$\det \begin{pmatrix} a_1 \cdot (\det(C)u_1) & a_1 \cdot u_2 \\ a_2 \cdot (\det(C)u_1) & a_2 \cdot u_2 \end{pmatrix} = \det(C) \det \begin{pmatrix} \gamma_{11} & \gamma_{12} \\ \gamma_{21} & \gamma_{22} \end{pmatrix}$$
$$= \det(C)^2 \geq 0.$$

This shows that a and $\det(C)u$ must have the same orientation; hence $a = \det(C)u$.

3.4.2   Reversion

Our next operation is one that does not exist for vectors in $\mathbb{R}^n$, but for simple k-vectors and, later on, for geometric algebra and geometric calculus, it turns out to be quite useful.

Definition 3.5. If a is the simple k-vector $a = a_1 \wedge \cdots \wedge a_k$, then the *reversion of a* is

$$a^\dagger \overset{\text{def.}}{=} a_k \wedge \cdots \wedge a_1.$$

The well-definedness of reversion is an easy consequence of Proposition 3.7. Here is another way a simple k-vector and its reversion are related:

Proposition 3.9. *For a simple k-vector,*

$$(a_1 \wedge \cdots \wedge a_k)^\dagger = (-1)^r (a_1 \wedge \cdots \wedge a_k)$$

where $r = k(k-1)/2$.

We leave it to the reader to justify this result by showing that one can pass from the increasing sequence $(1, \ldots, k)$ to the decreasing sequence $(k, \ldots, 1)$ by $k(k-1)/2$ interchanges (or at least by that number up to some integer multiple of 2).

3.4.3 Dot product

For vectors $a = \sum_{i=1}^{n} \alpha_i e_i$ and $b = \sum_{i=1}^{n} \beta_i e_i$ in $\mathbb{R}^n$, we take the dot product to be $a \cdot b = \sum_{i=1}^{n} \alpha_i \beta_i$. But beyond this formal definition, the fact that we can write $a \cdot b = |a| \, |b| \cos \theta$, where θ is the angle between a and b, is of the greatest importance. It brings the dot product to life with geometric significance and permits us to use it in applications.

We now show a way to extend the definition of dot product to simple k-vectors:

Definition 3.6. Let $a_1, \ldots, a_k$ and $b_1, \ldots, b_k$ all be vectors in $\mathbb{R}^n$. We define the *dot product* of two simple k-vectors by this formula:

$$(a_1 \wedge \cdots \wedge a_k) \cdot (b_k \wedge \cdots \wedge b_1) \stackrel{\text{def.}}{=} \det \begin{pmatrix} a_1 \cdot b_1 & \cdots & a_1 \cdot b_k \\ \cdots & & \\ a_k \cdot b_1 & \cdots & a_k \cdot b_k \end{pmatrix}.$$

(The dot product is well-defined; see Proposition A.2 of Appendix A.2.)

Notice that in the definition we define $a \cdot b^\dagger$ rather than $a \cdot b$. It will not be clear why we do it this way until much later when we introduce the idea of *geometric product*. The geometric product will be a binary operation that unifies both the dot product and the wedge product. When we see how it operates, it should be obvious that we want to write the reversion of $b_1 \wedge \cdots \wedge b_k$ in the definition of dot product.

A more important question is this: What is the dot product *about*? Has it some intuition behind it? Here is an example that may provide a partial answer:

Example 3.9. Let a_1, a_2 be orthogonal, nonzero vectors in $\mathbb{R}^n$. Let u_1 be a unit vector in the same direction as a_1; so $u_1 = a_1 / |a_1|$. Let u_2 be another unit vector that is orthogonal to u_1. See Figure 3.8. Then a_2 must make an angle θ with u_2 where

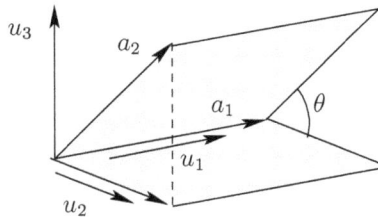

Figure 3.8: Simple 2-vectors with an angle of θ between them

$0 \le \theta \le \pi$. Let u_3 be a third unit vector such that u_1, u_2, u_3 are orthonormal and a_2 lies in the $u_2 u_3$-plane. Then $a_1 \wedge a_2$ and $u_1 \wedge u_2$ are two simple 2-vectors that meet at an angle of θ.

What does $(a_1 \wedge a_2) \cdot (u_2 \wedge u_1)$ represent?

We can write

$$a_1 = |a_1| u_1,$$

$$a_2 = |a_2| \left((\cos \theta) \, u_2 + (\sin \theta) \, u_3 \right),$$

and, because the vectors are orthogonal, $\text{vol}(a_1, a_2) = |a_1| \, |a_2|$ and $\text{vol}(u_1, u_2) = 1$. We then use the definition of dot product to compute

$$(a_1 \wedge a_2) \bullet (u_2 \wedge u_1) = \det \begin{pmatrix} a_1 \bullet u_1 & a_1 \bullet u_2 \\ a_2 \bullet u_1 & a_2 \bullet u_2 \end{pmatrix}$$

$$= |a_1| \, |a_2| \cos \theta.$$

Since the area of (u_1, u_2) is 1, we choose to write this as

$$(a_1 \wedge a_2) \bullet (u_2 \wedge u_1) = \text{vol}(a_1, a_2) \, \text{vol}(u_1, u_2) \, \cos \theta.$$

Compare the answer of this last example to the dot product of two vectors a and b of $\mathbb{R}^n$: $a \bullet b = |a| \, |b| \cos \theta$ where θ is the angle between a and b. It turns out that something of this nature is *always* true of the dot product of simple k-vectors. This is something about which we shall say more later.

Definition 3.7. If a is a simple k-vector, then its *magnitude* is $|a| \overset{\text{def.}}{=} \sqrt{a \bullet a^{\dagger}}$.

Notice the similarity of this definition to that of the magnitude of a vector in $\mathbb{R}^n$. We show in the proposition that follows that $a \bullet a^{\dagger}$ is always a nonnegative number, so there is no problem with taking its square root.

Proposition 3.10. *Suppose that a and b are simple k-vectors in $\mathbb{R}^n$ and λ is a scalar. Then the following are true:*

1. *$|a| = vol(a)$.*

2. *Suppose a and b lie in the same k-dimensional subspace and $a, b \neq 0$. Then a and b have the same orientation if and only if $a \bullet b^{\dagger} > 0$. They have opposite orientations if and only if $a \bullet b^{\dagger} < 0$.*

3. *$\lambda \, (a \bullet b) = (\lambda a) \bullet b = a \bullet (\lambda b)$.*

4. *$a \bullet b^{\dagger} = a^{\dagger} \bullet b$.*

5. *$a \bullet b = b \bullet a$.*

Proof. Part 1 follows from the definitions of dot product and volume.

Part 2 follows from the definitions of dot product and having the same orientation.

For part 3, take $a = a_1 \wedge \cdots \wedge a_k$ and $b = b_1 \wedge \cdots \wedge b_k$. We can take λa to be $(\lambda a_1) \wedge a_2 \wedge \cdots \wedge a_k$ and λb to be $(\lambda b_1) \wedge b_2 \wedge \cdots \wedge b_k$. Then

$$(\lambda a) \bullet b = \det \begin{pmatrix} (\lambda a_1) \bullet b_k & \cdots & (\lambda a_1) \bullet b_1 \\ & \cdots & \\ a_k \bullet b_k & \cdots & a_k \bullet b_1 \end{pmatrix}$$

$$= \lambda \det \begin{pmatrix} a_1 \bullet b_k & \cdots & a_1 \bullet b_1 \\ & \cdots & \\ a_k \bullet b_k & \cdots & a_k \bullet b_1 \end{pmatrix} = \lambda \, (a \bullet b)$$

$$= \det \begin{pmatrix} a_1 \cdot b_k & \cdots & a_1 \cdot (\lambda b_1) \\ \cdots & & \\ a_k \cdot b_k & \cdots & a_k \cdot (\lambda b_1) \end{pmatrix}$$

$$= a \cdot (\lambda b).$$

We leave the proof of part 4 as an exercise.

The proof of part 5 reduces to the fact that the determinant of a matrix is equal to the determinant of its transpose. $\qquad\square$

We now present a result that will, later on, turn out to be very useful.

If a is a vector in $\mathbb{R}^n$, we can write it in the form $a = \sum_{i=1}^n \alpha_i e_i$ where $\{e_i\}_{i=1}^n$ is the standard basis for $\mathbb{R}^n$. The scalars α_i are uniquely determined and are given by $\alpha_i = a \cdot e_i$. Because of this, given any two vectors a and b in $\mathbb{R}^n$, we know that they are equal if and only if $a \cdot e_i = b \cdot e_i$ for all i. More generally, we can say that $a = b$ if and only if $a \cdot c = b \cdot c$ for all vectors c in $\mathbb{R}^n$.

This result generalizes to the setting of simple k-vectors:

Proposition 3.11. *If a and b are simple k-vectors in $\mathbb{R}^n$, then $a = b$ if and only if $a \cdot c = b \cdot c$ for all simple k-vectors c in $\mathbb{R}^n$.*

We defer the proof to Appendix A.2.

3.4.4 Linear transformations acting on simple k-vectors

Now we consider how linear transformations can operate on simple k-vectors:

Definition 3.8. Suppose that $f : V \to \mathbb{R}^n$ is a linear transformation where V is a vector subspace of $\mathbb{R}^m$. If $a_1 \wedge \cdots \wedge a_k$ is a simple k-vector where each $a_i \in V$, then we set

$$\wedge^k f(a_1 \wedge \cdots \wedge a_k) \stackrel{\text{def.}}{=} f(a_1) \wedge \cdots \wedge f(a_k).$$

Thus $\wedge^k f$ is a transformation operating on simple k-vectors lying in V for $k = 1, 2, \ldots$. We take $\wedge^0 f$ to be the identity map on scalars; that is, $\wedge^0 f(\lambda) = \lambda$ whenever $\lambda \in \mathbb{R}$.

Of course there is the question of whether or not this operation is well-defined. That is, if we have two different ways to write the same simple k-vector, for example, $a_1 \wedge \cdots \wedge a_k = b_1 \wedge \cdots \wedge b_k$, then how do we know that $f(a_1) \wedge \cdots \wedge f(a_k) = f(b_1) \wedge \cdots \wedge f(b_k)$? We prove well-definedness in Proposition A.3 of Appendix A.2.

Example 3.10. Let $f : \mathbb{R}^2 \to \mathbb{R}^2$ be the linear transformation given by $f(\chi_1, \chi_2) = (\chi_1 + 2\chi_2, \chi_2 - \frac{1}{2}\chi_1)$. Then the result of applying $\wedge^2 f$ to $e_1 \wedge e_2$, namely,

$$\wedge^2 f(e_1 \wedge e_2) = (e_1 - \tfrac{1}{2}e_2) \wedge (2e_1 + e_2),$$

is displayed in Figure 3.9.

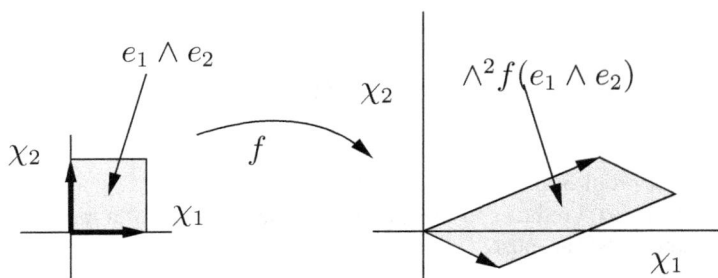

Figure 3.9: Linear transformation acting on a simple 2-vector

3.4.5 Wedge product

Definition 3.9. Suppose a and b are simple p- and q-vectors respectively in $\mathbb{R}^n$ where $p, q \geq 1$. If $a = a_1 \wedge \cdots \wedge a_p$ and $b = b_1 \wedge \cdots \wedge b_q$, then we define the *wedge product of a and b* by

$$a \wedge b \overset{\text{def.}}{=} a_1 \wedge \cdots \wedge a_p \wedge b_1 \wedge \cdots \wedge b_q.$$

In the event that either a or b is a simple 0-vector, that is a scalar λ, then the wedge product is defined by

$$\lambda \wedge b \overset{\text{def.}}{=} \lambda b \quad \text{and} \quad a \wedge \lambda \overset{\text{def.}}{=} \lambda a.$$

(The wedge product of simple p- and q-vectors is well-defined; see Proposition A.4 of Appendix A.2.)

The next results are now quite easy:

Proposition 3.12. *1. The wedge product is associative. That is, if a, b, c are simple p-, q-, and r-vectors respectively, then*

$$a \wedge (b \wedge c) = (a \wedge b) \wedge c.$$

2. *If a and b are simple p- and q-vectors respectively and λ is a scalar, then*

$$\lambda (a \wedge b) = (\lambda a) \wedge b = a \wedge (\lambda b).$$

3. *If a and b are simple p- and q-vectors respectively in $\mathbb{R}^n$ and $p + q > n$, then $a \wedge b = 0$.*

Proof. We prove only the last part. We write $a = a_1 \wedge \cdots \wedge a_p$ and $b = b_1 \wedge \cdots \wedge b_q$ where each a_i and b_j is a vector in $\mathbb{R}^n$. If $p + q > n$, then $\{a_1, \ldots, a_p, b_1, \ldots, b_q\}$ must be a linearly dependent set, thus $a \wedge b = 0$. $\qquad \square$

Exercises 3.4.

1. Suppose

$$a = 2e_1 + 3e_2, \qquad\qquad b = -4e_1 + e_2,$$
$$u_1 = e_1 + e_2, \qquad\qquad u_2 = e_1 - e_2.$$

Find scalars λ that satisfy each of the following:

(a) $a \wedge b = \lambda (e_1 \wedge e_2)$.

(b) $a \wedge b = \lambda (u_1 \wedge u_2)$.

(c) $e_1 \wedge e_2 = \lambda (u_1 \wedge u_2)$.

(d) $u_1 \wedge u_2 = \lambda (e_1 \wedge e_2)$.

2. Show that if $(a_1 \wedge \cdots \wedge a_k)^\dagger = (-1)^r (a_1 \wedge \cdots \wedge a_k)$, we can take r to be $k(k-1)/2$.

3. In $\mathbb{R}^4$, let us form all simple 2-vectors $e_i \wedge e_j$ where $i < j$. How many of these are there? Compute $(e_{i_1} \wedge e_{i_2}) \bullet (e_{j_1} \wedge e_{j_2})^\dagger$ for all $i_1 < i_2$ and $j_1 < j_2$.

4. Suppose we are given vectors $a_1, \ldots, a_k$ and $b_1, \ldots, b_k$ in $\mathbb{R}^n$. Show that if there is some b_i with the property that $a_j \bullet b_i = 0$ for $j = 1, \ldots, k$, then $(a_1 \wedge \cdots \wedge a_k) \bullet (b_1 \wedge \cdots \wedge b_k) = 0$.

5. Suppose a_1, a_2, a', b_1, b_2 are vectors in $\mathbb{R}^n$ and λ is a scalar.

(a) Show that

$$[(\lambda a_1) \wedge a_2] \bullet (b_1 \wedge b_2) = \lambda (a_1 \wedge a_2) \bullet (b_1 \wedge b_2).$$

(b) Show that

$$[(a_1 + a') \wedge a_2] \bullet (b_1 \wedge b_2) = (a_1 \wedge a_2) \bullet (b_1 \wedge b_2) + (a' \wedge a_2) \bullet (b_1 \wedge b_2).$$

(c) Recalling that $\{e_i\}_{i=1}^n$ is the standard basis for $\mathbb{R}^n$, show that $(a_1 \wedge a_2) \bullet (b_1 \wedge b_2)$ can be written in the form

$$\sum_{i_1 < i_2, j_1 < j_2} \lambda_{i_1, i_2, j_1, j_2} (e_{i_1} \wedge e_{i_2}) \bullet (e_{j_1} \wedge e_{j_2})^\dagger$$

where each $\lambda_{i_1, i_2, j_1, j_2}$ is a scalar and it is understood that the summation is over all ordered pairs of indices (i_1, i_2) and (j_1, j_2) where $i_1 < i_2$ and $j_1 < j_2$.

(d) Show that the only terms which occur in the last sum are those for which $(i_1, i_2) = (j_1, j_2)$.

6. Let $f : \mathbb{R}^2 \to \mathbb{R}^2$ be the rotation of $\mathbb{R}^2$ by an angle of θ. Show that $\wedge^2 f$ is the identity map on the set of simple 2-vectors.

7. Check the other cases in the proof of Proposition 3.6.

8. Suppose we are given a square matrix

$$A = \begin{pmatrix} \lambda_{11} & \cdots & \lambda_{n1} \\ \vdots & & \vdots \\ \lambda_{1n} & \cdots & \lambda_{nn} \end{pmatrix}.$$

Show how to choose vectors $a_1, \ldots, a_n$ in $\mathbb{R}^n$ in terms of the standard basis $e_1, \ldots, e_n$ such that

$$a_1 \wedge \cdots \wedge a_n = \det(A) \, e_1 \wedge \cdots \wedge e_n.$$

Chapter 4

$\Lambda^k \mathbb{R}^n$ and some applications

4.1 The space of k-vectors

We have been talking about simple k-vectors, yet they are not really vectors; we have no means to add them. We now remedy this.

Here is the trick: Given a simple k-vector a, we can identify it with the map

$$c \overset{a}{\mapsto} a \cdot c$$

where c ranges over all simple k-vectors in $\mathbb{R}^n$. Since these maps take values in the reals, we can add them and multiply them by scalars.

Definition 4.1. Let $\mathcal{S}^k$ be the set of simple k-vectors in $\mathbb{R}^n$ where $1 \leq k \leq n$. By $\Lambda^k \mathbb{R}^n$ we mean the set of maps $f : \mathcal{S}^k \to \mathbb{R}$ with the property that for each f there exist simple k-vectors $a_1, \ldots, a_p$ and scalars $\lambda_1, \ldots, \lambda_p$ such that

$$f(c) = \lambda_1 (a_1 \cdot c) + \cdots + \lambda_p (a_p \cdot c)$$

for all $c \in \mathcal{S}^k$. In this case, we write

$$f = \lambda_1 a_1 + \cdots + \lambda_p a_p.$$

We identify $\Lambda^1 \mathbb{R}^n$ with $\mathbb{R}^n$, and we set $\Lambda^0 \mathbb{R}^n \overset{\text{def.}}{=} \mathbb{R}$ and $\Lambda^k \mathbb{R}^n \overset{\text{def.}}{=} \{0\}$, the trivial vector space, when $k > n$. We call $\Lambda^k \mathbb{R}^n$ *the space of k-vectors over $\mathbb{R}^n$*.

It is easily seen that

Proposition 4.1. $\Lambda^k \mathbb{R}^n$ *is a vector space over* $\mathbb{R}$.

But are we really justified in saying that we have extended the simple k-vectors to a vector space? What does it mean to say that $\Lambda^k \mathbb{R}^n$ "contains" the simple k-vectors?

Here is where Proposition 3.11 comes to the rescue: In general, an f in $\Lambda^k \mathbb{R}^n$ will have different representations, for example $f = \sum_{i=1}^p \lambda_i a_i$ and $f = \sum_{j=1}^q \xi_j b_j$. However if we can write $f = a$ and $f = b$ where a and b are simple k-vectors, then for all simple k-vectors c, we have $a \cdot c = f(c) = b \cdot c$ and hence by Proposition 3.11, $a = b$. That is, the set of simple k-vectors is embedded in $\Lambda^k \mathbb{R}^n$ in a one-to-one manner; we may, by an abuse of notation, write "$\mathcal{S}^k \subseteq \Lambda^k \mathbb{R}^n$".

One other minor point: We have introduced the idea of multiplication by a scalar in two different senses. We know what it means to multiply a simple k-vector by a scalar, λa, and what it means to multiply a real-valued function by a scalar, λf. Fortunately, the two senses coincide. This is shown by the fact that if $f = a$ where a is simple, then $(\lambda f)(c) = \lambda (a \cdot c) = (\lambda a) \cdot c$.

We now have available to us the standard properties of a vector space such as

$$\lambda (a + b) = \lambda a + \lambda b,$$
$$a + b = b + a,$$
$$a + (b + c) = (a + b) + c,$$

where a, b, c are k-vectors (including the simple ones) and λ is a scalar.

If $\Lambda^k \mathbb{R}^n$ is vector space, we ought to be able to figure out its dimension. To do this, we look at a simple k-vector $a_1 \wedge \cdots \wedge a_k$ and think about the symbol $\wedge$. Before long we will show that we can think of $\wedge$ as a binary operation, a kind of generalization of the operation of vector product that one encounters in calculus and vector analysis.

However the way we introduced $\wedge$ does not really imply that it is a binary operation in the sense that, say, $a + b$ or $a \cdot b$ are binary operations. Rather, $\wedge$ is just part of the symbolism for a simple k-vector, and we could as easily have written **simple**$(a_1, \ldots, a_k)$ as $a_1 \wedge \cdots \wedge a_k$.

We do know two ways in which $\wedge$ begins to look suspiciously like a binary operation with vectors. First,

$$a_1 \wedge \cdots \wedge (\lambda a_i) \wedge \cdots \wedge a_k = \lambda (a_1 \wedge \cdots \wedge a_i \wedge \cdots \wedge a_k)$$

when λ is a scalar. Second,

$$a_1 \wedge \cdots \wedge a_k = \text{sgn}(i_1, \ldots, i_k)(a_{i_1} \wedge \cdots \wedge a_{i_k})$$

whenever $(i_1, \ldots, i_k)$ is a permutation of $(1, \ldots, k)$. This second property has the nice implication that if, in $a_1 \wedge \cdots \wedge a_k$, we can find a_i and a_j such that $a_i = a_j$ when $i \neq j$, then $a_1 \wedge \cdots \wedge a_k = 0$. This is because we can switch a_i and a_j and change the sign of $a_1 \wedge \cdots \wedge a_k$, yet $a_1 \wedge \cdots \wedge a_k$ does not change!

It is easily seen in the next proposition that $\wedge$ also exhibits the distributivity property under limited circumstances. (We do away with the limitations later.)

Proposition 4.2. *If $a_1, \ldots, a_k$ and a' are vectors in $\mathbb{R}^n$, then*

$$a_1 \wedge \cdots \wedge (a_i + a') \wedge \cdots \wedge a_k$$
$$= (a_1 \wedge \cdots \wedge a_i \wedge \cdots \wedge a_k) + (a_1 \wedge \cdots \wedge a' \wedge \cdots \wedge a_k).$$

Since we treat k-vectors as functions operating on simple k-vectors, we merely have to apply the two expressions in Proposition 4.2 to an arbitrary simple k-vector b and see that we get the same thing both times. We leave this as an exercise.

We now try our hand at a revealing calculation:

Example 4.1. Consider 2-vectors in $\mathbb{R}^3$. These are linear combinations of simple 2-vectors, and every simple 2-vector has the form

$$a = (\alpha_1 e_1 + \alpha_2 e_2 + \alpha_3 e_3) \wedge (\alpha_1' e_1 + \alpha_2' e_2 + \alpha_3' e_3).$$

If we multiply this out, taking full advantage of distributivity and the fact that

$$e_i \wedge e_j = \begin{cases} -e_j \wedge e_i & \text{if } i \neq j, \\ 0 & \text{if } i = j, \end{cases}$$

we see that we can write

$$a = \beta_{12}(e_1 \wedge e_2) + \beta_{13}(e_1 \wedge e_3) + \beta_{23}(e_2 \wedge e_3)$$

for appropriate scalars β_{ij}. The fact that we can do this suggests the possibility that $e_1 \wedge e_2$, $e_1 \wedge e_3$, and $e_2 \wedge e_3$ might constitute a basis for $\Lambda^2 \mathbb{R}^3$. This is a good guess. Look at the next result:

Proposition 4.3. *Suppose that $1 \leq k \leq n$ and $\{u_1, \ldots, u_n\}$ is an orthonormal basis for $\mathbb{R}^n$. Then first, whenever $i_1 < \cdots < i_k$ and $j_1 < \cdots < j_k$, we have*

$$(u_{i_1} \wedge \cdots \wedge u_{i_k}) \bullet (u_{j_1} \wedge \cdots \wedge u_{j_k})^\dagger = \begin{cases} 1 & \text{if } (i_1, \ldots, i_k) = (j_1, \ldots, j_k) \\ 0 & \text{otherwise.} \end{cases}$$

Second, the set of simple k-vectors of the form $u_{i_1} \wedge \cdots \wedge u_{i_k}$ where $i_1 < \cdots < i_k$ is a basis for $\Lambda^k \mathbb{R}^n$.

Proof. The fact that we can write any k-vector as a linear combination of simple k-vectors of the form $u_{i_1} \wedge \cdots \wedge u_{i_k}$ where $i_1 < \cdots < i_k$ is easily seen from what we did in Example 4.1.

Next notice that

$$(u_{i_1} \wedge \cdots \wedge u_{i_k}) \bullet (u_{j_1} \wedge \cdots \wedge u_{j_k})^\dagger = \det(u_{i_s} \bullet u_{j_t}).$$

Every $u_{i_s} \bullet u_{j_t}$ is either 0 or 1. If the sequences $(i_1, \ldots, i_k)$ and $(j_1, \ldots, j_k)$ differ in any entry, then we will have $\det(u_{i_s} \bullet u_{j_t}) = 0$ since the matrix will have a row and column of zeros. If, on the other hand, the two sequences are identical, then $\det(u_{i_s} \bullet u_{j_t}) = 1$ since our matrix is the identity matrix.

We need only show that elements of the form $u_{i_1} \wedge \cdots \wedge u_{i_k}$ where $i_1 < \cdots < i_k$ are linearly independent to see that they form a basis for $\Lambda^k \mathbb{R}^n$. Suppose we have

$$\sum_{i_1 < \cdots < i_k} \lambda_{i_1 \ldots i_k}(u_{i_1} \wedge \cdots \wedge u_{i_k}) = 0$$

where each $\lambda_{i_1 \ldots i_k}$ is a scalar. Recall that

$$f = \sum_{i_1 < \cdots < i_k} \lambda_{i_1 \ldots i_k} (u_{i_1} \wedge \cdots \wedge u_{i_k})$$

is a function that operates on simple k-vectors; in this case we assume it is the zero function. We also know that when a simple k-vector operates on another simple k-vector, the result is their dot product. Suppose we let f operate on the simple k-vector $(u_{j_1} \wedge \cdots \wedge u_{j_k})^\dagger$ where $j_1 < \cdots < j_k$. We obtain the following results:

$$f\big((u_{j_1} \wedge \cdots \wedge u_{j_k})^\dagger\big) = 0,$$
$$\sum_{i_1 < \cdots < i_k} \lambda_{i_1 \ldots i_k} (u_{i_1} \wedge \cdots \wedge u_{i_k}) \bullet (u_{j_1} \wedge \cdots \wedge u_{j_k})^\dagger = 0,$$
$$\lambda_{j_1 \ldots j_k} = 0.$$

Since $\lambda_{j_1 \ldots j_k} = 0$, we have independence. $\square$

Example 4.2. If $\{u_1, u_2, u_3, u_4\}$ is an orthonormal basis for $\mathbb{R}^4$, then an orthonormal basis for $\Lambda^2 \mathbb{R}^4$ consists of the elements

$$u_1 \wedge u_2, \quad u_1 \wedge u_3, \quad u_1 \wedge u_4, \quad u_2 \wedge u_3, \quad u_2 \wedge u_4, \quad u_3 \wedge u_4$$

while an orthonormal basis for $\Lambda^3 \mathbb{R}^4$ consists of

$$u_1 \wedge u_2 \wedge u_3, \quad u_1 \wedge u_2 \wedge u_4, \quad u_1 \wedge u_3 \wedge u_4, \quad u_2 \wedge u_3 \wedge u_4.$$

Remark 4.1. An immediate consequence of Proposition 4.3 is $\dim \Lambda^k \mathbb{R}^n = \binom{n}{k}$, a binomial coefficient.

Proposition 4.3 is more restrictive than it needs to be:

Proposition 4.4. *Let $\{u_1, \ldots, u_n\}$ be a basis for $\mathbb{R}^n$. Then the set of simple k-vectors $u_{i_1} \wedge \cdots \wedge u_{i_k}$ where $i_1 < \cdots < i_k$ is a basis for $\Lambda^k \mathbb{R}^n$.*

Proof. It is easily seen that any k-vector a can be written as a linear combination of $u_{i_1} \wedge \cdots \wedge u_{i_k}$ where $i_1 < \cdots < i_k$. There are exactly $\binom{n}{k}$ such simple k-vectors. Since this is the dimension of $\Lambda^k \mathbb{R}^n$, they must be a basis for the space of k-vectors. $\square$

It is convenient and very easy at this point to extend the idea of reversion to arbitrary k-vectors. We know that for a simple k-vector we have

$$(a_1 \wedge \cdots \wedge a_k)^\dagger = (-1)^r a_1 \wedge \cdots \wedge a_k \quad \text{where } r = \frac{k(k-1)}{2}.$$

Therefore we do the following:

Definition 4.2. For an arbitrary (not necessarily simple) k-vector, we set

$$a^\dagger \stackrel{\text{def.}}{=} (-1)^r a \quad \text{where } r = \frac{k(k-1)}{2}.$$

Given a vector subspace V of $\mathbb{R}^n$, it is often convenient to restrict one's attention to k-vectors that "live in V." We say that a is a *simple k-vector in V* provided we can write it in the form $a = a_1 \wedge \cdots \wedge a_k$ where each $a_i \in V$. By a *k-vector in V*, we mean a finite linear combination of simple k-vectors in V. We denote the space of such k-vectors as $\Lambda^k V$.

Of course the propositions we have established for $\Lambda^k \mathbb{R}^n$ have appropriate versions for $\Lambda^k V$. In particular, we have this:

Proposition 4.5. *If V is a vector subspace of $\mathbb{R}^n$ with dimension m, then*

$$\dim\left(\Lambda^k V\right) = \binom{m}{k}$$

where $0 \leq k \leq m$.

We leave the proof of this as an exercise.

Exercises 4.1.

1. Prove Proposition 4.2.

2. Show that every n-vector in $\mathbb{R}^n$ has the form $\lambda\left(e_1 \wedge \cdots \wedge e_n\right)$ where λ is a scalar.

3. Calculate each of the following in terms of $e_i \wedge e_j$ where $i < j$.

    (a) $(e_1 - e_2) \wedge (e_1 + e_2)$.
    (b) $(e_1 - 2e_2 + 3e_3) \wedge (e_1 - 2e_2 + 3e_3)$.
    (c) $(e_1 - e_2) \wedge (e_1 + e_2) \wedge (3e_1 + 4e_2)$.
    (d) $(2e_1 - e_3) \wedge (e_1 - 3e_2 + 4e_3)$.

4. Calculate $(\lambda_1 e_1 + \lambda_2 e_2 + \lambda_3 e_3) \wedge (\xi_1 e_1 + \xi_2 e_2 + \xi_3 e_3)$ in terms of λ_i, ξ_j, and $e_i \wedge e_j$ where $i < j$ in this last term. Show that each coefficient of $e_i \wedge e_j$ is a 2×2 determinant.

5. Show that if a and b are k-vectors and λ a scalar, then $(\lambda a)^\dagger = \lambda a^\dagger$ and $(a+b)^\dagger = a^\dagger + b^\dagger$.

6. Let V be an m-dimensional vector subspace of $\mathbb{R}^n$ and suppose that $\{u_i\}_{i=1}^m$ is a basis for V. Show that the set of k-vectors of the form $u_{i_1} \wedge \cdots \wedge u_{i_k}$ where $i_1 < \cdots < i_k$ is a basis for $\Lambda^k V$ and that therefore Proposition 4.5 holds.

4.2   The dot product of arbitrary k-vectors

We know how to take the dot and wedge products of *simple k-vectors*; we would like to extend these operations to arbitrary k-vectors.

Before showing how to do that, it is convenient to discuss some notation.

Working in $\mathbb{R}^n$, we find ourselves discussing objects α_i, u_j, f_k, etc. where $i, j, k, \ldots$ are *indices* from the set $\{1, 2, \ldots, n\}$. But when working with wedge products in $\Lambda^p \mathbb{R}^n$, it is often convenient to use *multi-indices*.

A *multi-index of length p* is an ordered sequence $I = (i_1, i_2, \ldots, i_p)$ where each i_k is an element of the index set $\{1, \ldots, n\}$. We write $|I| = p$ to indicate the length of I.

Notice that if we have a wedge product $u_{j_1} \wedge \cdots \wedge u_{j_p}$ that is nonzero, then the indices $\{j_1, \ldots, j_p\}$ must be distinct and it must be possible to reorder the sequence $(j_1, \ldots, j_p)$ into a sequence $(i_1, \ldots, i_p)$ such that $i_1 < \cdots < i_p$. In that case, we have

$$u_{j_1} \wedge \cdots \wedge u_{j_p} = (-1)^r u_{i_1} \wedge \cdots \wedge u_{i_p}$$

for some r of uniquely determined parity. If $I = (i_1, \ldots, i_p)$ has the property that $i_1 < i_2 < \cdots < i_p$, then we call I an *ordered multi-index*. We let $\mathcal{O}_p$ be the *set of ordered multi-indices of length p*.

If we have a p-vector of the form $u_{i_1} \wedge \cdots \wedge u_{i_p}$, then we feel free to write

$$u_I = u_{i_1} \wedge \cdots \wedge u_{i_p}$$

where I is the multi-index $(i_1, \ldots, i_p)$. We also feel free in this case to omit the parentheses on the multi-index and write, for example,

$$u_3 \wedge u_1 \wedge u_4 = u_{314} \quad \text{instead of} \quad u_3 \wedge u_1 \wedge u_4 = u_{(3,1,4)}.$$

In particular, if we have a basis $\{v_1, \ldots, v_m\}$ for an m-dimensional vector subspace V of $\mathbb{R}^n$, then every p-vector a in V has a unique expansion of the form

$$a = \sum_{I \in \mathcal{O}_p} \alpha_I v_I$$

where $\sum_{I \in \mathcal{O}_p}$ means the summation over all ordered multi-indices of length p and each α_I is a scalar.

In the event that we need it, we also permit ourselves to imagine that we have a multi-index $I = \emptyset$ of length zero, the empty multi-index. If $\{u_i\}_{i=1}^n$ is an orthonormal basis for $\mathbb{R}^n$, then we take

$$u_\emptyset \overset{\text{def.}}{=} 1.$$

We now turn our attention to the problem of constructing dot products of arbitrary k-vectors.

If a and b are k-vectors, we know we can expand them into finite linear sums $a = \sum_I \alpha_I a_I$ and $b = \sum_J \beta_J b_J$ where a_I and b_J are simple k-vectors and I and J are ordered multi-indices of length k. Would it make sense to set

$$a \cdot b \overset{?}{=} \sum_{I,J} \alpha_I \beta_J (a_I \cdot b_J) ? \tag{4.1}$$

We *do* know how to take dot products of simple k-vectors. There is a difficulty: The expansions of a and b are not unique. If we used different expansions of a, for example, could we be sure that Equation (4.1) would give us the same answer both times? However this problem is easily disposed of.

Suppose we have two expansions of a, namely $a = \sum_I \alpha_I a_I$ and $a = \sum_K \alpha'_K a'_K$. To say these are different expansions of the same k-vector means that

$$\sum_I \alpha_I (a_I \cdot c) = \sum_K \alpha'_K (a'_K \cdot c)$$

for all simple k-vectors c. We then observe that

$$\sum_{I,J} \alpha_I \beta_J (a_I \cdot b_J) = \sum_J \beta_J \left(\sum_I \alpha_I (a_I \cdot b_J) \right)$$
$$= \sum_J \beta_J \left(\sum_K \alpha'_K (a'_K \cdot b_J) \right)$$
$$= \sum_{K,J} \alpha'_K \beta_J (a'_K \cdot b_J).$$

One can play a similar trick using two different expansions of b. Thus our choice of expansions of a and b does not matter.

Definition 4.3. If $a, b \in \Lambda^k \mathbb{R}^n$, then

$$a \cdot b \stackrel{\text{def.}}{=} \sum_{I,J} \alpha_I \beta_J (a_I \cdot b_J)$$

where $a = \sum_I \alpha_I a_I$ and $b = \sum_J \beta_J b_J$ are expansions of a and b as finite linear combinations of simple k-vectors and I, J range over sets of multi-indices of length k.

This is clearly an extension of the definition of dot product of simple k-vectors. Here are easy consequences of the definition:

Proposition 4.6. *For $a, b, c \in \Lambda^k \mathbb{R}^n$ and λ a scalar, the dot product satisfies*

1. $\lambda (a \cdot b) = (\lambda a) \cdot b = a \cdot (\lambda b)$,

2. $a \cdot b = b \cdot a$.

3. $a \cdot (b + c) = a \cdot b + a \cdot c$,

4. $(a + b) \cdot c = a \cdot c + b \cdot c$,

where $\lambda \in \mathbb{R}$ and $a, b, c \in \Lambda^k \mathbb{R}^n$.

It turns out that computations of dot products of k-vectors take on a particularly nice form when carried out in an orthonormal basis. It is convenient to first extend the definition of the *magnitude* of a k-vector a thus:

$$|a| \stackrel{\text{def.}}{=} \sqrt{a \cdot a^\dagger}.$$

Now recall that by Proposition 4.3, if $\{u_i\}_{i=1}^n$ is an orthonormal basis for $\mathbb{R}^n$, then

$$u_I \cdot u_J{}^\dagger = \begin{cases} 1 & \text{if } I = J, \\ 0 & \text{if } I \neq J, \end{cases}$$

where I and J are ordered multi-indices of length k. The following result is then a matter of straightforward computation:

Proposition 4.7. *Let* $\{u_i\}_{i=1}^n$ *be an orthonormal basis for* $\mathbb{R}^n$. *If a and b are k-vectors with expansions* $a = \sum_{I \in \mathcal{O}_k} \alpha_I u_I$ *and* $b = \sum_{J \in \mathcal{O}_k} \beta_J u_J$, *then*

1. $a \cdot b^\dagger = a^\dagger \cdot b = \sum_{I \in \mathcal{O}_k} \alpha_I \beta_I$,

2. $|a| = \sqrt{\sum_{I \in \mathcal{O}_k} \alpha_I^2}$.

Corollary 4.1. *If a and b are k-vectors, then* $|a \cdot b^\dagger| \leq |a|\,|b|$.

Proof. Recall that if we have $(\chi_1, \dots, \chi_n)$ and $(\xi_1, \dots, \xi_n)$ in $\mathbb{R}^n$, then the Schwarz inequality says that

$$\left| \sum_{i=1}^n \chi_i \xi_i \right| \leq \sqrt{\sum_{i=1}^n \chi_i^2} \sqrt{\sum_{i=1}^n \xi_i^2}. \tag{4.2}$$

If we expand a and b in terms, then we see from Proposition 4.7 that $|a \cdot b^\dagger| \leq |a|\,|b|$ amounts to

$$\left| \sum_I \alpha_I \beta_I \right| \leq \sqrt{\sum_I \alpha_I^2} \sqrt{\sum_I \beta_I^2}$$

which is simply (4.2) with the sum over a different indexing set. $\square$

Corollary 4.2. *Suppose* $a, b \in \Lambda^k \mathbb{R}^n$ *and* λ *is a scalar. Then*

1. $|\lambda a| = |\lambda|\,|a|$,

2. $|a + b| \leq |a| + |b|$.

Exercises 4.2.

1. Prove Proposition 4.6.

2. Prove Proposition 4.7.

3. Prove that for k-vectors, $|a \cdot b^\dagger| = |a \cdot b|$.

4. Suppose that $\{u_i\}_{i=1}^n$ and $\{v_i\}_{i=1}^n$ are two orthonormal bases for $\mathbb{R}^n$. Let I, J, K be ordered multi-indices of length k and show that

$$u_I = \sum_{J \in \mathcal{O}_k} (u_I \cdot v_J^\dagger) v_J \quad \text{and} \quad \sum_{J \in \mathcal{O}_k} (u_I \cdot v_J^\dagger)(v_J \cdot u_K^\dagger) = \delta_{IK}.$$

It is to be understood here that δ_{IK} is a version of Kronecker's delta for multi-indices; that is,

$$\delta_{IK} = \begin{cases} 1 & \text{if } I = K, \\ 0 & \text{if } I \neq K. \end{cases}$$

5. Prove Corollary 4.2.

4.3   The wedge product of arbitrary k-vectors

As an aid to constructing an extension of the concept of wedge product to a binary operation on arbitrary p- and q-vectors, we temporarily choose and hold fixed an orthonormal basis $v = \{v_i\}_{i=1}^n$ for $\mathbb{R}^n$. In Appendix A.3, we show this apparent dependence on our choice of v is illusory, that all such choices lead to the same definition.

Recall that if $I = (i_1, \ldots, i_k)$, a multi-index, then $v_I = v_{i_1} \wedge \cdots \wedge v_{i_k}$. Let $\mathcal{O}_k$ be the set of ordered multi-indices of length k; that is, if $I \in \mathcal{O}_k$, then $i_1 < \cdots < i_k$. Let a and b be p- and q-vectors respectively in $\mathbb{R}^n$ and write their unique expansions in terms of the basis:

$$a = \sum_{I \in \mathcal{O}_p} \alpha_I v_I \quad \text{and} \quad b = \sum_{J \in \mathcal{O}_q} \beta_J v_J.$$

Notice that since we have defined the wedge product of simple k-vectors, the product $v_I \wedge v_J$ makes sense. Then we define the binary operation $\wedge : \Lambda^p \mathbb{R}^n \times \Lambda^q \mathbb{R}^n \to \Lambda^{p+q} \mathbb{R}^n$, defined for all p- and q-vectors, simple or not, by

$$a \wedge b \overset{\text{def.}}{=} \sum_{I,J} \alpha_I \beta_J (v_I \wedge v_J). \tag{4.3}$$

If a is a scalar, $a = \lambda$, then we set

$$\lambda \wedge b \overset{\text{def.}}{=} \lambda b = \sum_J \lambda \beta_J v_J.$$

A similar remark applies if b is a scalar.

The following is then trivial:

Proposition 4.8. *Let a be a p-vector, b and c be q-vectors, and λ a scalar. Then*

1. $1 \wedge a = a \wedge 1 = a.$

2. $a \wedge (b+c) = a \wedge b + a \wedge c.$

3. $(b+c) \wedge a = b \wedge a + c \wedge a.$

4. $\lambda (a \wedge b) = (\lambda a) \wedge b = a \wedge (\lambda b).$

The next result is almost as easy:

Proposition 4.9. *The wedge product is associative.*

Proof. The associativity of the wedge product (considered now as a binary operation) follows from the fact that $v_I \wedge (v_J \wedge v_K) = (v_I \wedge v_J) \wedge v_K$. (Proposition 3.12.) □

Now here are two results that we prove in Appendix A.3:

Proposition 4.10. *Given a simple k-vector $a_1 \wedge \cdots \wedge a_k$, we may regard each instance of $\wedge$ as the binary operation defined in (4.3).*

It may not be immediately clear what this last result is saying. Recall that when we originally defined $a_1 \wedge \cdots \wedge a_k$, it was an equivalence class of oriented parallelepipeds; there was not really any thought that $\wedge$ was a binary operation. This result is saying that if we write down each of the vectors $a_1, \ldots, a_k$ in terms of the basis $\{v_i\}_{i=1}^n$ and calculate their wedge product using (4.3), this gives us exactly the simple k-vector $a_1 \wedge \cdots \wedge a_k$ described in Definition 3.3.

Proposition 4.11. *The definition (4.3) of the wedge product is independent of our choice of the orthonormal basis $v = \{v_i\}_{i=1}^n$.*

That is, if we compute $a \wedge b$ using v and if we then replace v by another orthonormal basis $\{w_i\}_{i=1}^n$ in (4.3) and recompute $a \wedge b$, we will get the same thing both times.

Here are two useful computational results whose proofs we leave as exercises:

Proposition 4.12. *If a is a p-vector and b is a q-vector, then*

$$a \wedge b = (-1)^{pq} b \wedge a.$$

Proposition 4.13. *If $a \in \Lambda^p \mathbb{R}^n$ and $b \in \Lambda^q \mathbb{R}^n$, then*

$$(a \wedge b)^\dagger = b^\dagger \wedge a^\dagger.$$

Example 4.3. We have talked about simple k-vectors and arbitrary k-vectors. Let us finally give an example of a k-vector that is *not* simple. Let $a = (e_1 \wedge e_2) + (e_3 \wedge e_4)$, a 2-vector in $\mathbb{R}^4$. If a is simple, it can be written as $a = a_1 \wedge a_2$, and this implies that $a \wedge a = 0$. However

$$\begin{aligned} a \wedge a &= (e_1 \wedge e_2 \wedge e_3 \wedge e_4) + (e_3 \wedge e_4 \wedge e_1 \wedge e_2) \\ &= 2 e_1 \wedge e_2 \wedge e_3 \wedge e_4. \end{aligned}$$

Thus a cannot be simple.

Exercises 4.3.

1. Show that for $a \in \Lambda^p \mathbb{R}^n$ and $b \in \Lambda^q \mathbb{R}^n$ we have $ab = (-1)^{pq} ba$.

2. Show that for $a \in \Lambda^p \mathbb{R}^n$ and $b \in \Lambda^q \mathbb{R}^n$, we have $(a \wedge b)^\dagger = b^\dagger \wedge a^\dagger$.

3. Prove that all 2-vectors in $\mathbb{R}^3$ are simple.

4.4 Oriented hyperplanes, blades, and angles

Suppose that V is a k-dimensional subspace of $\mathbb{R}^n$ and $\{a_1, \ldots, a_k\}$ is a basis for V. A point x of $\mathbb{R}^n$ lies in V precisely when x is a linear combination of $a_1, \ldots, a_k$. But this is true if and only if $x \wedge a_1 \wedge \cdots \wedge a_k = 0$.

More generally, suppose that W is a hyperplane that is a translate of V and W passes through the point p_0. Then x lies in W precisely when the vectors $x - p_0, a_1, \ldots, a_k$ are linearly dependent; that is, when $(x - p_0) \wedge a_1 \wedge \cdots \wedge a_k = 0$. (See Figure 4.1.) This

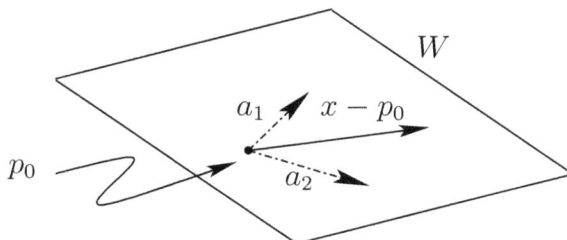

Figure 4.1: Point in a hyperplane

suggests the following concept:

Definition 4.4. By a *k-blade* in $\mathbb{R}^n$ we mean a nonzero simple k-vector $a_1 \wedge \cdots \wedge a_k$.

Our discussion above shows the following:

Proposition 4.14. *Suppose the k-dimensional hyperplane W in $\mathbb{R}^n$ passes through the point p_0 and is a translate of the vector subspace V. If a is a k-blade in V, then the equation of W is $(x - p_0) \wedge a = 0$.*

In this last result we talk of V as being the vector subspace *generated* or *determined* by the blade a, and we may refer to a as being *parallel* to V or W.

Example 4.4. Suppose we write down an equation of the form $(x - p_0) \wedge a = 0$ for the $\chi_1 \chi_2$-plane in $\mathbb{R}^3$. (Figure 4.2.) We know that e_1 and e_2 lie in the $\chi_1 \chi_2$-plane, so we can

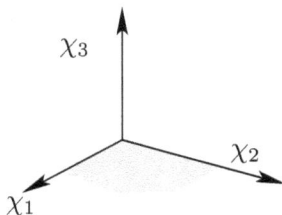

Figure 4.2: $\chi_1 \chi_2$-plane in $\mathbb{R}^3$

set $a = e_1 \wedge e_2$, and we can take p_0 to be $(0,0,0)$. Let us set $x = \chi_1 e_1 + \chi_2 e_2 + \chi_3 e_3$. The equation $(x - p_0) \wedge a = 0$ is

$$(\chi_1 e_1 + \chi_2 e_2 + \chi_3 e_3) \wedge (e_1 \wedge e_2) = 0.$$

Multiplying out the left-hand side, we obtain

$$\chi_3 (e_1 \wedge e_2 \wedge e_3) = 0$$

so that (not surprisingly) we can say that the equation of the $\chi_1 \chi_2$-plane is $\chi_3 = 0$.

If we consider a line in $\mathbb{R}^n$, we know that we can think of it as having an *orientation*, that is, a preferred direction. To be a little more formal, we specify a direction by choosing a vector w parallel to the line $\mathcal{L}$. (Figure 4.3.) Since it is only the direction

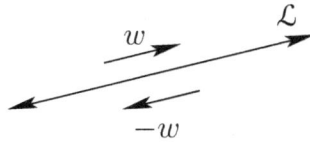

Figure 4.3: Line with orientations

of the vector that is important, not its magnitude, we may as well suppose w is a unit vector. That being the case, we see that $\mathcal{L}$ has precisely two possible orientations, w and $-w$.

We may, if we wish to sound important, describe a line as a 1-dimensional hyperplane. We generalize the idea of orientation to k-dimensional hyperplanes.

Let V_0 be a k-dimensional hyperplane in $\mathbb{R}^n$ passing through the point p_0. Let V be the vector subspace associated with V_0. This V is unique and $V_0 = V + p_0$. We define an *orientation* of V_0 to be a unit k-blade w that is parallel to V_0. (When we call w a *unit* blade, we mean of course that $|w| = 1$.)

To say that w is parallel to V_0 means we can write w in the form $w = a_1 \wedge \cdots \wedge a_k$ where $a_1, \ldots, a_k$ are vectors in V. If $b_1 \wedge \cdots \wedge b_k$ is any other simple k-vector in V, we know from Proposition 3.8 that $b_1 \wedge \cdots \wedge b_k = \lambda (a_1 \wedge \cdots \wedge a_k)$ for some scalar λ.

Suppose w_0 is a second orientation for V_0. We must have $w_0 = \lambda w$ and $|w_0| = |w| = 1$. This forces $\lambda = \pm 1$. It follows that V_0 can have only two orientations; if w is one of them, then the other one is $-w$.

We may, if we wish, think of an orientation of a hyperplane as a sense of *left-handedness* or *right-handedness*.

Example 4.5. The orientations of the $\chi_1 \chi_2$-plane in $\mathbb{R}^3$ are $\pm e_1 \wedge e_2$.

Example 4.6. Let us find the orientations of the plane V_0 in $\mathbb{R}^3$ given by $\chi_1 + \chi_2 + \chi_3 = 1$. (Technically, this is a 2-dimensional hyperplane.)

The associated vector subspace is V described by the equation $\chi_1 + \chi_2 + \chi_3 = 0$. Since $(1, -1, 0)$ and $(1, 0, -1)$ are solutions of the last equation, we see that $e_1 - e_2$ and

$e_1 - e_3$ are two independent vectors in V. Then $a = (e_1 - e_2) \wedge (e_1 - e_3)$ is a 2-blade parallel to V_0. This is not yet an orientation since it is not a unit 2-blade. If we replace a by $a/|a|$, we will have a unit 2-blade that is parallel to V_0.

We can compute $|a|^2$ either using the formula

$$|a|^2 \; = \; a \cdot a^\dagger \; = \; \det(a_i \cdot a_j)$$

where $a_1 = e_1 - e_2$ and $a_2 = e_1 - e_3$ or by rewriting a in terms of the form $e_{i_1} \wedge e_{i_2}$ and appealing to Proposition 4.3. In either case, we obtain $|a|^2 = 3$. Thus one orientation of V_0 is

$$w \; = \; \frac{1}{\sqrt{3}} \, (e_1 - e_2) \wedge (e_1 - e_3)$$

and the other is its negative.

We have seen in Example 3.9 how we can use the dot product of 2-blades to talk about the angle between them. We want to define the angle between oriented hyperplanes of the same dimension by using blades parallel to the hyperplanes.

Recall that according to Corollary 4.1, k-blades satisfy $|a \cdot b| \le |a| \, |b|$. This means that $-1 \le (a \cdot b)/|a| \, |b| \le 1$, and we can use this fact thus:

Definition 4.5. By the *angle θ between two k-blades a and b* we mean the unique value $0 \le \theta \le \pi$ defined by

$$\cos(\theta) \; = \; \frac{a \cdot b^\dagger}{|a| \, |b|}.$$

If A and B are k-dimensional hyperplanes in $\mathbb{R}^n$ with orientations w_A and w_B respectively, then we define the *angle between the oriented hyperplanes* by $\cos(\theta) = w_A \cdot w_B^\dagger$ where, as before, $0 \le \theta \le \pi$.

Notice that the angle between hyperplanes depends on their orientation and can change if the orientation changes.

Example 4.7. In $\mathbb{R}^3$, let A be the $\chi_1 \chi_2$-plane and assign it orientation $w_A = e_1 \wedge e_2$. Let B be the plane through the origin spanned by $e_1 + e_3$ and e_2; assign it orientation $w_B = (1/\sqrt{2}) (e_1 + e_3) \wedge e_2$. This is the plane $\chi_1 + \chi_3 = 0$. The angle between A and B is given by

$$\cos(\theta) \; = \; w_A \cdot w_B^\dagger \; = \; -\frac{1}{\sqrt{2}}$$

so that $\theta = 3\pi/4$. If, however, we switch the orientation of B to $-w_B$, then we have $\cos(\theta) = 1/\sqrt{2}$ so that $\theta = \pi/4$. See Figure 4.4.

We can use *any* blades parallel to oriented hyperplanes to compute the angles between the hyperplanes provided we know that the blades "agree with" the orientations. We make this idea explicit thus:

If a and b are k-vectors, then $a \overset{\text{ray}}{=} b$ means that $a = \lambda b$ for some scalar $\lambda > 0$. That is, a lies on the "ray" of k-vectors τb, $\tau > 0$, generated by b.

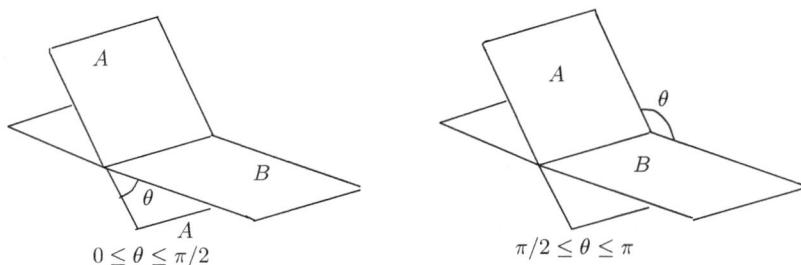

Figure 4.4: Angles between hyperplanes depend on orientation

If a hyperplane A has orientation w_A, then a blade a is parallel to A if and only if $a \overset{\text{ray}}{=} \pm w_A$. We say that *a agrees with orientation of A* provided $a \overset{\text{ray}}{=} w_A$; this amounts to $a = |a| w_A$. We will find this to be a useful idea later.

In the case where A and B are oriented hyperplanes, we then have

$$\cos(\theta) = \frac{a \cdot b^\dagger}{|a| \, |b|}$$

provided a and b agree with w_A and w_B respectively. This is because if $a \overset{\text{ray}}{=} w_A$ and $b \overset{\text{ray}}{=} w_B$, then

$$w_A = \frac{a}{|a|} \quad \text{and} \quad w_B = \frac{b}{|b|}.$$

Exercises 4.4.

1. Let $\mathcal{L}$ be the line in $\mathbb{R}^n$ given by the parametric equation $x(\tau) = p_0 + a\tau$ where τ is our parameter, p_0 is a point in $\mathbb{R}^n$, a is a nonzero vector, and $x(\tau)$ is an arbitrary point on the line. What are the orientations of $\mathcal{L}$? How do you know this?

2. Let P_i be the $(n-1)$-dimensional vector subspace $\chi_i = 0$ in $\mathbb{R}^n$. What are the orientations of P_i? How do you know this?

3. If P is the plane in $\mathbb{R}^3$ given by the equation $2\chi_1 + \chi_2 - 5\chi_3 = 2$, what are the orientations of P? How do you know this?

4. In $\mathbb{R}^3$, let A be the plane $\chi_1 + \chi_2 + \chi_3 = 1$ with orientation

$$w = (e_1 - e_2) \wedge (e_1 - e_3)/\sqrt{3}$$

and P_i be the plane $\chi_i = 0$ for $i = 1, 2, 3$. If θ_i is the angle between A and P_i, show that $\cos(\theta_i) = \pm 1/\sqrt{3}$. What orientations should we choose for P_i so that we always have $\cos(\theta_i) = 1/\sqrt{3}$?

5. In $\mathbb{R}^n$, let A_n be the hyperplane $\sum_{i=1}^n \chi_i = 1$ and P_i be the hyperplane $\chi_i = 0$ for $i = 1, \ldots, n$. Assign P_i the orientation

$$w_{i,n} = (-1)^i e_1 \wedge \cdots \wedge \widehat{e_i} \wedge \cdots \wedge e_n.$$

   (a) Show that

   $$a_n = \sum_{i=1}^n (-1)^i e_1 \wedge \cdots \wedge \widehat{e_i} \wedge \cdots \wedge e_n$$

   is a blade parallel to A_n.

   (b) Find an orientation for A_n.

   (c) If $\theta_{i,n}$ is the angle between A_n and P_i, show that $\theta_{i,n}$ is the same for all i.

   (d) Show that $\theta_{i,n} \to \pi/2$ as $n \to \infty$.

6. Suppose that $a = a_1 \wedge \cdots \wedge a_k$ and $b = b_1 \wedge \cdots \wedge b_k$ are k-blades parallel to the hyperplanes A and B respectively. Show that if there is an a_i that is orthogonal to each of $b_1, \ldots, b_k$, then the angle between A and B is $\pi/2$.

4.5 Tangent blades and oriented manifolds

We know what it means to have an orientation for a hyperplane. We would like to extend this idea to manifolds.

We can think of curves in $\mathbb{R}^n$ as having an orientation in the sense of a preferred direction. (Figure 4.5.) There are only two possibilities for such an orientation, "this

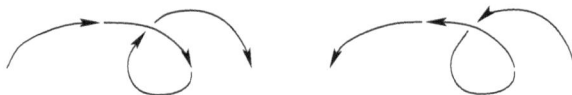

Figure 4.5: A curve admits two orientations

way" or "the opposite one."

   In more detail, we can attach a unit tangent vector to each point of the curve to indicate its orientation. (See Figure 4.6.) This must be done in a *continuous* fashion;

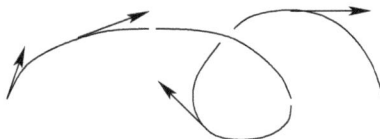

Figure 4.6: Curve orientation given by unit tangent vectors

that is, there cannot be any place where the unit tangent vector suddenly switches from one direction to its opposite. Since a curve $\mathcal{A}$ is usually described by a parametrization $\tau \mapsto x(\tau)$, that is, by a function x from some interval of real numbers to the points of $\mathcal{A}$ in $\mathbb{R}^n$, we see that we can take $x'(\tau)$ as a tangent vector. (If we think of τ as time and $x(\tau)$ as the position at time τ of a particle traversing $\mathcal{A}$, then $x'(\tau)$ is just the velocity vector.) In general, $x'(\tau)$ will not be a unit vector, but if the point p on $\mathcal{A}$ corresponds to τ via $p = x(\tau)$, then we can attach a unit tangent vector w to $\mathcal{A}$ at p by

$$w = w(p) = \frac{x'(\tau)}{|x'(\tau)|}.$$

We then refer to the map $p \mapsto w(p)$, the assignment of a unit tangent vector to every point of the curve $\mathcal{A}$, as the *orientation of $\mathcal{A}$ induced by the parametrization $x(\tau)$.*

It turns out that this same idea can be generalized to manifolds.

Recall that a tangent vector to a manifold $\mathcal{M}$ at the point x_0 is an element of the tangent space $T_{x_0}\mathcal{M}$ and is described in Definitions 2.13 and 2.15.

Definition 4.6. Let x_0 be a point on the k-manifold $\mathcal{M}$ where, as always, the manifold is at least $\mathcal{C}^1$. Assuming $\mathcal{M}$ lies in $\mathbb{R}^n$, we say that a simple q-vector a in $\mathbb{R}^n$ is a simple *tangent q-vector to $\mathcal{M}$ at x_0* provided a can be written in the form $a = a_1 \wedge \cdots \wedge a_q$ where each a_i is a tangent vector to $\mathcal{M}$ at x_0. By the *space of tangent q-vectors to $\mathcal{M}$ at x_0, $\Lambda^q T_{x_0}\mathcal{M}$,* we mean the set of all finite linear combinations of simple tangent q-vectors to $\mathcal{M}$ at the point x_0.

By $\Lambda^1 T_{x_0}\mathcal{M}$ we mean, of course, $T_{x_0}\mathcal{M}$, and we take $\Lambda^0 T_{x_0}\mathcal{M}$ to be $\mathbb{R}$, the set of scalars.

Remark 4.2. If $\mathcal{M}$ is a k-manifold, then $\Lambda^q T_{x_0}\mathcal{M}$ must be the trivial vector space consisting of only 0 whenever $q > k$. Also, by Proposition 4.5, we have

$$\dim\left(\Lambda^q T_{x_0}\mathcal{M}\right) = \binom{k}{q}$$

for $0 \leq q \leq k$.

Example 4.8. Let $\mathcal{S}$ be the manifold $\chi_3 = \phi(\chi_1, \chi_2)$ in $\mathbb{R}^3$ where ϕ is a $\mathcal{C}^1$ real-valued function. We take $x : \mathbb{R}^2 \to \mathcal{S}$ to be the $\mathcal{C}^1$ parametrization

$$x = x(\tau_1, \tau_2) = \tau_1 e_1 + \tau_2 e_2 + \phi(\tau_1, \tau_2) e_3.$$

See Figure 4.7. We can also think of x as the point

$$x = \chi_1 e_1 + \chi_2 e_2 + \chi_3 e_3 = \tau_1 e_1 + \tau_2 e_2 + \phi(\tau_1, \tau_2) e_3$$

where (χ_1, χ_2, χ_3) are the coordinates of the point $x \in \mathcal{S}$ in $\mathbb{R}^3$. We then calculate tangent vectors to $\mathcal{S}$ at the point x:

$$\frac{\partial x}{\partial \tau_i}(x) = e_i + \frac{\partial \phi}{\partial \tau_i} e_3 = e_i + \frac{\partial \chi_3}{\partial \chi_i} e_3 \quad \text{for } i = 1, 2.$$

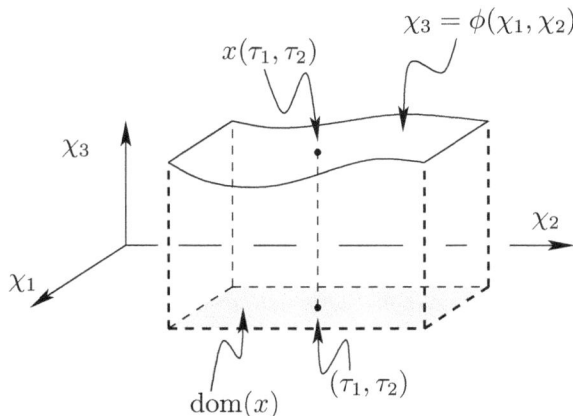

Figure 4.7: The graph of $\chi_3 = \phi(\chi_1, \chi_2)$

$T_x S$ is a two-dimensional vector space and every tangent vector at x has the form

$$v = \lambda_1 \left(e_1 + \frac{\partial \chi_3}{\partial \chi_1} e_3 \right) + \lambda_2 \left(e_2 + \frac{\partial \chi_3}{\partial \chi_2} e_3 \right)$$

where λ_1, λ_2 are arbitrary scalars. By Remark 4.2, we know that $\Lambda^2 T_x S$ is a one-dimensional vector space, so every tangent 2-vector at x must be of the form

$$a = \lambda \left(e_1 + \frac{\partial \chi_3}{\partial \chi_1} e_3 \right) \wedge \left(e_2 + \frac{\partial \chi_3}{\partial \chi_2} e_3 \right)$$

where λ is an arbitrary scalar.

We now extend the idea of orientation to manifolds:

Definition 4.7. Let $\mathcal{M}$ be a k-dimensional $\mathcal{C}^1$ manifold. By an *orientation of* $\mathcal{M}$ we mean a continuous function $x \mapsto w(x)$ where $x \in \mathcal{M}$ and $w(x)$ is a unit tangent k-blade to $\mathcal{M}$ at x.

The intuition behind the definition is something like this: Consider a 2-cell $\mathcal{M}$ in $\mathbb{R}^3$. We imagine a unit tangent 2-blade $w(x)$ attached to $\mathcal{M}$ at successive points, $x_0, x_1,$ x_2, etc. (See Figure 4.8.) Each 2-blade has a certain "handedness" or orientation. We think of moving from point to point, being careful not to "change" the orientation as we do so. (This is the reason for the requirement of continuity.) This process can be thought of as imparting a certain "handedness" at each point of $\mathcal{M}$ and hence to the cell as a whole.

Example 4.9. Let us return to the surface $\chi_3 = \phi(\chi_1, \chi_2)$ of Example 4.8. If $x = (\chi_1, \chi_2, \chi_3)$ is an arbitrary point on the manifold, we can assign an orientation $w(x)$ at

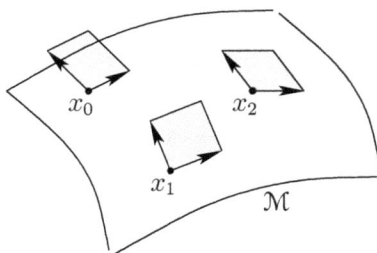

Figure 4.8: Orientation of a 2-cell

that point by setting

$$w(x) = \lambda(x) \left[\left(e_1 + \frac{\partial \chi_3}{\partial \chi_1}(x) e_3 \right) \wedge \left(e_2 + \frac{\partial \chi_3}{\partial \chi_2}(x) e_3 \right) \right]$$

where $\lambda(x)$ is chosen to force $w(x)$ to be a unit blade. That is,

$$\lambda(x) = \frac{1}{\left| \left(e_1 + \frac{\partial \chi_3}{\partial \chi_1}(x) e_3 \right) \wedge \left(e_2 + \frac{\partial \chi_3}{\partial \chi_2}(x) e_3 \right) \right|}$$

$$= \frac{1}{\sqrt{1 + \left(\frac{\partial \chi_3}{\partial \chi_1} \right)^2 + \left(\frac{\partial \chi_3}{\partial \chi_2} \right)^2}}.$$

Proposition 4.15. *If* $\mathcal{M}$ *is a* $\mathcal{C}^1$ *k-manifold with a coordinate patch* $x(t) = x(\tau_1, \ldots, \tau_k)$ *defined on some neighborhood of* $p_0 \in \mathcal{M}$, *then we can define an orientation on that neighborhood of* p_0 *by*

$$w(p) = \lambda(p) \left(\frac{\partial x}{\partial \tau_1}(t) \wedge \cdots \wedge \frac{\partial x}{\partial \tau_k}(t) \right) \tag{4.4}$$

where $p = x(t)$ *and* $\lambda(p)$ *is chosen so that* $w(p)$ *is a unit blade.*

Proof. Suppose that $p = x(t)$. Since $\{(\partial x / \partial \tau_i)(t)\}_{i=1}^{k}$ is a basis for $T_p \mathcal{M}$, we see that

$$\frac{\partial x}{\partial \tau_1}(t) \wedge \cdots \wedge \frac{\partial x}{\partial \tau_k}(t) \neq 0$$

and it is a tangent k-blade to $\mathcal{M}$ at p. We set

$$\lambda(p) = \frac{1}{\left| \frac{\partial x}{\partial \tau_1}(t) \wedge \cdots \wedge \frac{\partial x}{\partial \tau_k}(t) \right|}$$

in (4.4). Finally, Notice that since x and x^{-1} are $\mathcal{C}^1$, the map

$$p \mapsto w(p) = (w \circ x \circ x^{-1})(p)$$

must be continuous on $\mathcal{M}$. This gives us the desired orientation. $\square$

Sometimes we begin with a given $\mathcal{C}^1$ k-manifold $\mathcal{M}$ and a given orientation w. We can then ask whether or not the orientation induced by a chart $x = x(\tau_1, \ldots, \tau_k)$ is the same as w.

Definition 4.8. We say that the orientation w of $\mathcal{M}$ *agrees* with the coordinatization $(\tau_1, \ldots, \tau_k)$ at $x \in \mathcal{M}$ provided

$$w(x) \;=\; \lambda(x) \left(\frac{\partial x}{\partial \tau_1}(x) \wedge \cdots \wedge \frac{\partial x}{\partial \tau_k}(x) \right)$$

for $\lambda(x) > 0$.

Of course this condition of *agreement* between w and the coordinatization can be written

$$w \;\overset{\text{ray}}{=}\; \frac{\partial x}{\partial \tau_1} \wedge \cdots \wedge \frac{\partial x}{\partial \tau_k}.$$

Now here is an important generalization of a property of orientations on curves to orientations on cells:

Proposition 4.16. *A $\mathcal{C}^1$ cell has precisely two orientations, and each is the negative of the other.*

Proof. Let $\mathcal{M}$ be a $\mathcal{C}^1$ k-cell. $\mathcal{M}$ must have a $\mathcal{C}^1$ parametrization $x : \mathcal{I}^k \to \mathcal{M}$, so it must have an orientation w.

Let p be a point on $\mathcal{M}$ and consider the tangent space $T_p\mathcal{M}$.

We must have $p = x(t)$ for some $t \in \mathcal{I}^k$. We know that the vectors $u_i = (\partial x / \partial \tau_i)(t)$, where $i = 1, \ldots, k$, constitute a basis for $T_p\mathcal{M}$. By Proposition 3.8, every k-vector tangent to $\mathcal{M}$ at p must be of the form $\lambda (u_1 \wedge \cdots \wedge u_k)$. In particular,

$$w(p) \;=\; \lambda (u_1 \wedge \cdots \wedge u_k) \quad \text{where} \quad \lambda \;=\; \pm 1 / |u_1 \wedge \cdots \wedge u_k|. \tag{4.5}$$

Let w_0 be a second orientation of $\mathcal{M}$. Equation (4.5) must hold for $w_0(p)$ as it does for $w(p)$, hence we see that $w_0(p) = \pm w(p)$. It follows that $|w(p) - w_0(p)|$ can only have the values 0 or 2. Now the map $p \mapsto |w(p) - w_0(p)|$, considered as map of $\mathcal{M}$ into the reals, is continuous and can only have the values 0 or 2. Thus either $|w(p) - w_0(p)| = 0$ for all p, in which case $w = w_0$, or $|w(p) - w_0(p)| = 2$ for all p, in which case $w = -w_0$. $\qquad\square$

More generally, one can show that a manifold that is $\mathcal{C}^1$ and connected will have precisely two orientations, assuming that it has any orientations at all. It is well-known that the Möbius strip is connected and, though it can be $\mathcal{C}^1$, it *cannot* have an orientation which is defined on the whole manifold.

Exercises 4.5.

1. Let H be the helix in $\mathbb{R}^3$ with parametrization $x : \mathbb{R} \to H$ given by

$$x(\tau) \;=\; \cos(\tau) e_1 + \sin(\tau) e_2 + \tau e_3.$$

Find an orientation $w = w(\chi_1, \chi_2, \chi_3)$ of H in terms of the point (χ_1, χ_2, χ_3) of H at which the orientation is evaluated and also such that w agrees with the parametrization $x(\tau)$.

2. Find an orientation $w = w(\xi_1, \xi_2, \xi_3)$ of the surface S in $\mathbb{R}^3$ given by the equation $\xi_3 = \xi_1^2 - \xi_2^2$. (Hint: A chart of this surface is given by

$$x(\xi_1, \xi_2) \;=\; \xi_1 e_1 + \xi_2 e_2 + \xi_3 e_3$$

where $\xi_3 = \xi_1^2 - \xi_2^2$.)

3. Construct an orientation $w = w(\xi_1, \xi_2, \xi_3)$ of the upper hemisphere of the unit sphere in $\mathbb{R}^3$,

$$\xi_1^2 + \xi_2^2 + \xi_3^2 \;=\; 1 \quad \text{restricted to } \xi_3 > 0,$$

where (ξ_1, ξ_2, ξ_3) is an arbitrary point on the upper hemisphere. Does this expression give an orientation for the *whole* sphere?

4. If a k-dimensional manifold S has an orientation w that agrees with a coordinate patch x, then show that

$$w(p) \;=\; \frac{\left(\wedge^k x'(t)\right)(e_1 \wedge \cdots \wedge e_k)}{\left|\left(\wedge^k x'(t)\right)(e_1 \wedge \cdots \wedge e_k)\right|}$$

at the point $p = x(t)$.

5. It is not always true that orientable manifolds have only two orientations. If a manifold S consists of two disjoint circles in $\mathbb{R}^2$, then how many orientations can S have?

4.6   Integration over manifolds

We assume the reader is familiar with integrals of real-valued functions over n-dimensional regions, at least in the cases $n = 2, 3$. Typically we use the notation $\int_U \phi$ for such an integral where U is some "nice" subset of $\mathbb{R}^n$ and ϕ is a "well-behaved" real-valued function on U.

If U is, for example, a 2-dimensional region, we might write the integral thus:

$$\int_U \phi \;=\; \int_U \phi(\chi_1, \chi_2)\, d\chi_1\, d\chi_2.$$

In the event that U is a rectangle, say, $U = [\alpha_0, \alpha_1] \times [\beta_0, \beta_1]$, we might evaluate the integral by turning it into an *iterated* integral:

$$\int_U \phi \;=\; \int_{\beta_0}^{\beta_1} \left(\int_{\alpha_0}^{\alpha_1} \phi(\chi_1, \chi_2)\, d\chi_1 \right) d\chi_2.$$

More generally, we might need to resort to some sort of numerical technique to evaluate the integral of ϕ.

The knowledgable reader may assume we use either Riemann or Lebesgue integration; it makes no difference for the functions we consider.

In this chapter, we introduce integration of functions—whether real-valued or k-vector-valued—over *unoriented* manifolds lying in some $\mathbb{R}^n$, usually an $\mathbb{R}^n$ of higher dimension than the manifold.

We shall briefly indicate how orientation might enter into such integrals, but in order to really explain integration over oriented manifolds, we need the machinery of geometric algebra which comes later. Even so, without touching on orientation, the notion of tangent blades is very helpful in seeing the right way to define integration.

4.6.1 The change-of-variables formula

Before we can start on manifolds, we need the change-of-variables formula for multiple integrals over regions in $\mathbb{R}^n$. Unfortunately, this is a fairly difficult thing to prove and would take us too far afield. We will state it and discuss the intuition behind it; simple k-vectors are helpful here. A full proof can be found in [24] or [35].

Proposition 4.17 (Change-of-variables formula). *Suppose U and V are open subsets of $\mathbb{R}^n$ and $x : U \to V$ is a one-to-one onto map such that both x and x^{-1} are $\mathcal{C}^1$. If $\phi : V \to \mathbb{R}$ and ϕ is integrable over V, then $(\phi \circ x) |\det x'|$ is integrable over U and*

$$\int_U (\phi \circ x) |\det x'| = \int_V \phi. \tag{4.6}$$

(*See Figure 4.9.*)

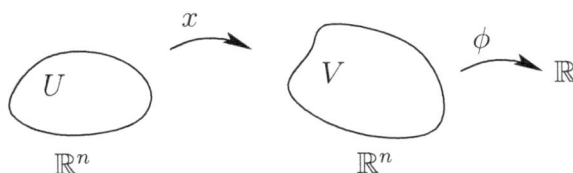

Figure 4.9: The setup for change-of-variables

This formula also applies to certain n-dimensional subsets of $\mathbb{R}^n$ such as cells, simplexes, solid tori, etc. because our definition of $\mathcal{C}^1$ requires this property to extend beyond the boundary of the cell, hence onto an open set containing the cell.

Example 4.10. Let $U = [0,1] \times [\frac{1}{2}, 1]$ and let V be the image in $\mathbb{R}^2$ of U under the map

$$x(t) = x(\tau_1, \tau_2) = (\tau_1 + \tau_2) e_1 + \tau_2^2 e_2.$$

See Figure 4.10. We think of x as being written in the form $x(t) = \chi_1(t) e_1 + \chi_2(t) e_2$ and recall that the matrix of $x'(t)$ is $(\partial \chi_i / \partial \tau_j)_{i,j=1}^2$. It is easily calculated that $|\det(x'(t))| = 2\tau_2$. So for a real-valued function ϕ on V, the change-of-variables formula amounts to

$$\int_0^1 \int_0^1 \phi(\tau_1 + \tau_2, \tau_2^2) 2\tau_2 \, d\tau_1 \, d\tau_2 = \int_V \phi(\chi_1, \chi_2) \, d\chi_1 \, d\chi_2.$$

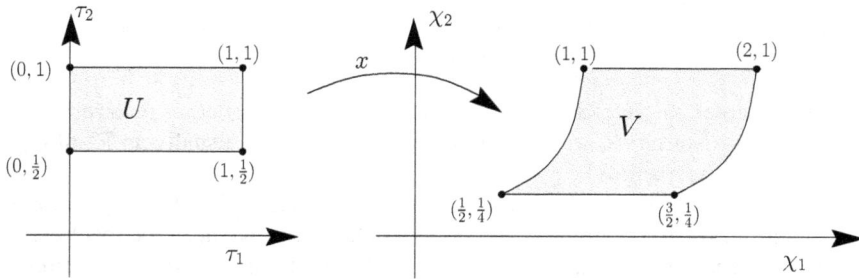

Figure 4.10: A transformation of the unit square

Let us try to see the intuition behind the change-of-variables formula. We draw some pictures in two dimensions:

First we slice U into a large number of "small" squares by an imposed grid. This induces a division of V into small curved regions, the images under x of the squares. (Figure 4.11.)

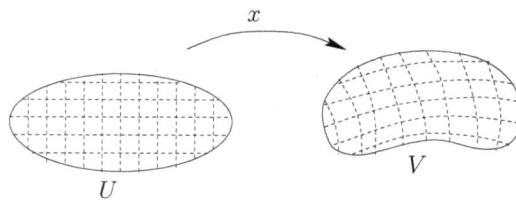

Figure 4.11: A partition of U induces one of V

For the sake of simplicity, we suppose that each small square in U has height and width λ. From each square we choose a point t_i and we set $x_i = x(t_i)$ (where we take i to be an index with which we tag the small square). Consider a small square U_i and its image V_i. (Figure 4.12.) We can approximate the integral of ϕ over V by a Riemann

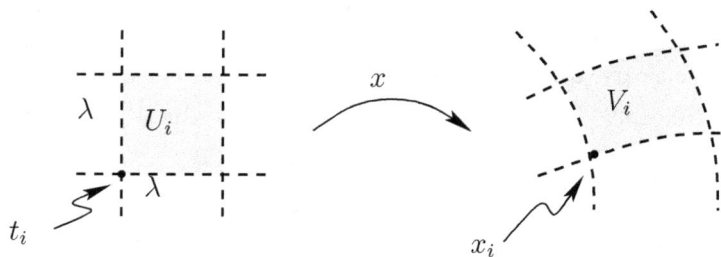

Figure 4.12: U_i and its image

sum:

$$\int_V \phi \approx \sum_i \phi(x_i) \, \text{area}(V_i).$$

Let us try to construct a Riemann sum over U that should give us (approximately) $\int_V \phi$. It should look something like this:

$$\sum_i \phi\left(x(t_i)\right)(?) \approx \sum_i \phi(x_i) \, \text{area}(V_i). \tag{4.7}$$

We cannot plug $\text{area}(U_i)$ in for the question mark; in general, $\text{area}(U_i)$ and $\text{area}(V_i)$ are very different. How do these areas compare?

Consider Figure 4.13. We may take the sides of U_i to be the vectors λe_1 and λe_2. If

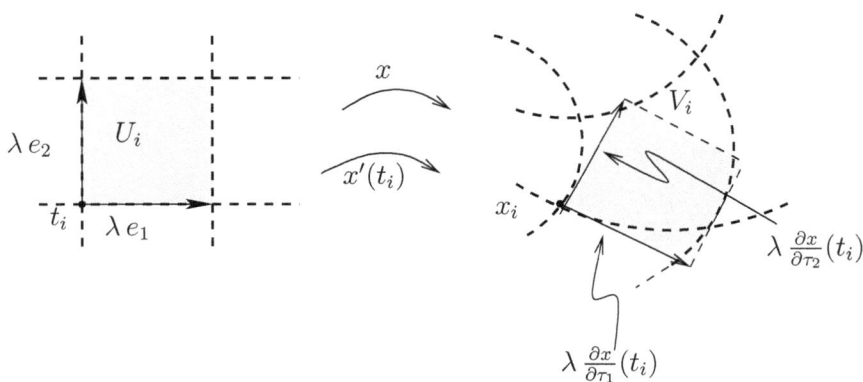

Figure 4.13: Approximating V_i using $x'(t_i)$

we apply the linear transformation $x'(t_i)$, we see that the sides of V_i are approximated by the vectors $\lambda \left[(\partial x)/(\partial \tau_i)\right](t_i)$ for $i = 1, 2$. The region V_i is, in some sense, approximated by the oriented parallelogram

$$\left(\lambda \frac{\partial x}{\partial \tau_1}(t_i)\right) \wedge \left(\lambda \frac{\partial x}{\partial \tau_2}(t_i)\right),$$

and the area of this parallelogram is simply the magnitude of the simple 2-vector. Therefore

$$\text{area}(U_i) = \lambda^2,$$

$$\text{area}(V_i) \approx \lambda^2 \left| \frac{\partial x}{\partial \tau_1}(t_i) \wedge \frac{\partial x}{\partial \tau_2}(t_i) \right|.$$

So we can replace Equation (4.7) by

$$\sum_i \phi\left(x(t_i)\right) \left| \frac{\partial x}{\partial \tau_1}(t_i) \wedge \frac{\partial x}{\partial \tau_2}(t_i) \right| \left(\text{area}(U_i)\right) \approx \sum_i \phi(x_i) \left(\text{area}(V_i)\right). \tag{4.8}$$

The left-hand side of (4.8) is a Riemann sum for an integral over U. Thus what we expect to be true for this 2-dimensional integral is

$$\int_U (\phi \circ x)(\tau_1, \tau_2) \left| \frac{\partial x}{\partial \tau_1}(\tau_1, \tau_2) \wedge \frac{\partial x}{\partial \tau_2}(\tau_1, \tau_2) \right| d\tau_1 \, d\tau_2$$

$$= \int_V \phi(\chi_1, \chi_2) \, d\chi_1 \, d\chi_2,$$

or, more compactly,

$$\int_U (\phi \circ x) \left| \frac{\partial x}{\partial \tau_1} \wedge \frac{\partial x}{\partial \tau_2} \right| = \int_V \phi. \tag{4.9}$$

This is not quite the change-of-variables formula; we still need to introduce $\det\left(x'(t)\right)$. This is easy to do: Write $x(t) = \chi_1(t) e_1 + \chi_2(t) e_2$. Then

$$\frac{\partial x}{\partial \tau_i} = \frac{\partial \chi_1}{\partial \tau_i} e_1 + \frac{\partial \chi_2}{\partial \tau_i} e_2, \quad i = 1, 2,$$

and by Proposition 3.8 and Example 3.8, we have

$$\frac{\partial x}{\partial \tau_1} \wedge \frac{\partial x}{\partial \tau_2} = \det\left(\frac{\partial \chi_i}{\partial \tau_j}\right)_{2 \times 2} e_1 \wedge e_2 = \left(\det(x'(t))\right) e_1 \wedge e_2.$$

Thus

$$\left| \frac{\partial x}{\partial \tau_1} \wedge \frac{\partial x}{\partial \tau_2} \right| = \left| \det(x'(t)) \right|.$$

This gives us the change-of-variables formula in Proposition 4.17 for $n = 2$.

The heuristic argument above can be made for integrals of ϕ defined on n-dimensional regions; one replaces squares by n-dimensional cubes and areas by volumes.

We now restate our formula with a change that is useful when we define integrals over k-dimensional manifolds in $\mathbb{R}^n$:

Restatement of the change-of-variables formula:

Given the hypotheses of Proposition 4.17,

$$\int_U (\phi \circ x) \left| \frac{\partial x}{\partial \tau_1} \wedge \cdots \wedge \frac{\partial x}{\partial \tau_n} \right| = \int_V \phi. \tag{4.10}$$

4.6.2   Integrals over manifolds

Let $\mathcal{M}$ be a k-dimensional manifold in $\mathbb{R}^n$ and ϕ a real-valued function defined on $\mathcal{M}$. We want to define $\int_{\mathcal{M}} \phi$, the integral of ϕ over $\mathcal{M}$.

Suppose that $x : U \to \mathcal{M}$ is a parametrization of $\mathcal{M}$ and it assigns coordinates $(\tau_1, \ldots, \tau_k)$ to points of $\mathcal{M}$. Recall this means that a point $p \in \mathcal{M}$ has coordinates $(\tau_1, \ldots, \tau_k)$ with respect to the parametrization (or coordinate patch) x if $p = x(\tau_1, \ldots, \tau_k)$

where $(\tau_1, \ldots, \tau_k) \in U \subseteq \mathbb{R}^k$. We intend U to be a "nice" subset of $\mathbb{R}^k$, for example, a k-cell, rectangle, simplex, open set, etc. We set

$$\int_{\mathcal{M}} \phi \overset{\text{def.}}{=} \int_U (\phi \circ x) \left| \frac{\partial x}{\partial \tau_1} \wedge \cdots \wedge \frac{\partial x}{\partial \tau_k} \right|. \tag{4.11}$$

The integral on the right is that of a real-valued function over a "nice" subset of $\mathbb{R}^k$, so we expect it to be integrable using standard techniques of calculus.

The intuitive justification for (4.11) is essentially the same as for the change-of-variables formula:

A "small" rectangle U_j in $\mathbb{R}^k$ with sides λe_i is mapped by the linear transformation $x'(t)$ to a rectangle V_j with sides $\lambda (\partial x/\partial \tau_i)(t)$ that approximates a small portion of a manifold. See Figure 4.14. We know that

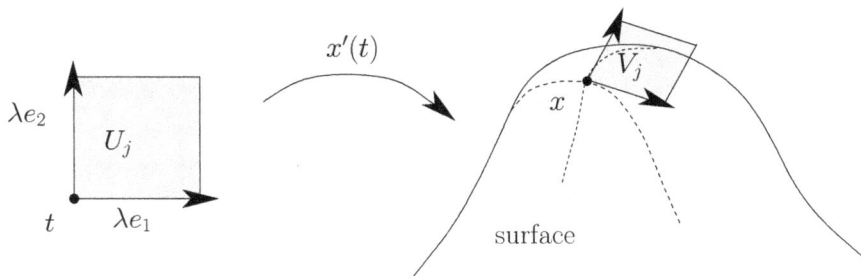

Figure 4.14: Tangent blade approximation

$$\text{vol}(V_j) = \left| \frac{\partial x}{\partial \tau_1}(t) \wedge \cdots \wedge \frac{\partial x}{\partial \tau_k}(t) \right| \lambda^k,$$

so that the factor by which $x'(t)$ changes the volume of "small" k-rectangles is

$$\left| \frac{\partial x}{\partial \tau_1}(t) \wedge \cdots \wedge \frac{\partial x}{\partial \tau_k}(t) \right|.$$

This suggests definition (4.11).

One difference between definition (4.11) and the change-of-variables formula is that $\det(x'(t))$ usually does not play a role in (4.11). This is because, in general, the matrix for $x'(t)$ is not square so the determinant is not there.

Example 4.11. Suppose that $\mathcal{M}$ is an arc in $\mathbb{R}^n$. It should have a parametrization $x : [\alpha, \beta] \to \mathcal{M}$, and we will assume all functions involved are at least $\mathcal{C}^1$. We must be able to write the parametrization in the form

$$x(\tau) = \chi_1(\tau) e_1 + \cdots + \chi_n(\tau) e_n$$

where $\alpha \leq \tau \leq \beta$. So

$$\frac{dx}{d\tau}(\tau) \;=\; \frac{d\chi_1}{d\tau}(\tau)e_1 + \cdots + \frac{d\chi_n}{d\tau}(\tau)e_n.$$

Therefore

$$\int_{\mathcal{M}} \phi \;=\; \int_{[\alpha,\beta]} (\phi \circ x)\left|\frac{dx}{d\tau}\right| \;=\; \int_\alpha^\beta \phi(x(\tau)) \left(\sqrt{\sum_{i=1}^n \left(\frac{d\chi_i}{d\tau}(\tau)\right)^2}\right) d\tau.$$

This is a standard formula for the integral of ϕ "with respect to arclength" and is often written in the form

$$\int_{\mathcal{M}} \phi \;=\; \int_{\mathcal{M}} \phi(x)\, ds$$

where s is thought of as arclength as measured along $\mathcal{M}$.

Example 4.12. Let $\mathcal{M}$ be the upper hemisphere of the unit sphere $\chi_1^2 + \chi_2^2 + \chi_3^2 = 1$; that is, we restrict ourselves to points $x = (\chi_1, \chi_2, \chi_3)$ for which $\chi_3 \geq 0$. See Figure 4.15. Let us calculate $\int_{\mathcal{M}} 1$. This has to be the area of the hemisphere which we already

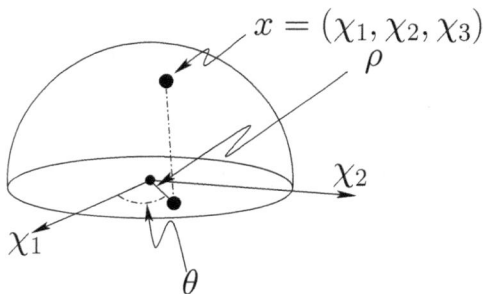

Figure 4.15: Upper hemisphere of the unit sphere

know from elementary geometry is 2π.

We parametrize $\mathcal{M}$ using polar coordinates in the $\chi_1\chi_2$-plane and then projecting straight up. By θ we mean an angle measured in the $\chi_1\chi_2$-plane from the χ_1-axis, and ρ is distance from the origin measured in the $\chi_1\chi_2$-plane. This leads to a parametrization

$$x(\rho,\theta) \;=\; \rho\cos(\theta)e_1 + \rho\sin(\theta)e_2 + \sqrt{1-\rho^2}\,e_3$$

where $0 \leq \theta \leq 2\pi$ and $0 \leq \rho \leq 1$. (There is a technical difficulty here in that the parametrization is not one-to-one everywhere. However when we integrate, the region where the equations are not well-behaved is a "small" set, what is known as "a set of measure zero," and its contribution to the integral is zero. A more careful justification of our flouting the rules depends on convergence theorems which are not available to us.)

It is straightforward to calculate that

$$\left| \frac{\partial x}{\partial \rho} \wedge \frac{\partial x}{\partial \theta} \right| = \frac{\rho}{\sqrt{1 - \rho^2}}.$$

It then follows that

$$\int_{\mathcal{M}} 1 = \int_0^{2\pi} \int_0^1 \frac{\rho}{\sqrt{1 - \rho^2}} \, d\rho \, d\theta = 2\pi$$

which is the desired answer.

Example 4.13. Let S^1 be the unit circle in $\mathbb{R}^2$, that is, the set with equation $\chi_1^2 + \chi_2^2 = 1$. We will take $\mathcal{T}$ to be $S^1 \times S^1$. $\mathcal{T}$ is a torus; we can see this from the fact that it is obtained by taking a circle, S^1, and rotating it in a circle, $S^1 \times S^1$. It lies in $\mathbb{R}^4$ and satisfies the two equations $\chi_1^2 + \chi_2^2 = 1$ and $\chi_3^2 + \chi_4^2 = 1$. We consider the question of integrating a real-valued function ϕ that is defined on $\mathcal{T}$.

We know that we can parametrize S^1 thus:

$$\theta \mapsto \cos(\theta) e_1 + \sin(\theta) e_2.$$

So we take

$$x(\psi, \xi) = \cos(\psi) e_1 + \sin(\psi) e_2 + \cos(\xi) e_3 + \sin(\xi) e_4,$$

where $(\psi, \xi) \in [0, 2\pi] \times [0, 2\pi]$, as our parametrization of $\mathcal{T}$.

(There is a slight technical difficulty here since this parametrization is not one-to-one on the edges of the square. However we can break $[0, 2\pi]^2$ into smaller squares as in Figure 4.16, and x will be a proper parametrization on each of the smaller squares.)

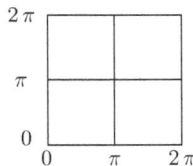

Figure 4.16: Partition of $[0, 2\pi]^2$

We see that

$$\frac{\partial x}{\partial \psi} = -\sin(\psi) e_1 + \cos(\psi) e_2,$$

$$\frac{\partial x}{\partial \xi} = -\sin(\xi) e_3 + \cos(\xi) e_4,$$

and we readily calculate that

$$\left| \frac{\partial x}{\partial \psi} \wedge \frac{\partial x}{\partial \xi} \right| = 1.$$

We finally set up the integral of the function ϕ over $\mathcal{T}$ thus:

$$\int_{\mathcal{T}} \phi = \int_{[0,2\pi]^2} (\phi \circ x) \left| \frac{\partial x}{\partial \psi} \wedge \frac{\partial x}{\partial \xi} \right|$$

$$= \int_0^{2\pi} \int_0^{2\pi} \phi\big(\cos(\psi), \sin(\psi), \cos(\xi), \sin(\xi)\big)\, d\psi\, d\xi.$$

Here is a fundamental property which must be checked before we can consider our definition of integral over a manifold satisfactory:

Proposition 4.18. *Suppose that ϕ is an integrable, real-valued function defined on the k-dimensional $\mathcal{C}^1$ manifold $\mathcal{M}$. Then the integral of ϕ as defined by (4.11) is independent of our choice of parametrization.*

We prove this in Appendix A.4.

If we can integrate real-valued functions over manifolds, it is not hard to see that we can also integrate p-vector fields:

Suppose $\mathcal{M}$ is a k-manifold in $\mathbb{R}^n$ and f is a p-vector field defined over $\mathcal{M}$. Choose a basis $\{u_i\}_{i=1}^n$ for $\mathbb{R}^n$ and expand f in terms of that basis:

$$f = \sum_{I \in \mathcal{O}_p} \phi_I u_I$$

where the summation is over ordered multi-indices of length p. Then set

$$\int_{\mathcal{M}} f \overset{\text{def.}}{=} \sum_{I \in \mathcal{O}_p} \left(\int_{\mathcal{M}} \phi_I \right) u_I. \tag{4.12}$$

(It should be kept in mind here that $u_1, \dots, u_n$ are fixed and constant vectors as would be true for example of $\{e_i\}_{i=1}^n$.) It is not hard to see (Exercise 7 this section) that this definition is independent of the choice of basis $\{u_i\}_{i=1}^n$. It is also clear that it must satisfy linearity properties:

$$\int_{\mathcal{M}} (f+g) = \int_{\mathcal{M}} f + \int_{\mathcal{M}} g \quad \text{and} \quad \int_{\mathcal{M}} \lambda f = \lambda \int_{\mathcal{M}} f$$

where λ is a real number constant.

Example 4.14. Let $\mathcal{A}$ be a $\mathcal{C}^1$ 1-cell, an arc, in $\mathbb{R}^n$ with orientation w. Suppose that $x : \mathcal{I} \to \mathcal{A}$ is a parametrization of $\mathcal{A}$ that agrees with the orientation and that p and q are the initial and terminal points; thus $x(0) = p$ and $x(1) = q$. Of course we have

$$w(x) = \frac{\frac{dx}{d\tau}(\tau)}{\left|\frac{dx}{d\tau}(\tau)\right|} = \frac{1}{\left|\frac{dx}{d\tau}(\tau)\right|} \left(\sum_{i=1}^n \frac{d\chi_i}{d\tau}(\tau) e_i \right)$$

whenever $x = x(\tau) = \sum_{i=1}^n \chi_i(\tau) e_i$. We integrate the orientation over $\mathcal{A}$ and obtain

$$\int_{\mathcal{A}} w = \int_0^1 (w \circ x)(\tau) \left| \frac{dx}{d\tau}(\tau) \right| d\tau$$

$$= \sum_{i=1}^{n} \left(\int_0^1 \frac{d\chi_i}{d\tau}(\tau)\, d\tau \right) e_i$$
$$= q - p,$$

the vector from the initial to the terminal point.

Example 4.15. Let $\mathcal{M}$ be the cylinder in $\mathbb{R}^3$ defined by $\chi_1^2 + \chi_2^2 = 1$ and $0 \leq \chi_3 \leq 1$. See Figure 4.17. We would like to evaluate $\int_{\mathcal{M}} e_1 \wedge e_2 \wedge e_3$.

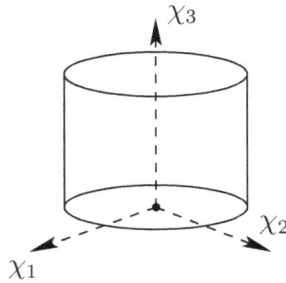

Figure 4.17: Cylinder

We do not even need a parametrization for this:

$$\int_{\mathcal{M}} e_1 \wedge e_2 \wedge e_3 = \left(\int_{\mathcal{M}} 1 \right) e_1 \wedge e_2 \wedge e_3$$
$$= \left(\mathrm{area}(\mathcal{M}) \right) e_1 \wedge e_2 \wedge e_3$$
$$= 2\pi \left(e_1 \wedge e_2 \wedge e_3 \right).$$

A concluding remark: The integral introduced here, $\int_{\mathcal{M}} \phi$, is an *unoriented* integral. However integrals of a single variable in introductory calculus are *oriented* integrals. This is reflected in the equation

$$\int_\alpha^\beta \phi(\tau)\, d\tau = -\int_\beta^\alpha \phi(\tau)\, d\tau$$

in which we integrate over $[\alpha, \beta]$ in two different directions. Once the machinery of geometric algebra is in place, we shall show how we can take into account the orientation not merely of an interval but of a manifold as well when we integrate over it.

Exercises 4.6.

1. Let H be the helix in $\mathbb{R}^3$ given by the parametrization

$$x(\tau) = \cos(\tau) e_1 + \sin(\tau) e_2 + \tau e_3, \quad 0 \leq \tau \leq 2\pi.$$

Let f be the 2-vector field defined on H by

$$f(x) = f(\chi_1, \chi_2, \chi_3) = (e_1 \wedge e_3) - \chi_1(e_1 \wedge e_3) - \chi_2(e_2 \wedge e_3)$$

where $x = (\chi_1, \chi_2, \chi_3) \in H$. Evaluate $\int_H f$.

2. We construct a "helix" H in $\mathbb{R}^n$ (where $n \geq 3$) by means of the parametrization

$$x(\tau) = \cos(\tau) e_1 + \sin(\tau) e_2 + \tau e_3 + \cdots + \tau e_n.$$

Compute the arclength of H.

3. Let S^1 be the unit circle in R^2, namely, the graph of the equation $\chi_1^2 + \chi_2^2 = 1$. Let $\mathcal{T} = S^1 \times S^1$, a torus in $\mathbb{R}^4$. $\mathcal{T}$ is the graph of $x = (\chi_1, \chi_2, \chi_3, \chi_4)$ that satisfies $\chi_1^2 + \chi_2^2 = 1$ and $\chi_3^2 + \chi_4^2 = 1$. Compute the 2-volume (area) of $\mathcal{T}$.

4. Let us construct a torus in $\mathbb{R}^3$. Suppose that ρ_0 and ρ_1 are constants such that $0 < \rho_0 < \rho_1$. We take a circle S in the $\chi_1 \chi_3$-plane centered at $(\rho_1, 0, 0) = \rho_1 e_1$

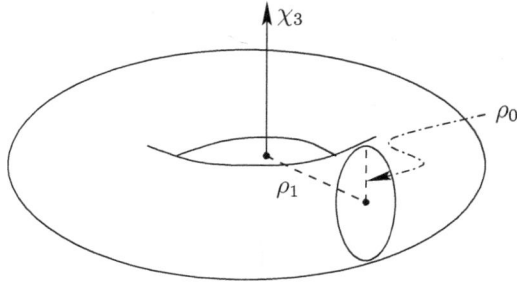

Figure 4.18: Torus in $\mathbb{R}^3$

with radius ρ_0 and rotate it about the χ_3-axis to sweep out a torus $\mathcal{T}$ in $\mathbb{R}^3$. See Figure 4.18. Compute the area of $\mathcal{T}$.

5. Let S^{2+} be the upper hemisphere of the unit 2-sphere in $\mathbb{R}^3$; that is,

$$\chi_1^2 + \chi_2^2 + \chi_3^2 = 1 \quad \text{where } \chi_3 > 0.$$

Assume that an orientation of S^{2+} is

$$w(x) = w(\chi_1, \chi_2, \chi_3) = \chi_3 (e_1 \wedge e_2) - \chi_2 (e_1 \wedge e_3) + \chi_1 (e_2 \wedge e_3)$$

where $x = (\chi_1, \chi_2, \chi_3) \in S^{2+}$. Evaluate $\int_{S^{2+}} w$.

6. Choose positive numbers ρ_0 and ρ_1 and form the two circles $\rho_0 S^1$ and $\rho_1 S^1$. That is, $\rho_0 S^1$ is the set of points (χ_1, χ_2) satisfying $\chi_1^2 + \chi_2^2 = \rho_0^2$ while $\rho_1 S^1$ is the set of (ξ_1, ξ_2) such that $\xi_1^2 + \xi_2^2 = \rho_1^2$. Now form the 2-dimensional torus $\mathcal{T} = (\rho_0 S^1) \times (\rho_1 S^1)$ in $\mathbb{R}^4$. Construct an orientation $w(x) = w(\chi_1, \chi_2, \xi_1, \xi_2)$ for $\mathcal{T}$ and evaluate $\int_{\mathcal{T}} w$.

7. We would like to verify that the definition given in (4.12) is indeed independent of our choice of $\{u_i\}_{i=1}^n$.

   Suppose $\mathcal{M}$ is a k-manifold in $\mathbb{R}^n$ and f is a p-vector field defined over $\mathcal{M}$. Choose two bases $\{u_i\}_{i=1}^n$ and $\{v_i\}_{i=1}^n$ for $\mathbb{R}^n$ and expand f in terms of those bases:

   $$f = \sum_{I \in \mathcal{O}_p} \phi_I\, u_I \quad \text{and} \quad f = \sum_{J \in \mathcal{O}_p} \psi_J\, v_J$$

   where ϕ_i and ψ_j are real-valued functions. Show that

   $$\sum_{I \in \mathcal{O}_p} \left(\int_{\mathcal{M}} \phi_I \right) u_I = \sum_{J \in \mathcal{O}_p} \left(\int_{\mathcal{M}} \psi_J \right) v_J.$$

Chapter 5

The geometric algebra $\mathbb{G}^n$

Before we return to doing calculus on manifolds, we want to construct a new product, the *geometric product*, that, in a sense, includes both the dot and the wedge product.

To do this, we first construct a single new vector space, $\mathbb{G}^n$, that contains the real numbers, $\mathbb{R}^n$, and all the k-vectors in $\mathbb{R}^n$. In this space, it will be perfectly legitimate to write $a + b$ where a is a p-vector and b is a q-vector with $p \neq q$. This space will then be endowed with our new operation ab, the geometric product of a and b, which will enjoy the standard linearity, distributivity, and associativity properties of products. We will call $\mathbb{G}^n$ endowed with this product the *geometric algebra over $\mathbb{R}^n$*.

5.1 $\mathbb{G}^n$ as a vector space

It is very easy to construct $\mathbb{G}^n$ simply as a vector space and nothing more. (We come back to the geometric product in a moment.) We set

$$\mathbb{G}^n = \mathbb{R} \times \mathbb{R}^n \times \Lambda^2 \mathbb{R}^n \times \cdots \times \Lambda^n \mathbb{R}^n$$
$$= \Lambda^0 \mathbb{R}^n \times \Lambda^1 \mathbb{R}^n \times \cdots \times \Lambda^n \mathbb{R}^n.$$

That is, $\mathbb{G}^n$ is the set of all $(n+1)$-tuples $(a_0, a_1, \ldots, a_n)$ where a_0 is a 0-vector (scalar), a_1 is a $1 - vector$ (vector in $\mathbb{R}^n$), etc. We define multiplication by a scalar and addition by

$$\lambda (a_0, a_1, \ldots, a_n) = (\lambda a_0, \lambda a_1, \ldots, \lambda a_n),$$
$$(a_0, \ldots, a_n) + (b_0, \ldots, b_n) = (a_0 + b_0, \ldots, a_n + b_n).$$

It is easily checked that $\mathbb{G}^n$ satisfies the axioms of a vector space. But we cannot call the elements of $\mathbb{G}^n$ vectors (we reserve that name for the inhabitants of $\mathbb{R}^n$) or k-vectors. They are now hybrid creatures and we call them *multivectors*.

By an abuse of notation, we identify any k-vector $a \in \Lambda^k \mathbb{R}^n$ with an obvious element of $\mathbb{G}^n$ by setting

$$a \text{ "="} (0, \ldots, 0, a, 0, \ldots, 0). \tag{5.1}$$

In this way we can say that $\Lambda^k \mathbb{R}^n \subseteq \mathbb{G}^n$ for $k = 0, 1, \ldots, n$.

If $a = (a_0, \ldots, a_n) \in \mathbb{G}^n$, we set

$$\langle a \rangle_k \overset{\text{def.}}{=} a_k.$$

We call $\langle a \rangle_k$ the *grade k part of a*. If $\langle a \rangle_k = a$, then we say that *a has grade k*. (The zero element of $\mathbb{G}^n$ can be considered to be any grade whatsoever.)

Notice that if we appeal to the "notational abuse" defined by (5.1), it follows that we can rewrite $a = (a_0, a_1, \ldots, a_n) \in \mathbb{G}^n$ as

$$a = a_0 + a_1 + \cdots + a_n.$$

We therefore abandon the notation $(a_0, \ldots, a_n)$ for elements of the geometric algebra in favor of summations. In connection with this, we can write

$$a = \langle a \rangle_0 + \cdots \langle a \rangle_n$$

with the expansion being unique.

Here is an obvious property of the grade:

Proposition 5.1. *The map $a \mapsto \langle a \rangle_k$ is linear.*

In algebraic terms, $\mathbb{G}^n$ is the direct sum of the spaces, $\Lambda^k \mathbb{R}^n$:

$$\mathbb{G}^n = \Lambda^0 \mathbb{R}^n \oplus \cdots \oplus \Lambda^n \mathbb{R}^n.$$

That is, $\mathbb{G}^n$ is a vector space consisting of finite linear combinations of the elements of $\Lambda^k \mathbb{R}^n$, $k = 0, \ldots, n$, and whenever $p \neq q$, then $\Lambda^p \mathbb{R}^n$ and $\Lambda^q \mathbb{R}^n$ have only the zero element in common.

Example 5.1. If we form the element

$$a = 3 + 4(e_1 \wedge e_3) - (e_2 \wedge e_4) - 16(e_1 \wedge e_2 \wedge e_4)$$

of $\mathbb{G}^4$, then

$$\begin{aligned}
\langle a \rangle_0 &= 3, \\
\langle a \rangle_1 &= 0, \\
\langle a \rangle_2 &= 4(e_1 \wedge e_3) - (e_2 \wedge e_4), \\
\langle a \rangle_3 &= -16(e_1 \wedge e_2 \wedge e_4), \\
\langle a \rangle_4 &= 0.
\end{aligned}$$

Recall that $\mathcal{O}_k$ is the set of ordered multi-indices of length k; and if $\{u_i\}_{i=1}^n$ is a basis for $\mathbb{R}^n$, then the set of u_I such that $|I| = k$ (the length of I is k) is a basis for $\Lambda^k \mathbb{R}^n$. The following is then an easy result:

Proposition 5.2. *If* $\{u_i\}_{i=1}^n$ *is a basis for* $\mathbb{R}^n$, *then*

$$\{u_I \,:\, I \in \mathcal{O}_k \text{ and } k = 0, 1, \ldots, n\}$$

is a basis for $\mathbb{G}^n$. *Thus every element* $a \in \mathbb{G}^n$ *has a unique expansion*

$$a = \sum_{k=0}^n \sum_{I \in \mathcal{O}_k} \alpha_I u_I$$

where $\alpha_I \in \mathbb{R}$.

Corollary 5.1. *dim* $\mathbb{G}^n = 2^n$.

Proof. We simply add the dimensions of the spaces $\Lambda^k \mathbb{R}^n$:

$$\binom{n}{0} + \binom{n}{1} + \cdots + \binom{n}{n} = 2^n. \qquad \square$$

Exercises 5.1.

1. Utilizing the standard bases, write out bases for $\mathbb{G}^2$ and $\mathbb{G}^3$.

2. Show that the map $\mathbb{G}^n \to \Lambda^k \mathbb{R}^n$ defined by $a \mapsto \langle a \rangle_k$ is a vector space homomorphism onto $\Lambda^k \mathbb{R}^n$.

5.2   Construction of the geometric product

We will go rather slowly and carefully in this section because we want it to be clear that the definition we make is logically sound and that it has the nice properties we want.

First, some informal considerations:

Given a vector $a \in \mathbb{R}^n$, we would like the geometric product aa to be the dot product $a \cdot a = |a|^2$. On the other hand, if a and b are orthogonal vectors, then we would like their geometric product ab to be $a \wedge b$. We can capture both these behaviors by saying that if $a, b \in \mathbb{R}^n$, then we want to have

$$ab = a \cdot b + a \wedge b.$$

This is not a sufficient condition to define the geometric product because it deals only with 1-vectors. We want to define ab when a and b are arbitrary multivectors.

We know that if $v = \{v_i\}_{i=1}^n$ is an orthonormal basis for $\mathbb{R}^n$, then $\{v_I\}_I$, where I ranges over all ordered multi-indices, is a basis for $\mathbb{G}^n$. Every $a \in \mathbb{G}^n$ can be expanded uniquely in the form $a = \sum_I \alpha_I v_I$. If we define the geometric product $v_I v_J$ where I and J are ordered multi-indices and then extend the definition linearly, then this will give us a way to define an arbitrary geometric product. (Of course, we expect and hope that our definition will turn out not to depend on our particular choice of v.)

Before giving a formal construction/definition of the geometric product, we turn to an examination of multi-indices.

5.2.1 Proper transformations of multi-indices

Suppose that $I = (i_1, \ldots, i_p)$ is a multi-index, not necessarily ordered and possibly with repetitions. We permit ourselves to transform I into another multi-index by the repeated application of two elementary operations:

1. An *interchange* is an operation in which we switch the order of two adjacent and distinct indices. For example, $(3,1,4,7) \mapsto (3,4,1,7)$.

2. An *annihilation/creation* is an operation in which two adjacent and identical indices are either removed from or introduced into a multi-index. Example:

$$\text{annihilation:} \quad (1,5,5,4) \mapsto (1,4),$$
$$\text{creation:} \quad (7,6) \mapsto (2,2,7,6).$$

If we say that F is a *proper transformation* of the multi-index I into J, we mean that F is a sequence of interchanges and annihilation/creations that carries I into J, and we write $J = F(I)$. It is perhaps not appropriate to think of F as a function. This is because there are, in general, many proper transformations that carry a given I into a given J. For example,

$$F_1: \quad (1,1,2,3) \mapsto (2,3) \mapsto (3,2),$$
$$F_2: \quad (1,1,2,3) \mapsto (1,1,3,2) \mapsto (3,2),$$

but $F_1 \neq F_2$. On the other hand, F always has an inverse, F^{-1}; just reverse the elementary operations of F.

We say that a proper transformation F is *even* or has *even parity* provided it has an even number of interchanges. Otherwise F is *odd* or has *odd parity*. Thus

$$F: \quad (1,2,3) \mapsto (1,3,2) \mapsto (3,1,2)$$

is even and

$$G: \quad (1,2,1,2) \mapsto (1,1,2,2) \mapsto (1,1) \mapsto \emptyset$$

(where it is understood that $\emptyset$ is the empty multi-index) is odd. It is obvious that F and F^{-1} always have the same parity.

Lemma 5.1. *If I and J are given multi-indices, then all proper transformations that carry I to J have the same parity.*

Proof. Suppose F is a proper transformation such that $J = F(I)$. For any multi-index K, let $E(K)$ be the number of times an index in K is greater than an index to its right. Thus $E(3,1,2) = 2$. The interchange of any two adjacent distinct indices in K changes $E(K)$ by ± 1. A creation or annihilation operation performed on K will change $E(K)$ by an even number. It follows then that $E(I)$ and $E(J)$ will have the same parity if and only if F is even; and they will have different parities if and only if F is odd. However whether or not $E(I)$ and $E(J)$ have the same parity is independent of which proper transformation was used to get from I to J. $\square$

Let us now introduce the notations

$$I \overset{+1}{\sim} J$$

to mean there is an even proper transformation that carries I to J and

$$I \overset{-1}{\sim} J$$

to mean there is an odd proper transformation of I to J. By $I \sim J$ we shall simply mean there is some proper transformation of I to J.

We easily see that

Lemma 5.2. *If $I \overset{\sigma_1}{\sim} J \overset{\sigma_2}{\sim} K$ where $\sigma_1, \sigma_2 = \pm 1$, then $I \overset{\sigma_1 \sigma_2}{\sim} K$.*

It should also be clear that we have the following:

Lemma 5.3. *Given any multi-index $I = (i_1, \ldots, i_p)$, there is a unique ordered multi-index $R = (j_1, \ldots, j_q)$ such that $I \sim R$.*

Of course the entries of R will simply be those indices of I which appear an odd number of times rearranged in increasing order. We call R the *root* of I.

We need one other idea here: If $I = (i_1, \ldots, i_p)$ and $J = (j_1, \ldots, j_q)$, then the *concatenation* of these two multi-indices is the multi-index

$$IJ \overset{\text{def.}}{=} (i_1, \ldots, i_p, j_1, \ldots, j_q).$$

If we concatenate with the empty multi-index, then, of course, $I\emptyset = \emptyset I = I$.

The following should be obvious:

Lemma 5.4. *If $I_0 \overset{\sigma_0}{\sim} J_0$ and $I_1 \overset{\sigma_1}{\sim} J_1$, then*

$$I_0 I_1 \overset{\sigma_0 \sigma_1}{\sim} J_0 J_1.$$

5.2.2 The geometric product

We are now ready to define the geometric product. First we choose and fix an orthonormal basis $\{v_i\}_{i=1}^n$ for $\mathbb{R}^n$.

If I and J are ordered multi-indices, we want to define the geometric product $v_I v_J$. Let R be the root of the concatenated multi-index IJ. Then

$$v_I v_J \overset{\text{def.}}{=} \sigma v_R \quad \text{where } IJ \overset{\sigma}{\sim} R. \tag{5.2}$$

That is, $v_I v_J = v_R$ if the number of interchanges required to change IJ to R is even; if it is odd, then $v_I v_J = -v_R$. We now extend the definition of geometric product to all of $\mathbb{G}^n$ by taking arbitrary $a, b \in \mathbb{G}^n$, writing their unique expansions in terms of our given basis,

$$a = \sum_I \alpha_I v_I \quad \text{and} \quad b = \sum_J \beta_J v_J$$

(where it is understood that the summations are over all ordered multi-indices) and setting

$$ab \overset{\text{def.}}{=} \sum_{I,J} \alpha_I \beta_J \, (v_I v_J).$$

This definition easily (exercise) yields the following result:

Proposition 5.3. *If $a, b, c \in \mathbb{G}^n$ and $\lambda \in \mathbb{R}$, then*

1. $1a = a1 = a$.

2. $a(b+c) = ab + ac$ *and* $(a+b)c = ac + bc$.

3. $\lambda(ab) = (\lambda a)b = a(\lambda b)$.

Example 5.2. Using the results above and Equation (5.2), we compute a few geometric products:

1. Since $(i,i) \overset{+1}{\sim} \emptyset$, we see that $v_i v_i = v_\emptyset = 1$.

2. Consider the geometric product $v_i v_j$ where $i \neq j$. Taking $I = (i)$ and $J = (j)$ as length one multi-indices, we have

$$(i)(j) \overset{+1}{\sim} (i,j),$$

   So looking back to Equation (5.2), we see that

$$v_i v_j = v_{ij} = v_i \wedge v_j.$$

3. Suppose i, j, k are distinct indices. We know from the last example that $v_j v_k = v_j \wedge v_k = v_{jk}$. Since $(i)(j,k) \overset{+1}{\sim} (i,j,k)$, we have

$$v_i(v_j v_k) = v_i(v_j \wedge v_k) = v_i v_{jk} = v_{ijk} = v_i \wedge v_j \wedge v_k.$$

4. $v_2(v_1 \wedge v_2) = -v_1$ since $(2,1,2) \overset{-1}{\sim} (1)$.

5. $(v_2 \wedge v_3)(v_1 \wedge v_2) = -v_1 \wedge v_3$ since $(2,3,1,2) \overset{-1}{\sim} (1,3)$.

6. $(v_1 \wedge v_2)(v_1 \wedge v_2)^\dagger = 1$.

7. $\left(\alpha + \beta\,(v_1 \wedge v_2)\right)\left(\alpha - \beta\,(v_1 \wedge v_2)\right)$
$$= \alpha^2 - \alpha\beta\,(v_1 \wedge v_2) + \alpha\beta\,(v_1 \wedge v_2) - \beta^2\,(v_1 \wedge v_2)(v_1 \wedge v_2)$$
$$= \alpha^2 + \beta^2$$
   where α and β are arbitrary scalars.

We now state some important properties of geometric algebra with the proofs being mostly relegated to an appendix.

Proposition 5.4. *The geometric product is associative.*

Proof. See Appendix A.5. □

We state identities that show the relation between the dot, wedge, and geometric products. For the proof, see Appendix A.5. Some simple and useful properties follow from these identities, including a proof that the definition of the geometric product is independent of our choice of the orthonormal basis $\{v_i\}_{i=1}^n$.

Proposition 5.5. *Let $a, b_1, \ldots, b_p \in \mathbb{R}^n$. Then*

$$a(b_1 \wedge \cdots \wedge b_p) = \sum_{r=1}^p (-1)^{r-1} (a \cdot b_r)(b_1 \wedge \cdots \wedge \widehat{b_r} \wedge \cdots \wedge b_p)$$
$$+ a \wedge b_1 \wedge \cdots \wedge b_p \quad and$$

$$(b_1 \wedge \cdots \wedge b_p)a = \sum_{r=1}^p (-1)^{p-r} (a \cdot b_r)(b_1 \wedge \cdots \wedge \widehat{b_r} \wedge \cdots \wedge b_p)$$
$$+ b_1 \wedge \cdots \wedge b_p \wedge a.$$

For $p = 1$ in the proposition and using the fact that a and b are orthogonal if and only if $a \cdot b = 0$, we obtain

Corollary 5.2. *If a and b are vectors in $\mathbb{R}^n$, then $ab = a \cdot b + a \wedge b$. In particular, $aa = |a|^2$, and $ab = a \wedge b$ if and only if a and b are orthogonal.*

Corollary 5.3. *If $a_1, \ldots, a_p$ are orthogonal vectors in $\mathbb{R}^n$, then*

$$a_1 \cdots a_p = a_1 \wedge \cdots \wedge a_p.$$

Proof. Since $a_i \cdot a_j = 0$ if $i \neq j$, Proposition 5.5 tells us the following:

$$a_{p-1} a_p = a_{p-1} \wedge a_p,$$
$$a_{p-2} a_{p-1} a_p = a_{p-2}(a_{p-1} \wedge a_p) = a_{p-2} \wedge a_{p-1} \wedge a_p,$$
$$\text{etc.} \qquad\qquad \square$$

For the next corollary, the proof will be found in Appendix A.5.

Corollary 5.4. *The definition of the geometric product does not depend on our choice of the orthonormal basis $v = \{v_i\}_{i=1}^n$.*

5.2.3 Fundamental properties of $\mathbb{G}(V)$ and $\mathbb{G}^n$

We would like to do two things here:

Suppose V is a vector subspace of $\mathbb{R}^n$. We know how to construct $\Lambda^k V$, the space of k-vectors in V. This is a vector subspace of $\Lambda^k \mathbb{R}^n$ and every element of $\Lambda^k V$ has the form

$$\sum_{i_1 < \cdots < i_k} \lambda_{i_1 \ldots i_k} v_{i_1} \wedge \cdots \wedge v_{i_k}$$

where each v_{i_j} is a vector in V. The first thing we would like to do is extend this process and construct the *geometric algebra over V, $\mathbb{G}(V)$*.

The second thing we would like to do is gather into one spot the most basic properties of $\mathbb{G}(V)$ and $\mathbb{G}^n$, those that determine everything else.

The first project is straightforward: For a multivector a in $\mathbb{G}^n$, we declare it to be a member of $\mathbb{G}(V)$ if and only if it can be written in the form

$$a = \sum_{k \geq 0} \sum_{i_1 < \cdots < i_k} \lambda_{i_1 \ldots i_k} \, v_{i_1} \wedge \cdots \wedge v_{i_k} \tag{5.3}$$

where each $v_{i_j} \in V$. As in $\mathbb{G}^n$, we can write $a = \sum_k \langle a \rangle_k$ where each term $\langle a \rangle_k$ is uniquely determined and $\langle a \rangle_k \in \Lambda^k V$. If V is a p-dimensional space, it makes sense to write

$$\mathbb{G}(V) = \Lambda^0 V \oplus \Lambda^1 V \oplus \cdots \oplus \Lambda^p V.$$

It is clear from (5.3) that $\mathbb{G}(V)$ is a vector subspace of $\mathbb{G}^n$. The only thing we might want to check is that $\mathbb{G}(V)$ is closed under the geometric product. That is, if $a, b \in \mathbb{G}(V)$, we want to see that $ab \in \mathbb{G}(V)$. This is easy to check. Let $\{v_i\}_{i=1}^p$ be an orthonormal basis for V and extend it to an orthonormal basis $\{v_i\}_{i=1}^n$ for $\mathbb{R}^n$. We can expand a, b thus:

$$a = \sum_{0 \leq k \leq p} \sum_{i_1 < \cdots < i_k} \alpha_{i_1 \ldots i_k} \, v_{i_1 \ldots i_k} \quad \text{and} \quad b = \sum_{0 \leq k \leq p} \sum_{j_1 < \cdots < j_k} \beta_{j_1 \ldots j_k} \, v_{j_1 \ldots j_k}.$$

It suffices to show that if we have v_I and v_J where I, J are ordered multi-indices whose terms come from the index subset $\{1, \ldots, p\}$, then $v_I v_J = \pm v_R$ where R is a multi-index whose terms are also in $\{1, \ldots, p\}$. But this is true by Equation (5.2) which defines the geometric product.

Example 5.3. A particularly interesting instance of this construction arises when we consider a manifold $\mathcal{M}$ and form $\mathbb{G}(T_x\mathcal{M})$, the set of multivectors tangent to $\mathcal{M}$ at the point x.

Notice that for every proposition we have about $\mathbb{G}^n$, we can construct a similar proposition for $\mathbb{G}(V)$ since the fact that we were working in the particular vector space $\mathbb{R}^n$ never played a crucial role.

For the second project, we consider V to be a p-dimensional vector subspace of $\mathbb{R}^n$. Here is a summary of important properties of $\mathbb{G}^n$ and $\mathbb{G}(V)$; there is a certain amount of overlap in the statements.

Properties

1. $\mathbb{G}(V) = \Lambda^0 V \oplus \cdots \oplus \Lambda^p V$, hence every element $a \in \mathbb{G}(V)$ has a unique expansion

$$a = \langle a \rangle_0 + \cdots + \langle a \rangle_p.$$

Of course, $\mathbb{G}^n = \mathbb{G}(\mathbb{R}^n)$.

2. If $a, b, c \in \mathbb{G}^n$ and $\lambda \in \mathbb{R}$, then

(a) $1a = a1 = a$.

(b) $a(b+c) = ab + ac$ and $(a+b)c = ac + bc$.

(c) $\lambda(ab) = (\lambda a)b = a(\lambda b)$.

(d) $a(bc) = (ab)c$.

3. If $a, b \in \mathbb{R}^n$, then
$$ab = a \cdot b + a \wedge b.$$

In particular, $aa = |a|^2$ and $ab = a \wedge b$ if and only if a and b are orthogonal. More generally, if $a, b_1, \ldots, b_p \in \mathbb{R}^n$, then

$$a(b_1 \wedge \cdots \wedge b_p) = \sum_{r=1}^{p} (-1)^{r-1} (a \cdot b_r)(b_1 \wedge \cdots \wedge \widehat{b_r} \wedge \cdots \wedge b_p)$$
$$+ a \wedge b_1 \wedge \cdots \wedge b_p \quad \text{and}$$

$$(b_1 \wedge \cdots \wedge b_p)a = \sum_{r=1}^{p} (-1)^{p-r} (a \cdot b_r)(b_1 \wedge \cdots \wedge \widehat{b_r} \wedge \cdots \wedge b_p)$$
$$+ b_1 \wedge \cdots \wedge b_p \wedge a.$$

4. If $u_1, \ldots, u_k$ are orthogonal vectors, then
$$u_1 \wedge \cdots \wedge u_k = u_1 \cdots u_k = \langle u_1 \cdots u_k \rangle_k.$$

5. If $\{u_i\}_{i=1}^{p}$ is a basis for V, then the multivectors

$$1,$$
$$u_1, \ldots, u_p,$$
$$\cdots$$
$$\{u_{i_1} \wedge \cdots \wedge u_{i_k} : i_1 < \cdots < i_k\},$$
$$\cdots$$
$$u_1 \wedge \cdots \wedge u_p$$

constitute a basis for $\mathbb{G}(V)$. Hence $\dim \mathbb{G}(V) = 2^p$.

Exercises 5.2.

1. Give an example of $a, b \in \mathbb{G}^n$ such that a and b have grades p and q respectively but the geometric product ab does not have a single grade.

2. Prove Proposition 5.3.

3. Suppose that $\{u_i\}_{i=1}^{n}$ is an orthonormal basis for $\mathbb{R}^n$.

    (a) Show that
    $$u_i u_j = \begin{cases} -u_j u_i & \text{if } i \neq j, \\ 1 & \text{if } i = j. \end{cases}$$

(b) Show that if $i \neq j$, then $u_i u_j$ is a square root of -1.

(c) If $i_1 < \cdots < i_k$ and $v = u_{i_1} \cdots u_{i_k}$, show that $v^{-1} = u_{i_k} \cdots u_{i_1}$. (We mean an inverse with respect to the geometric product.)

4. Suppose that $\{u_i\}_{i=1}^n$ is an orthonormal basis for $\mathbb{R}^n$. Compute each of the following geometric products and express the result as a linear combination of 1 and terms of the form $u_{i_1} \cdots u_{i_k}$ where $i_1 < \cdots < i_k$.

(a) $(u_1 u_2)(u_1 u_2 u_3)$.

(b) $(1 + u_i)^2$.

(c) $(u_1 + \cdots + u_n)^2$.

(d) $u_i (u_1 \cdots u_n)$.

(e) $u_j (u_1 \cdots \widehat{u}_j \cdots u_n)$.

(f) $u_i \sum_{j=1}^n (-1)^{j-1} u_1 \cdots \widehat{u}_j \cdots u_n$.

5.3   Reversion and the wedge product in $\mathbb{G}^n$

We have at this point defined reversion and the wedge product only on k-vectors, citizens of $\Lambda^k \mathbb{R}^n$. We now extend these operations to arbitrary multivectors. (Later we shall do the same for the dot product.)

Definition 5.1. For $a \in \mathbb{G}^n$ we define the *reversion* of a by

$$a^\dagger = \sum_{k=0}^n (-1)^{\frac{k(k-1)}{2}} \langle a \rangle_k.$$

Notice that $\alpha^\dagger = \alpha$ for $\alpha \in \mathbb{R}$ and $a^\dagger = a$ for $a \in \mathbb{R}^n$.

Example 5.4. Using the standard basis $\{e_i\}_{i=1}^4$ for $\mathbb{R}^4$, we can write the basis vectors for $\mathbb{G}^4$ in the form $e_{i_1} \cdots e_{i_k}$ rather than $e_{i_1} \wedge \cdots \wedge e_{i_k}$. Assuming $i_1 < \cdots < i_k$, we have

$$
\begin{aligned}
1^\dagger &= 1 \\
e_i^\dagger &= e_i \\
(e_{i_1} e_{i_2})^\dagger &= -e_{i_1} e_{i_2} \\
(e_{i_1} e_{i_2} e_{i_3})^\dagger &= -e_{i_1} e_{i_2} e_{i_3} \\
(e_1 e_2 e_3 e_4)^\dagger &= e_1 e_2 e_3 e_4.
\end{aligned}
$$

We would then have, for instance,

$$\left(3 - \frac{1}{2} e_3 + 7 (e_1 e_2) + e_1 e_2 e_3 e_4\right)^\dagger = 3 - \frac{1}{2} e_3 - 7 (e_1 e_2) + e_1 e_2 e_3 e_4.$$

Proposition 5.6. *Given $a, b \in \mathbb{G}^n$, reversion has the following properties:*

1. $a \mapsto a^\dagger$ *is a linear transformation.*

2. $a^{\dagger\dagger} = a.$

3. $\langle a^\dagger \rangle_k = \langle a \rangle_k^\dagger.$

4. $(ab)^\dagger = b^\dagger a^\dagger.$

Proof. Properties 1, 2 and 3 are easy. We prove only 4.

Let $\{u_1, \ldots, u_n\}$ be an orthonormal basis for $\mathbb{R}^n$. Notice that calculations and simplifications that involve only $u_1, \ldots, u_n$ depend on switching and cancelling adjacent elements, and because of this, a sort of "mirror image" principle arises: If, for example,

$$u_{i_1} \cdots u_{i_k} = (-1)^r u_{p_1} \cdots u_{p_s},$$

then

$$u_{i_k} \cdots u_{i_1} = (-1)^r u_{p_s} \cdots u_{p_1}.$$

Similarly, if

$$(u_{i_1} \cdots u_{i_k})(u_{j_1} \cdots u_{j_m}) = (-1)^r u_{p_1} \cdots u_{p_s},$$

then

$$(u_{j_m} \cdots u_{j_1})(u_{i_k} \cdots u_{i_1}) = (-1)^r u_{p_s} \cdots u_{p_1}.$$

It is easily seen that if $i_1, \ldots, i_k$ are distinct, then

$$(u_{i_1} \cdots u_{i_k})^\dagger = (-1)^{\frac{k(k-1)}{2}} u_{i_1} \cdots u_{i_k} = u_{i_k} \cdots u_{i_1}. \tag{5.4}$$

Next suppose that $i_1 < \cdots < i_k$ and $j_1 < \cdots j_m$. There exists r and $p_1 < \cdots < p_s$ such that

$$(u_{i_1} \cdots u_{i_k})(u_{j_1} \cdots u_{j_m}) = (-1)^r u_{p_1} \cdots u_{p_s}.$$

Then by (5.4) and the mirror image principle,

$$\begin{aligned}((u_{i_1} \cdots u_{i_k})(u_{j_1} \cdots u_{j_m}))^\dagger &= (-1)^r (u_{p_1} \cdots u_{p_s})^\dagger = (-1)^r u_{p_s} \cdots u_{p_1}\\ &= (u_{j_m} \cdots u_{j_1})(u_{i_k} \cdots u_{i_1}) = (u_{j_1} \cdots u_{j_m})^\dagger (u_{i_1} \cdots u_{i_k})^\dagger.\end{aligned} \tag{5.5}$$

Finally, given $a, b \in \mathbb{G}^n$, if we expand them in terms of the blades $u_{i_1} \cdots u_{i_k}$ and apply (5.5), then it is straightforward to see that $(ab)^\dagger = b^\dagger a^\dagger$. $\qquad\square$

Definition 5.2. We extend the wedge product from k-vectors to all of $\mathbb{G}^n$ by the following formula:

$$a \wedge b \overset{\text{def.}}{=} \sum_{p,q=0}^{n} \left\langle \langle a \rangle_p \langle b \rangle_q \right\rangle_{p+q} \tag{5.6}$$

where $a, b \in \mathbb{G}^n$.

It may not be immediately evident that Equation (5.6) has anything to do with the wedge product that we studied earlier or that it is an extension of that operation, so we look at an example:

Example 5.5. Recall that $\{e_i\}_{i=1}^n$ is the standard basis. We know that the wedge product of e_1 and $e_2 e_3$ should be $e_1 \wedge e_2 \wedge e_3$ (or, equivalently, $e_1 e_2 e_3$). Set $a = e_1$ and $b = e_2 e_3$ in Equation (5.6). Notice that

$$\langle a \rangle_p \;=\; \begin{cases} e_1 & \text{if } p = 1 \\ 0 & \text{otherwise,} \end{cases}$$

and

$$\langle b \rangle_q \;=\; \begin{cases} e_2 e_3 \;=\; e_2 \wedge e_3 & \text{if } q = 2 \\ 0 & \text{otherwise.} \end{cases}$$

Hence (5.6) becomes

$$a \wedge b \;=\; \langle e_1(e_2 e_3) \rangle_3 \;=\; e_1 e_2 e_3$$

which is the desired answer.

Suppose, on the other hand, that we want the wedge product of e_1 and $e_1 e_2$. We know that we should get 0 since e_1 occurs in both terms. If we appeal to Equation (5.6), we obtain

$$a \wedge b \;=\; \langle e_1(e_1 e_2) \rangle_3 \;=\; \langle e_2 \rangle_3 \;=\; 0$$

which is again the right answer.

More generally, to show that (5.6) extends our previous definition of $\wedge$, we choose $a \in \Lambda^k \mathbb{R}^n$ and $b \in \Lambda^m \mathbb{R}^n$ and check that the map

$$(a, b) \;\mapsto\; \sum_{p,q=0}^{n} \left\langle \langle a \rangle_p \, \langle b \rangle_q \right\rangle_{p+q} \tag{5.7}$$

gives us $a \wedge b$. If (5.7) is linear in both a and b, it will suffice to check this for $a = u_{i_1} \cdots u_{i_k}$ and $b = u_{j_1} \cdots u_{j_m}$ where $I = (i_1, \ldots, i_k)$ and $J = (j_1, \ldots, j_m)$ are ordered multi-indices and $\{u_i\}_{i=1}^n$ is an orthonormal basis for $\mathbb{R}^n$. We leave this step as an exercise.

The linearity of (5.7) in a and b (and other things as well) follow from the next result:

Proposition 5.7. *For all $a, b, c \in \mathbb{G}^n$ and all $\lambda \in \mathbb{R}$, if we use (5.6) as our definition of $\wedge$, then the following hold:*

1. $a \wedge (b + c) \;=\; (a \wedge b) + (a \wedge c).$

2. $(a + b) \wedge c \;=\; (a \wedge c) + (b \wedge c).$

3. $\lambda(a \wedge b) \;=\; (\lambda a) \wedge b \;=\; a \wedge (\lambda b).$

4. $(a \wedge b)^\dagger \;=\; b^\dagger \wedge a^\dagger.$

We leave the proof of this proposition to the reader and turn instead to a computational example.

Example 5.6. Here is a wedge product of multivectors which, before, it would not have made sense to compute. We appeal to Equation (5.6) and Proposition 5.7 to justify the computation:

$$(\alpha_0 + \alpha_1(e_2 e_3)) \wedge (\beta_0 e_2 + \beta_1 e_4)$$
$$= \alpha_0 \wedge (\beta_0 e_2) + \alpha_0 \wedge (\beta_1 e_4) + (\alpha_1(e_2 e_3)) \wedge (\beta_0 e_2) + (\alpha_1(e_2 e_3)) \wedge (\beta_1 e_4)$$
$$= \alpha_0 \beta_0 e_2 + \alpha_0 \beta_1 e_4 + \alpha_1 \beta_1 (e_2 e_3 e_4).$$

Remark 5.1. It is noteworthy that if we were somehow given the geometric product and its properties without any knowledge of the wedge product, then we could use Equation (5.6) to define and introduce the wedge product. A similar comment can be made about the dot product and the variants of the dot product that we shall meet later. So we see that we could, if we wished, treat the geometric product and geometric algebra as our fundamental setting and derive everything else from them.

A clear construction of the geometric algebra directly from $\mathbb{R}^n$ can be found in [30].

Exercises 5.3.

1. Show that if $a = u_{i_1} \cdots u_{i_k}$ and $b = u_{j_1} \cdots u_{j_m}$ where $I = (i_1, \ldots, i_k)$ and $J = (j_1, \ldots, j_m)$ are ordered multi-indices and $\{u_i\}_{i=1}^n$ is an orthonormal basis for $\mathbb{R}^n$, then the map defined by (5.7) yields $a \wedge b$.

2. Use Equation (5.6) to give a proof that if $\lambda \in \mathbb{R}$ and $a \in \mathbb{G}^n$, then $\lambda \wedge a = \lambda a$.

3. Prove Proposition 5.6. (Hint: It is sometimes helpful to expand the multivectors in terms of an orthonormal basis.)

4. Let u_1 and u_2 be two orthogonal unit vectors and set $u = u_1 u_2$.

   (a) Set $a = \lambda + \xi u$ where λ and ξ are scalars. Show that a^k has the form $\lambda_k + \xi_k u$ for $k = 2, 3, \ldots$. Find λ_k and ξ_k as polynomials in λ and ξ.

   (b) Set $a = \cos(\theta) + \sin(\theta) u$. Show that $a^k = \cos(k\theta) + \sin(k\theta) u$ for $k = 2, 3, \ldots$.

   (c) If $a = \lambda + \xi u$, show that for $k = 2, 3, \ldots$, we have

   $$a^k = |a|^k \left(\cos(k\theta) + \sin(k\theta) u \right)$$

   provided θ is properly introduced.

5.4 Division by multivectors

Before proceeding further, it useful to bring out the fact that sometimes multivectors have inverses; in particular, one can divide by blades.

Proposition 5.8. *If a is a blade, then a has an inverse (with respect to the geometric product), and $a^{-1} = a^\dagger / |a|^2$.*

Proof. Let a be a k-blade. It determines a k-dimensional subspace of $\mathbb{R}^n$. We let $\{u_i\}_{i=1}^k$ be an orthonormal basis for this subspace, and we know that we can write $a = \alpha(u_1 \wedge \cdots \wedge u_k) = \alpha(u_1 \cdots u_k)$ where the scalar α is $\pm|a|$. Then

$$a \frac{1}{|a|^2} a^\dagger \;=\; \frac{\alpha^2}{|a|^2} \, (u_1 \cdots u_k)(u_k \cdots u_1) \;=\; 1.$$

We get the same result when we calculate $(1/|a|^2)a^\dagger a$, so $a^{-1} = (1/|a|^2)a^\dagger$. $\square$

Of course there are multivectors that are not blades. Some of them have inverses and some of them do not. Examples occur in the exercises, and a discussion of this feature of geometric algebra can be found in [12].

Remark 5.2. A fact worth noting is that if u is a unit blade, then $u^{-1} = u^\dagger$. This is the case when we deal with orientations w of manifolds.

Remark 5.3. One must exercise a certain amount of caution when dividing since, in general, ab^{-1} and $b^{-1}a$ are not the same.

A useful application of division is finding orthogonal complements of blades and subspaces. More precisely, given a k-blade a in a p-dimensional vector subspace V, we may wish to construct a $(p-k)$-dimensional orthogonal complement b in V.

We proceed by choosing first an orientation w for V and then an orthonormal basis $\{u_i\}_{i=1}^p$ for V such that $a = \lambda\,(u_1 \cdots u_k)$ and $w = u_1 \cdots u_p$. If we divide the orientation by a or a by the orientation, then we obtain blades such as

$$b_0 \;=\; a^{-1}w \;=\; (-1)^{r_0} \frac{1}{\lambda}\, u_{k+1} \cdots u_p,$$

$$b_1 \;=\; aw^{-1} \;=\; (-1)^{r_1} \lambda\, u_{k+1} \cdots u_p,$$

where r_0, r_1 are integers to be determined, and b_0, b_1 are clearly blades orthogonal to a.

Of course there are other ways to calculate orthogonal complements. For example, wa^{-1} and $w^{-1}a$. And it is easily seen we do not need to appeal to division: aw, wa, $a^\dagger w$, etc. will also give us what we want.

Now here are some other uses of division by blades:

Example 5.7. Recall that in $\mathbb{R}^3$, one denotes the standard basis vectors as $\mathbf{i}, \mathbf{j}, \mathbf{k}$ rather than e_1, e_2, e_3. The cross product or vector product is then defined by setting

$$\begin{aligned}
\mathbf{i} \times \mathbf{j} &= \mathbf{k}, \\
\mathbf{j} \times \mathbf{k} &= \mathbf{i}, \\
\mathbf{k} \times \mathbf{i} &= \mathbf{j}
\end{aligned} \tag{5.8}$$

and extending linearly to arbitrary combinations of the basis vectors. The effect of this is that if a and b are vectors in $\mathbb{R}^3$, then $a \times b$ will be orthogonal to the plane determined by a and b, will have magnitude $|a|\,|b|\sin(\theta)$ (where θ is the angle between a and b), and will have the direction determined by the "right-hand rule."

We get the same effect using geometric algebra thus: Let $e = e_1 e_2 e_3 = \mathbf{ijk}$; we think of this as the standard orientation of $\mathbb{R}^3$. We set

$$a \times b \overset{\text{def.}}{=} (a \wedge b)e^{-1}.$$

The result must always be a 1-vector orthogonal to the plane determined by a and b, and from our knowledge of the wedge product, it gives the correct magnitude. The product is linear in both a and b, and it is easily checked that it gives the values in (5.8) for products of $\mathbf{i}$, $\mathbf{j}$, and $\mathbf{k}$.

Example 5.8. In $\mathbb{R}^3$, the unit sphere that is centered at the origin, S^2, has the equation $\chi_1^2 + \chi_2^2 + \chi_3^2 = 1$. The unit outward normal vector to S^2 at the point $x = (\chi_1, \chi_2, \chi_3)$ is

$$n(x) = x = \chi_1 e_1 + \chi_2 e_2 + \chi_3 e_3.$$

If we compute a unit 2-blade $w(x)$ that is orthogonal to $n(x)$ at every $x \in S^2$, then this should be tangent to the sphere at x and hence an orientation for S^2. Let $e = e_1 e_2 e_3$ be the orientation for $\mathbb{R}^3$. We construct an example of the desired w thus:

$$w(x) = n(x)e = (\chi_1 e_1 + \chi_2 e_2 + \chi_3 e_3)(e_1 e_2 e_3)$$
$$= \chi_1(e_2 e_3) - \chi_2(e_1 e_3) + \chi_3(e_1 e_2).$$

Example 5.9. An "orthogonal complementation" operation that is useful in the theory of differential forms is the *Hodge star operator* $a \mapsto {}^*a$. If it is carried out in a vector subspace V with orientation w, then it can be defined using geometric algebra by

$$a \mapsto {}^*a = a^\dagger w.$$

It is defined, of course, for all vectors in V (actually for all multivectors in $\mathbb{G}^n$) and only gives an orthogonal complement when a is a blade in V.

As a final point, we notice that multiplying or dividing by the orientation of a vector space gives a nice way of translating back and forth between k-vectors of "complementary dimensions":

Proposition 5.9. *Let V be a p-dimensional vector subspace of $\mathbb{R}^n$ with orientation w. Let $f \colon \mathbb{G}(V) \to \mathbb{G}^n$ be the map $f(x) = xw^{-1}$. Then the following hold:*

1. *If f is restricted to $\Lambda^k V$, then it acts as a vector space isomorphism of $\Lambda^k V$ onto $\Lambda^{p-k} V$.*

2. *f takes blades to blades.*

3. $\dim\left(\Lambda^k V\right) = \binom{p}{k} = \dim\left(\Lambda^{p-k} V\right).$

We leave the proof as an exercise.

Exercises 5.4.

1. Find $(e_1 - e_2)^{-1}$.

2. Some elements of $\mathbb{G}^n$ that are not blades also have inverses. Let $e = e_1 e_2 \in \mathbb{G}^2$. Show that if $x = \alpha + \beta\,e \neq 0$, then x has an inverse in $\mathbb{G}^2$. What is it?

3. There are nonzero elements of $\mathbb{G}^n$ that do not have inverses.

   (a) Let $e = e_1 e_2 \in \mathbb{G}^2$. Show that $x \in \mathbb{G}^2$ has the property that $x \neq 0$ and $x^2 = 0$ if and only if x has the form

   $$x = \rho\left((\cos\theta)\,e_1 + (\sin\theta)\,e_2 + e\right) \quad \text{where } \rho \neq 0.$$

   (b) Show that for every $n \geq 2$, there exists $x \in \mathbb{G}^n$ such that $x \neq 0$ and $x^2 = 0$.

   (c) Show that if $x \in \mathbb{G}^n$ has the property that $x \neq 0$ and $x^2 = 0$, then x does not have an inverse.

4. In $\mathbb{R}^{m+1}$, the unit sphere S^m, which is centered at the origin, satisfies the equation $\sum_{i=1}^{m+1} \chi_i^2 = 1$. Find an orientation $w(x)$ for S^m where $x = (\chi_1, \ldots, \chi_{m+1})$ is an arbitrary point on S^m.

5. Show that if a is a blade, then $(a^\dagger)^{-1} = (a^{-1})^\dagger$.

6. Suppose that a is a k-blade in $\mathbb{R}^n$ and $x \in \mathbb{R}^n$. Show that the geometric product axa^{-1} is a 1-vector and that $f(x) = axa^{-1}$ is a linear transformation of $\mathbb{R}^n$.

7. Prove Proposition 5.9.

5.5   The dot product in $\mathbb{G}^n$

There is more than one useful way to extend the dot product on $\Lambda^k \mathbb{R}^n$ to $\mathbb{G}^n$. We call the first of these the *scalar product*.

5.5.1   The scalar product

We begin with two easily proved results:

Proposition 5.10. *For all $a, b \in \mathbb{G}^n$, we have*

$$\langle ab \rangle_0 = \langle ba \rangle_0 \quad \text{and} \quad \langle ab^\dagger \rangle_0 = \langle a^\dagger b \rangle_0.$$

Definition 5.3. For $a, b \in \mathbb{G}^n$, call the product

$$(a, b) \;\mapsto\; \langle ab^\dagger \rangle_0 = \langle a^\dagger b \rangle_0$$

the *scalar product* of a and b.

Of course the scalar product is linear in both a and b. We leave it to the reader to show that if a, b are k-blades in $\mathbb{G}^n$, then $\langle ab^\dagger \rangle_0 = a \cdot b^\dagger$. Once we know that fact, it is straightforward to prove the following:

Proposition 5.11. *Let $a, b \in \mathbb{G}^n$. If $\{u_1, \ldots, u_n\}$ is an orthonormal basis for $\mathbb{R}^n$ in terms of which*

$$a = \sum_{k=0}^{n} \sum_{i_1 < \cdots < i_k} \alpha_{i_1 \cdots i_k} \left(u_{i_1} \cdots u_{i_k} \right)$$

and

$$b = \sum_{k=0}^{n} \sum_{i_1 < \cdots < i_k} \beta_{i_1 \cdots i_k} \left(u_{i_1} \cdots u_{i_k} \right),$$

then

$$\langle ab^\dagger \rangle_0 = \sum_{k=0}^{n} \sum_{i_1 < \cdots < i_k} \alpha_{i_1 \cdots i_k} \, \beta_{i_1 \cdots i_k}.$$

In particular,

$$\langle aa^\dagger \rangle_0 = \sum_{k=0}^{n} \sum_{i_1 < \cdots < i_k} \alpha_{i_1 \cdots i_k}^2.$$

Notice that the formula for the "dot product" $\langle ab^\dagger \rangle_0$ is just the same as the more familiar $\sum_{i=1}^{n} \alpha_i \beta_i$ for vectors in $\mathbb{R}^n$ except that it must now be expressed in terms of ordered multi-indices.

In what follows, when we say that a map of a vector space into the reals, $a \mapsto |a|$, is a *norm*, we mean that $|a| \geq 0$, that $|a| = 0$ precisely when $a = 0$, that $|\lambda \, a| = |\lambda| \, |a|$ where λ is a scalar, and that $|a + b| \leq |a| + |b|$.

We now construct a very natural norm for $\mathbb{G}^n$. Later, when we talk about $a_k \to b$ or $f(x) \to y$ in $\mathbb{G}^n$, we will mean that the norms $|a_k - b|$, $|f(x) - y|$ converge to 0. It will be useful to have this sort of machinery in the background when we discuss derivatives and integrals.

Definition 5.4. For $a \in \mathbb{G}^n$, we define the *magnitude* of a by

$$|a| = \sqrt{\langle aa^\dagger \rangle_0} = \sqrt{\langle a^\dagger a \rangle_0} = \sqrt{\sum_{k=0}^{n} \left(\langle a \rangle_k \cdot \langle a^\dagger \rangle_k \right)}.$$

Remark 5.4. Notice that if a is a k-vector, then the earlier definition of $|a|$ agrees with our new definition of $|a|$ because, in that case, we have $\langle aa^\dagger \rangle_0 = a \cdot a^\dagger$.

The fact that $\langle aa^\dagger \rangle_0 \geq 0$ and that the definition of $|a|$ makes sense follows from this proposition:

Proposition 5.12. *Let $a, b \in \mathbb{G}^n$.*

1. *The map $a \mapsto |a|$ is a norm on $\mathbb{G}^n$; further, for every k, it is an extension of the norm previously defined on $\Lambda^k \mathbb{R}^n$.*

2. $|a|^2 = \sum_{k=0}^{n} |\langle a \rangle_k|^2$.

3. *If either a or b is a simple k-vector, then* $|ab| = |a| \, |b|$.

Proof. 1. Given an orthonormal basis $\{u_1, \ldots, u_n\}$ of $\mathbb{R}^n$, we set up a one-to-one correspondence between the elements $u_{i_1} \cdots u_{i_k}$, where $i_1 < \cdots < i_k$ and $k = 0, \ldots, n$, of $\mathbb{G}^n$ and the vectors e_j, where $j = 1, \ldots, 2^n$, of $\mathbb{R}^{2^n}$. This correspondence induces a vector space isomorphism of $\mathbb{G}^n$ onto $\mathbb{R}^{2^n}$. This isomorphism carries $|a|$ to the euclidean norm of the element of $\mathbb{R}^{2^n}$ that corresponds to a. Thus $|a|$ must be a norm. It is easily checked that for $a \in \Lambda^k \mathbb{R}^n$, $|a|$ as defined in $\Lambda^k \mathbb{R}^n$ and in $\mathbb{G}^n$ are the same.

2. This follows trivially from expanding a in terms of an orthonormal basis.

3. We consider the case where a is a simple k-vector. We may suppose that we have chosen the orthonormal basis $\mathcal{U} = \{u_1, \ldots, u_n\}$ in such a way that $a = \alpha u_1 \cdots u_k$ where α is a scalar. If b has the expansion

$$b = \sum_{m=0}^{n} \sum_{i_1 < \cdots < i_m} \beta_{i_1 \ldots i_m} u_{i_1} \cdots u_{i_m},$$

then

$$|a|^2 \, |b|^2 = \sum_{m=0}^{n} \sum_{i_1 < \cdots < i_m} \alpha^2 \beta_{i_1 \ldots i_m}^2.$$

Now

$$ab = \sum_{m=0}^{n} \sum_{i_1 < \cdots < i_m} \alpha \beta_{i_1 \ldots i_m} (u_1 \cdots u_k)(u_{i_1} \cdots u_{i_m}).$$

Let $\mathcal{B}$ be the basis for $\mathbb{G}^n$ generated by $\mathcal{U}$ and notice that the map

$$u_{i_1} \cdots u_{i_m} \mapsto (u_1 \cdots u_k)(u_{i_1} \cdots u_{i_m})$$

is a one-to-one map of $\mathcal{B}$ onto itself. (The inverse is obtained by dividing by $u_1 \cdots u_k$.) This means that if we write

$$ab = \sum_{m=0}^{n} \sum_{j_1 < \cdots < j_m} \gamma_{j_1 \ldots j_m} u_{j_1} \cdots u_{j_m},$$

then the numbers $\gamma_{j_1 \ldots j_m}$ are simply a rearrangement of the numbers $\alpha \beta_{i_1 \ldots i_m}$. Therefore,

$$|ab|^2 = \sum_{m=0}^{n} \sum_{j_1 < \cdots < j_m} \gamma_{j_1 \ldots j_m}^2 = \sum_{m=0}^{n} \sum_{i_1 < \cdots < i_m} \alpha^2 \beta_{i_1 \ldots i_m}^2.$$

This establishes the result.   □

5.5.2   Three variations on a theme

The scalar product was an extension of the dot product on $\Lambda^k \mathbb{R}^n$ to $\mathbb{G}^n$. We now consider three other extensions of the dot product from $\Lambda^k \mathbb{R}^n$ to $\mathbb{G}^n$ that are sometimes useful in different circumstances.

Definition 5.5. Let $a, b \in \mathbb{G}^n$. We define thus:

1. Left-hand dot product:

$$a \bullet_L b \overset{\text{def}}{=} \sum_{k,m=0}^{n} \left\langle \langle a \rangle_k \langle b \rangle_m \right\rangle_{m-k}.$$

2. Right-hand dot product:

$$a \bullet_R b \overset{\text{def}}{=} \sum_{k,m=0}^{n} \left\langle \langle a \rangle_k \langle b \rangle_m \right\rangle_{k-m}.$$

3. Two-sided dot product:

$$a \bullet b \overset{\text{def}}{=} \sum_{k,m=0}^{n} \left\langle \langle a \rangle_k \langle b \rangle_m \right\rangle_{|k-m|}.$$

It may not be clear that these definitions have anything to do with the dot product in the form with which we previously worked. All of these definitions are linear with respect to both a and b, so it should suffice to check what happens when we take products using orthonormal basis vectors.

In [9], these operations are referred to as *contractions*, and the symbols $\rfloor$ and $\lfloor$ are used for the left-hand and right-hand dot products:

$$a \rfloor b = a \bullet_L b \quad \text{and} \quad a \lfloor b = a \bullet_R b.$$

Example 5.10. Working with the standard basis vectors, we see that

$$(e_1 e_3) \bullet_L (e_1 e_2 e_3) = \langle (e_1 e_3)(e_1 e_2 e_3) \rangle_1 = e_2.$$

On the other hand,

$$(e_1 e_2 e_3) \bullet_L (e_1 e_3) = \langle (e_1 e_2 e_3)(e_1 e_2) \rangle_{-1} = 0$$

since there is no grade of -1. We try two more products:

$$(e_1 e_3) \bullet_L (e_1 e_3)^\dagger = \langle (e_1 e_3)(e_3 e_1) \rangle_0 = 1$$

and

$$(e_1 e_3) \bullet_L (e_2 e_3) = \langle (e_1 e_3)(e_2 e_3) \rangle_0 = \langle -e_1 e_2 \rangle_0 = 0.$$

These are exactly the answers we would expect using our previous definition of dot product on $\Lambda^k \mathbb{R}^n$.

Knowing how the left-hand dot product works essentially tells us everything we need to know about the right-hand dot product.

Proposition 5.13. *For all* $a, b \in \mathbb{G}^n$ *we have* $(a \bullet_L b)^\dagger = b^\dagger \bullet_R a^\dagger$.

Proof. It suffices to prove this result for blades that are products of the standard basis vectors $e_1, \ldots, e_n$ (or similar blades in any other orthonormal basis). So consider $e_{i_1} \cdots e_{i_k}$ and $e_{j_1} \cdots e_{j_{k+m}}$ where $1 \leq k$ and $0 \leq m$, the indices $i_1, \ldots, i_k$ are distinct, and the indices $j_1, \ldots, j_{k+m}$ are distinct. We see that if $\{i_1, \ldots, i_k\}$ is not a subset of $\{j_1, \ldots, j_{k+m}\}$, then

$$(e_{i_1} \cdots e_{i_k}) \cdot_L (e_{j_1} \cdots e_{j_{k+m}})$$
$$= \left\langle (e_{i_1} \cdots e_{i_k})(e_{j_1} \cdots e_{j_{k+m}}) \right\rangle_m$$
$$= 0$$

and, similarly,

$$(e_{j_{k+m}} \cdots e_{j_1}) \cdot_R (e_{i_k} \cdots e_{i_1}) = 0,$$

in which case the desired result is trivial. So suppose that $\{i_1, \ldots, i_k\}$ is a subset of $\{j_1, \ldots, j_{k+m}\}$. Set

$$I = (i_1, \ldots, i_k) \quad \text{and} \quad I + J = (j_1, \ldots, j_{k+m})$$

and let J be an ordered sequence of the indices which occur in $I + J$ but not in I. Then we may write

$$e_{i_1} \cdots e_{i_k} = e_I \quad \text{and} \quad e_{j_1} \cdots e_{j_{k+m}} = e_{I+J}.$$

By permuting the factors of e_{I+J} that have indices $i_1, \ldots, i_k$ to the left, we can write

$$e_{I+J} = (-1)^r e_I e_J$$

where the parity of r is determined by I and $I + J$. Then

$$e_I \cdot_L e_{I+J} = e_I e_{I+J} = e_I (-1)^r e_I e_J = (-1)^r (-1)^{\frac{k(k-1)}{2}} e_J.$$

On the other hand,

$$e_{I+J}^\dagger \cdot_R e_I^\dagger = (-1)^r (e_I e_J)^\dagger e_I^\dagger = (-1)^r (-1)^{\frac{k(k-1)}{2}} e_J^\dagger.$$

Thus $e_{I+J}^\dagger \cdot_R e_I^\dagger = (e_I \cdot_L e_{I+J})^\dagger$, and the desired result is established. □

We see now that if we talk about the left-hand dot product of a and b, we can expect a nonzero result only if the factor of smallest grade is on the left-hand side. With the right-hand dot product, it is the other way around.

Remark 5.5. We have introduced dot products of multivectors in several different places starting in Chapter 3. This is perhaps a good place to say something about how they are all related.

Looking back at Definition 3.6, the dot product of two simple k-vectors is given by

$$(a_1 \wedge \cdots \wedge a_k) \cdot (b_k \wedge \cdots \wedge b_1) = \det \begin{pmatrix} a_1 \cdot b_1 & \cdots & a_1 \cdot b_k \\ & \cdots & \\ a_k \cdot b_1 & \cdots & a_k \cdot b_k \end{pmatrix}.$$

The extension of this in Definition 4.3 to arbitrary k-vectors, not just simple ones, is a simple linear extension.

If we now look at Definition 4.3, at right-hand, left-hand, and two-sided dot products and scalar products, so long as a and b are k-vectors, that is, so long as they have the same grade, all the definitions yield the same value:

$$a \cdot_R b = a \cdot_L b = a \cdot b = \langle ab \rangle_0. \tag{5.9}$$

We conclude this section with a result that will be useful later and which extends the relation

$$ab = a \cdot b + a \wedge b$$

that holds for vectors.

Proposition 5.14. *If a is a vector in $\mathbb{R}^n$ and $b \in \mathbb{G}^n$, then*

$$ab = a \cdot_L b + a \wedge b \quad and \quad ba = b \cdot_R a + b \wedge a.$$

Proof. The second statement follows from the first via Propositions 5.7 and 5.13, so we verify only the first equation. We also note that if b is a scalar, $b = \beta$, then $a \cdot_L \beta = 0$ and $a \wedge \beta = a\beta = \beta a$; thus the first equation holds trivially. By the linearity of the geometric and dot products and the wedge, it suffices to establish the result when b is a simple p-vector with $p \geq 1$.

Let $\{u_i\}_{i=1}^n$ be an orthonormal basis for $\mathbb{R}^n$. Observe that

$$u_i \cdot_L (u_1 \cdots u_p) = \sum_{r=1}^p (-1)^{r-1} (u_i \cdot u_r)(u_1 \cdots \widehat{u_r} \cdots u_p)$$

where the product on the right-hand side is the standard dot product of vectors. It follows easily from this that if a is a vector, then

$$a \cdot_L (u_1 \cdots u_p) = \sum_{r=1}^p (-1)^{r-1} (a \cdot u_r)(u_1 \cdots \widehat{u_r} \cdots u_p).$$

We may, without loss of generality, assume b has the form $b = \beta (u_1 \cdots u_p)$ where β is a scalar. Then

$$a \cdot_L b = \beta \sum_{r=1}^p (-1)^{r-1} (a \cdot u_r)(u_1 \cdots \widehat{u_r} \cdots u_p).$$

We then consult Proposition 5.5 and notice that $a \cdot_L b$ appears as a term on the right-hand side of the first equation there. We conclude from this proposition that

$$ab = a \cdot_L b + a \wedge b$$

which was the desired conclusion. $\square$

Exercises 5.5.

1. Prove Proposition 5.10.

2. Show that if a, b are k-vectors in $\mathbb{G}^n$, then $\langle a b^\dagger \rangle_0 = a \cdot b^\dagger$.

3. Prove Proposition 5.11.

4. Show that the scalar product and each of the dot products is linear in both a and b. For the left dot product, for example, you should show that

   (a) $a \cdot_L (b + c) = a \cdot_L b + a \cdot_L c$,

   (b) $(a + b) \cdot_L c = a \cdot_L c + b \cdot_L c$,

   (c) $(\lambda a) \cdot_L b = a \cdot_L (\lambda b) = \lambda (a \cdot_L b)$ where λ is a scalar.

5. Give an example for which $\langle ab \rangle_0 \neq \langle a b^\dagger \rangle_0$.

6. Show that if $a, b_1, \ldots, b_p \in \mathbb{R}^n$, then

$$a \cdot_L (b_1 \wedge \cdots \wedge b_p) = \sum_{r=1}^{p} (-1)^{r-1} (a \cdot b_r)(b_1 \wedge \cdots \wedge \widehat{b_r} \wedge \cdots \wedge b_p),$$

$$(b_1 \wedge \cdots \wedge b_p) \cdot_R a = \sum_{r=1}^{p} (-1)^{p-r} (a \cdot b_r)(b_1 \wedge \cdots \wedge \widehat{b_r} \wedge \cdots \wedge b_p).$$

7. Justify Equation (5.9) in Remark 5.5. (Hint: Check this for an orthonormal basis.)

5.6   Reciprocal frames

The next piece of machinery will be useful in doing geometric calculus.

5.6.1   Construction and properties of reciprocal frames

We say that a set of vectors $\{a_1, \ldots, a_p\}$ in $\mathbb{R}^n$ is a *frame* provided $a_1 \wedge \cdots \wedge a_p \neq 0$.

Definition 5.6. We say that two frames $\{u_1, \ldots, u_p\}$ and $\{v_1, \ldots, v_p\}$ in $\mathbb{R}^n$ are *reciprocal* provided

$$\text{span}\{u_1 \ldots u_p\} = \text{span}\{v_1, \ldots, v_p\},$$

and

$$u_i \cdot v_j = \begin{cases} 1 & \text{if } i = j, \\ 0 & \text{if } i \neq j. \end{cases}$$

(The definition also applies if it holds under some rearrangement of the indices.) If W is the common span of the two sets of vectors, we say that $\{u_1, \ldots, u_p\}$ and $\{v_1, \ldots, v_p\}$ are reciprocal bases of W. We also say that $\{u_1, \ldots, u_p\}$ is the reciprocal basis (or frame) for $\{v_1, \ldots, v_p\}$ and vice versa.

Example 5.11. Every orthonormal basis is its own reciprocal basis.

Example 5.12. If $\{u_1, \ldots, u_p\}$ is a frame whose vectors are orthogonal, then the reciprocal frame is $\{v_1, \ldots, v_p\}$ where each $v_i = u_i/|u_i|^2$.

Example 5.13. If we take $\{e_1, e_1 + e_2\}$ as a basis for $\mathbb{R}^2$, the reciprocal basis is $\{e_1 - e_2, e_2\}$. This is easily checked by computing the dot products.

It turns out that we can use geometric algebra to construct reciprocal frames.

Proposition 5.15. *Let V be a p-dimensional vector subspace of $\mathbb{R}^n$. Suppose $\{u_i\}_{i=1}^p$ is a basis for V and set $u = u_1 \wedge \cdots \wedge u_p$. Then the reciprocal basis $\{m_i\}_{i=1}^p$ is given by*

$$m_i = (-1)^{i-1}(u_1 \wedge \cdots \wedge \widehat{u_i} \wedge \cdots \wedge u_p)\, u^{-1}$$
$$= (-1)^{p-i} u^{-1}(u_1 \wedge \cdots \wedge \widehat{u_i} \wedge \cdots \wedge u_p).$$

Proof. We establish only the first form for the reciprocal basis. The proof for $p = 1$ is trivial, so the reader may safely assume p is at least 2.

Set $m_i = (-1)^{i-1}(u_1 \wedge \cdots \wedge \widehat{u_i} \wedge \cdots \wedge u_p)\, u^{-1}$. By this definition,

$$m_i u = (-1)^{i-1}(u_1 \wedge \cdots \wedge \widehat{u_i} \wedge \cdots \wedge u_p)$$

while Proposition 5.5 gives us

$$m_i u = \sum_{j=1}^p (-1)^{j-1}(m_i \cdot u_j)(u_1 \wedge \cdots \wedge \widehat{u_j} \wedge \cdots \wedge u_p) + m_i \wedge u_1 \wedge \cdots \wedge u_p.$$

Since $\{u_j\}_{j=1}^p$ is a basis for V, the vectors $m_i, u_1, \ldots, u_p$ must be linearly dependent, so we have $m_i \wedge u_1 \wedge \cdots \wedge u_p = 0$. It follows that

$$(-1)^{i-1}(u_1 \wedge \cdots \wedge \widehat{u_i} \wedge \cdots \wedge u_p) = \sum_{j=1}^p (-1)^{j-1}(m_i \cdot u_j)(u_1 \wedge \cdots \wedge \widehat{u_j} \wedge \cdots \wedge u_p).$$

We know from Exercise 6 of Section 4.1 or from Propositions 4.4 and 4.5 and the discussions surrounding them that the $(p-1)$-vectors of the form $u_1 \wedge \cdots \wedge \widehat{u_j} \wedge \cdots \wedge u_p$ constitute a basis for $\Lambda^{p-1}V$ and hence are linearly independent in $\mathbb{G}^n$. It follows that the corresponding coefficients on either side of the last equation must be equal, and this means that

$$m_i \cdot u_j = \begin{cases} 1 & \text{if } i = j, \\ 0 & \text{if } i \neq j. \end{cases}$$

This is the defining property of the reciprocal basis. $\qquad\square$

Here are sometimes useful properties of reciprocal frames:

Proposition 5.16. *Let V be a p-dimensional vector subspace of $\mathbb{R}^n$. Suppose $\{u_i\}_{i=1}^p$ is a frame in V and $\{m_i\}_{i=1}^p$ is the reciprocal frame. Then the following hold:*

1. *Suppose that $I = (i_1, \ldots, i_k)$ and $J = (j_1, \ldots, j_k)$, where $1 \leq k \leq p$, are ordered multi-indices of the same length. Then*

$$u_I \cdot m_J^\dagger = \begin{cases} 1 & \text{if } I = J, \\ 0 & \text{if } I \neq J. \end{cases}$$

2. $(u_1 \wedge \cdots \wedge u_p)(m_1 \wedge \cdots \wedge m_p)^\dagger = 1$.

Proof. For the first part of the proposition, we appeal to Definition 3.6:

$$u_I \cdot m_J^\dagger = \det \begin{pmatrix} u_{i_1} \cdot m_{j_1} & \cdots & u_{i_1} \cdot m_{j_k} \\ \vdots & & \vdots \\ u_{i_k} \cdot m_{j_1} & \cdots & u_{i_k} \cdot m_{j_k} \end{pmatrix}.$$

If there is either a row or column vector in the matrix which is all zeros, then $u_I \cdot m_J^\dagger = 0$. From the definition of reciprocal frame and the fact that the multi-indices are ordered, we see that the following hold for the matrix in the last equation:

1. Each entry is either 0 or 1.

2. In each row or column vector of the matrix, there is at most a single 1.

3. If there are occurrences of 1 in two different row vectors, then the 1 in the lower row vector lies to the right of the 1 in the higher row vector.

It is easily seen from this that the only way the matrix can avoid having a row or column of all zeros is if the main diagonal is all ones while all other entries are zero. This is the case where $I = J$, so in this case, $u_I \cdot m_J^\dagger = 1$.

For the second part of the proposition, set $u = u_1 \wedge \cdots \wedge u_p$ and $m = m_1 \wedge \cdots \wedge m_p$. We are, in effect, about to show that $m^\dagger$ is the inverse of u; however we show only that $u m^\dagger = 1$ and leave it to the reader to show that $m^\dagger u = 1$.

We know that u and m are p-blades in the p-dimensional space V and we know that $\dim \wedge^p V = 1$, so we must have $m = \lambda_0 u$ for some nonzero scalar λ_0. If we multiply both sides of the last equation by $m^\dagger$ and divide by λ_0, it follows that $u m^\dagger = \lambda_1$, a scalar. Therefore we must have

$$u m^\dagger = \langle \langle u \rangle_p \langle m^\dagger \rangle_p \rangle_0.$$

If we consult Definition 5.5, we find this means that $u m^\dagger = u \cdot m^\dagger$. It then follows from the first part of this proposition that $u m^\dagger = 1$. $\square$

One nice application of this proposition is to compute the coefficients of a k-vector: Suppose we are given V, $\{u_i\}_{i=1}^p$, and $\{m_i\}_{i=1}^p$ as in the statement of Proposition 5.16. If a is a k-vector in V, we know that the set of u_I such that I is an ordered multi-index of length k (that is, $I \in \mathcal{O}_k$) is a basis for $\wedge^k V$. Then it must be possible to write $a = \sum_{I \in \mathcal{O}_k} \lambda_I u_I$ where each λ_I is a scalar. We would like to know the value of λ_J, and this can be computed by dotting both sides of the last equation by $m_J^\dagger$. Since $u_I \cdot m_J^\dagger$ is 1 if $I = J$ and otherwise 0, we find that $\lambda_J = a \cdot m_J^\dagger$.

Another application—and a very important one when doing calculus on manifolds—is to find the reciprocal frame for a frame of tangent vectors on a manifold.

Definition 5.7. If a manifold $\mathcal{M}$ has coordinates $\tau_1, \ldots, \tau_p$ induced on it by a chart x, we denote the associated set of tangent vectors $\{\partial x / \partial \tau_i\}_{i=1}^p$. Let us denote the reciprocal frame by $\{d\tau_i\}_{i=1}^p$. Thus $(\partial x / \partial \tau_i) \cdot d\tau_j = \delta_{ij}$, Kronecker's delta. Expressions of the form $d\tau_i$ are sometimes called *differentials*.

Example 5.14. Let us take as our manifold the radius 1 upper half-sphere S^{2+} in $\mathbb{R}^3$ with the parametrization given in Example 4.12:

$$
\begin{aligned}
x(\rho, \theta) &= \chi_1 e_1 + \chi_2 e_2 + \chi_3 e_3 \\
&= \rho \cos(\theta) e_1 + \rho \sin(\theta) e_2 + \sqrt{1 - \rho^2} e_3
\end{aligned}
$$

where $0 < \theta < 2\pi$ and $0 < \rho < 1$. At any given point $x = \chi_1 e_1 + \chi_2 e_2 + \chi_3 e_3$ of S^{2+}, we calculate the tangent vectors to be

$$
\begin{aligned}
\frac{\partial x}{\partial \rho} &= \frac{\chi_1}{\rho} e_1 + \frac{\chi_2}{\rho} e_2 - \frac{\rho}{\chi_3} e_3 \\
\frac{\partial x}{\partial \theta} &= -\chi_2 e_1 + \chi_1 e_2
\end{aligned}
$$

where we are using the notation $\partial x / \partial \chi_i$ for the tangent vectors that was introduced in Equation (2.6). Let the reciprocal frame for $\{\partial x / \partial \rho, \partial x / \partial \theta\}$ be denoted $\{d\rho, d\theta\}$. Since our pair of of tangent vectors is seen to be orthogonal, we need merely multiply $\partial x / \partial \rho$ and $\partial x / \partial \theta$ by appropriate scalars to ensure that

$$
d\rho \cdot \frac{\partial x}{\partial \rho} = d\theta \cdot \frac{\partial x}{\partial \theta} = 1.
$$

The solution for our reciprocal vectors is this:

$$
d\rho = \frac{1}{1 - \rho^2} \frac{\partial x}{\partial \rho} \quad \text{and} \quad d\theta = \frac{1}{\rho^2} \frac{\partial x}{\partial \theta}.
$$

5.6.2 Adjoint linear transformations

We have had occasion to make use of transposed matrices of linear transformations. There is an interesting connection between such transposed matrices, reciprocal frames, and the *adjoints* of linear transformations.

Definition 5.8. Suppose U and V are vector subspaces of $\mathbb{R}^m$ and $\mathbb{R}^n$ respectively and that $f: U \to V$ and $g: V \to U$ are linear transformations such that for all $u \in U$ and $v \in V$ we have

$$
f(u) \cdot v = u \cdot g(v).
$$

We then say that g is the *adjoint* of f and we write $g = f^t$.

There is an implication in this definition that the adjoint is unique, assuming that it exists. The existence and uniqueness are both evident from the following proposition:

Proposition 5.17. *We assume the following:*

1. *U and V are p- and q-dimensional vector subspaces of $\mathbb{R}^m$ and $\mathbb{R}^n$ respectively.*

2. *$f: U \to V$ and $g: V \to U$ are linear transformations.*

3. *U has a basis $\{u_i\}_{i=1}^p$ and reciprocal basis $\{m_i\}_{i=1}^p$ while V has a basis $\{v_j\}_{j=1}^q$ and reciprocal basis $\{n_j\}_{j=1}^q$.*

4. *A is the matrix of f calculated with respect to the bases $\{u_i\}_{i=1}^p$ and $\{v_j\}_{j=1}^q$, and B the matrix of g calculated with respect to the bases $\{n_j\}_{j=1}^q$ and $\{m_i\}_{i=1}^p$.*

Then $B = A^T$ if and only if $f(u) \cdot v = u \cdot g(v)$ for all $u \in U$ and $v \in V$.

Proof. We write how f acts on basis vectors:

$$f(u_i) = \sum_{j=1}^q \lambda_{ij} v_j \quad \text{where } \lambda_{ij} \in \mathbb{R}.$$

If we dot this last equation with n_j, we discover that $\lambda_{ij} = f(u_i) \cdot n_j$. Thus

$$f(u_i) = \sum_{j=1}^q \left(f(u_i) \cdot n_j \right) v_j. \tag{5.10}$$

Similarly,

$$g(n_j) = \sum_{i=1}^p \left(u_i \cdot g(n_j) \right) m_i. \tag{5.11}$$

If we write our vectors as column vectors, we have

$$f(u_i) = \begin{pmatrix} f(u_1) \cdot n_1 & \cdots & f(u_p) \cdot n_1 \\ \vdots & & \\ f(u_1) \cdot n_q & \cdots & f(u_p) \cdot n_q \end{pmatrix} \begin{pmatrix} 0 \\ \vdots \\ 1 \\ \vdots \\ 0 \end{pmatrix} = A u_i \tag{5.12}$$

where the column matrix with 1 in the ith location is u_i and A at the end of the equation is the $q \times p$ matrix for f. Similarly,

$$g(n_j) = \begin{pmatrix} u_1 \cdot g(n_1) & \cdots & u_1 \cdot g(n_q) \\ \vdots & & \\ u_p \cdot g(n_1) & \cdots & u_p \cdot g(n_q) \end{pmatrix} \begin{pmatrix} 0 \\ \vdots \\ 1 \\ \vdots \\ 0 \end{pmatrix} = B n_j \tag{5.13}$$

where B is the $p \times q$ matrix for g. We see from Equations (5.12) and (5.13) that $B = A^T$ if and only if $f(u_i) \cdot n_j = u_i \cdot g(n_j)$ for all i, j. Since $\{u_i\}_{i=1}^p$ and $\{n_j\}_{j=1}^q$ are bases for U and V respectively, the last condition is equivalent to $f(u) \cdot v = u \cdot g(v)$ for all $u \in U$, $v \in V$. $\qquad \square$

Example 5.15. Let $f \colon \mathbb{R}^n \to U$ be the orthogonal projection onto a p-dimensional vector subspace of $\mathbb{R}^n$. There must exist an orthonormal basis $\{u_i\}_{i=1}^n$ for $\mathbb{R}^n$ having the property that $\{u_i\}_{i=1}^p$ is a basis for U. The projection map must satisfy

$$f(u_j) = \sum_{i=1}^p \delta_{ij} u_j \quad \text{where } 1 \le j \le n$$

and where δ_{ij} is Kronecker's delta, that is, $\delta_{ij} = 1$ if $i = j$ and 0 if $i \ne j$. We can easily show (as in the proof of Proposition 5.17) that

$$f^t(u_i) = \sum_{j=1}^n \left(u_j \cdot f^t(u_i) \right) u_j.$$

Thus

$$f^t(u_i) = \sum_{j=1}^n \left(f(u_j) \cdot u_i \right) u_j = \sum_{j=1}^n \sum_{k=1}^p \delta_{kj} \left(u_i \cdot u_k \right) u_j = u_i$$

for $1 \le i \le p$. Since $\{u_i\}_{i=1}^p$ is a basis for U, we see that

$$f^t(x) = x \quad \text{for all } x \in U.$$

Example 5.16. Let $f \colon \mathbb{R}^n \to \mathbb{R}$ be a linear map. (For example, we might have $f(\chi_1, \ldots, \chi_n) = \chi_k$ or $f(\chi_1, \ldots, \chi_n) = \sum_{i=1}^n \chi_i$.) What is f^t?

We treat $\{1\}$ as the basis (and reciprocal basis) for $\mathbb{R}$ considered as a vector space. Notice that in $\mathbb{R}$, we have $1 \cdot \lambda = \lambda$. Then

$$f^t(1) = \sum_{i=1}^n \left(f^t(1) \cdot e_i \right) e_i = \sum_{i=1}^n \left(1 \cdot f(e_i) \right) e_i = \sum_{i=1}^n f(e_i) e_i.$$

Since f^t is linear, we conclude that

$$f^t(\lambda) = \lambda \sum_{i=1}^n f(e_i) e_i.$$

Example 5.17. Choose a constant θ and let $f \colon \mathbb{R}^2 \to \mathbb{R}^2$ be the rotation of $\mathbb{R}^2$ by an angle θ. Thus the matrix of f (with respect to the standard basis) is

$$[f] = \begin{pmatrix} \cos(\theta) & -\sin(\theta) \\ \sin(\theta) & \cos(\theta) \end{pmatrix}.$$

By Proposition 5.17, the matrix of f^t is

$$[f^t] = \begin{pmatrix} \cos(\theta) & \sin(\theta) \\ -\sin(\theta) & \cos(\theta) \end{pmatrix}.$$

So if we think of $x = \chi_1 e_1 + \chi_2 e_2$ as a column vector, we see that

$$f'(x) = \begin{pmatrix} \cos(\theta) & \sin(\theta) \\ -\sin(\theta) & \cos(\theta) \end{pmatrix} \begin{pmatrix} \chi_1 \\ \chi_2 \end{pmatrix},$$

the rotation of $\mathbb{R}^2$ by an angle of $-\theta$.

Exercises 5.6.

1. Given the frame $u_1 = e_1$ and $u_2 = \cos(\theta) e_1 + \sin(\theta) e_2$ where $0 < \theta < \pi/2$, find the reciprocal frame $\{m_1, m_2\}$.

2. Given the frame $\{u_i\}_{i=1}^n$ in $\mathbb{R}^n$ where $u_i = \sum_{k=1}^i e_k$, find the reciprocal frame $\{m_i\}_{i=1}^n$.

3. Let $\{u_1, u_2\}$ be a frame and suppose that it has a reciprocal frame $\{m_1, m_2\}$. If $D = |u_1|^2 |u_2|^2 - (u_1 \cdot u_2)^2$, show that $D = |u_1 \wedge u_2|^2 \neq 0$. Then show that

$$m_1 = \frac{|u_2|^2}{D} u_1 - \frac{u_1 \cdot u_2}{D} u_2.$$

   What is m_2?

4. Let V be the subspace $\chi_1 + \cdots + \chi_n = 0$ of $\mathbb{R}^n$ where $n \geq 2$. A basis for V is $\{e_1 - e_n, \ldots, e_{n-1} - e_n\}$. Show that the reciprocal basis is $\{m_1, \ldots, m_{n-1}\}$ where

$$m_i = e_i - \frac{1}{n}(e_1 + \cdots + e_n).$$

5. Given the hypotheses of Proposition 5.16, show that $m^\dagger u = 1$.

6. Suppose that U is a p-dimensional vector subspace of the q-dimensional subspace V where both lie in some $\mathbb{R}^n$. Given a basis $\{v_i\}_{i=1}^p$ for U and reciprocal basis $\{n_i\}_{i=1}^p$, show that it is possible to extend these to $\{v_i\}_{i=1}^q$ and $\{n_i\}_{i=1}^q$ in such a way that $\{v_i\}_{i=1}^q$ is a basis for V and $\{n_i\}_{i=1}^q$ is the reciprocal basis.

7. Define $\{u_i\}_{i=1}^p$ in $\mathbb{R}^{p+1}$ by

$$u_i = e_i + \lambda_i e_{p+1} \quad \text{where } i = 1, \ldots, p$$

   and each λ_i is a given scalar constant. Show that the reciprocal frame $\{m_i\}_{i=1}^p$ is given by

$$m_i = u_i - \frac{1}{1 + \sum_{j=1}^p \lambda_j^2} \sum_{k=1}^p \lambda_i \lambda_k u_k.$$

5.7 Outermorphisms and innermorphisms

We saw earlier that if $f : \mathbb{R}^m \to \mathbb{R}^n$ is a linear transformation, then the operation on *simple k-vectors* defined by $\wedge^k f(a_1 \wedge \cdots \wedge a_k) = f(a_1) \wedge \cdots \wedge f(a_k)$ is well-defined. We now extend $\wedge^k f$ from an operation on simple k-vectors to an actual linear transformation on all k-vectors.

Proposition 5.18. *If $f : \mathbb{R}^m \to \mathbb{R}^n$ is a linear transformation, then for $1 \leq k$ there is a unique linear transformation $g : \Lambda^k \mathbb{R}^m \to \Lambda^k \mathbb{R}^n$ such that for every simple k-vector we have*

$$g(a_1 \wedge \cdots \wedge a_k) = f(a_1) \wedge \cdots \wedge f(a_k). \tag{5.14}$$

Proof. If $a \in \mathbb{G}^m$, then it must have a unique expansion, using ordered multi-indices, of the form $a = \sum_{I \in \mathcal{O}_k} \lambda_I e_I$. We define a linear transformation $g : \Lambda^k \mathbb{R}^m \to \Lambda^k \mathbb{R}^n$ by

$$g(a) = \sum_{I \in \mathcal{O}_k} \lambda_I \left(\wedge^k f(e_I) \right).$$

If I is the ordered multi-index $(i_1, \ldots, i_k)$, then we know that $\wedge^k f(e_I) = f(e_{i_1}) \wedge \cdots \wedge f(e_{i_k})$.

We need to see that g satisfies Equation (5.14). Let $a_1, \ldots, a_k$ be vectors in $\mathbb{R}^n$ and expand them in terms of the standard basis:

$$a_i = \sum_{i=1}^{m} \alpha_{ij} e_j.$$

Then

$$a_1 \wedge \cdots \wedge a_k = \left(\sum_i \alpha_{1i} e_i \right) \wedge \cdots \wedge \left(\sum_j \alpha_{kj} e_j \right) = \sum_{I \in \mathcal{O}_k} \lambda_I e_I$$

where the coefficients λ_I are uniquely determined by the properties of the wedge product. Notice that

$$f(a_1) \wedge \cdots \wedge f(a_k) = \left(\sum_i \alpha_{1i} f(e_i) \right) \wedge \cdots \wedge \left(\sum_j \alpha_{kj} f(e_j) \right)$$

$$= \sum_{I \in \mathcal{O}_k} \lambda_I \left(\wedge^k f(e_I) \right)$$

where the coefficients λ_I must be the same as before. This amounts to saying

$$f(a_1) \wedge \cdots \wedge f(a_k) = g(a_1 \wedge \cdots \wedge a_k).$$

If $g_1 : \Lambda^k \mathbb{R}^m \to \Lambda^k \mathbb{R}^n$ is another linear transformation that satisfies Equation (5.14), then we see that

$$g_1(e_{i_1} \wedge \cdots \wedge e_{i_k}) = f(e_{i_1}) \wedge \cdots \wedge f(e_{i_k}) = g(e_{i_1} \wedge \cdots \wedge e_{i_k})$$

which ensures that $g_1 = g$. $\qquad \square$

We now extend the linear transformation $f : \mathbb{R}^m \to \mathbb{R}^n$ to a linear transformation $\underline{f} : \mathbb{G}^m \to \mathbb{G}^n$.

Definition 5.9. Suppose we are given a linear transformation $f : \mathbb{R}^m \to \mathbb{R}^n$. Every $a \in \mathbb{G}^m$ has a unique expansion

$$a = \langle a \rangle_0 + \langle a \rangle_1 + \cdots + \langle a \rangle_m,$$

so we define $\underline{f} : \mathbb{G}^m \to \mathbb{G}^n$ by

$$\underline{f}(a) = \wedge^0 f(\langle a \rangle_0) + \wedge^1 f(\langle a \rangle_1) + \cdots + \wedge^m f(\langle a \rangle_m)$$

or, equivalently,

$$\langle \underline{f}(a) \rangle_k = \wedge^k f(\langle a \rangle_k).$$

(Recall in this definition that $\wedge^0 f$ is the identity map on $\mathbb{R}$.) We call the linear transformation $\underline{f}$ generated by f an *outermorphism*.

Example 5.18. Let $f : \mathbb{R}^2 \to \mathbb{R}^2$ be the linear transformation satisfying $f(e_1) = e_1$ and $f(e_2) = e_1 + e_2$. Then

$$\underline{f}(e_1 e_2) = e_1 \wedge (e_1 + e_2) = e_1 e_2.$$

More generally,

$$\underline{f}(\lambda_0 + \lambda_1 e_1 + \lambda_2 e_2 + \lambda_3 e_1 e_2) = \lambda_0 + (\lambda_1 + \lambda_2) e_1 + \lambda_2 e_2 + \lambda_3 \, e_1 e_2.$$

Proposition 5.19. *If $f : \mathbb{R}^m \to \mathbb{R}^n$ is a linear transformation, then the outermorphism $\underline{f} : \mathbb{G}^m \to \mathbb{G}^n$ is the unique linear transformation satisfying $\underline{f}(a) = f(a)$ for all $a \in \mathbb{R}^m$ and $\underline{f}(b \wedge c) = \underline{f}(b) \wedge \underline{f}(c)$ for all $b, c \in \mathbb{G}^m$.*

The proof is quite straightforward and we leave it to the reader.

Of course this discussion still makes sense if we start with a linear transformation between two finite-dimensional vector subspaces of $\mathbb{R}^n$, $f : U \to V$, and extend it to a linear transformation $\underline{f}$ between the associated geometric algebras, $\underline{f} : \mathbb{G}(U) \to \mathbb{G}(V)$.

Though outermorphisms distribute over the wedge product, they do *not* distribute over the geometric product.

Example 5.19. Let $f : \mathbb{R}^2 \to \mathbb{R}^2$ be the linear transformation for which

$$f(e_1) = e_1 + e_2 \quad \text{and} \quad f(e_2) = e_2.$$

Then

$$\underline{f}(e_1 e_2) = (e_1 + e_2) \wedge e_2 = e_1 e_2$$

and

$$\underline{f}(e_1) \underline{f}(e_2) = (e_1 + e_2) e_2 = e_1 e_2 + 1.$$

So $\underline{f}(e_1 e_2) \neq \underline{f}(e_1) \underline{f}(e_2)$.

Definition 5.10. Given a linear transformation $f : U \to V$ between subspaces of some $\mathbb{R}^m$ and $\mathbb{R}^n$, we can, in a fashion similar to what we have done for $\underline{f}$, extend the adjoint transformation $f^t : V \to U$ to a linear transformation $\overline{f} : \mathbb{G}(V) \to \mathbb{G}(U)$ in such a way that $\overline{f}(a \wedge b) = \overline{f}(a) \wedge \overline{f}(b)$ for all $a, b \in \mathbb{G}(V)$. We call $\overline{f}$ an *innermorphism*.

Provided we think of the scalar product as our dot product on $\mathbb{G}^n$, we see that $\overline{f}$ plays the role of the adjoint to $\underline{f}$.

Proposition 5.20. *Given the outermorphism $\underline{f}$ and corresponding innermorphism $\overline{f}$, we have*

$$\langle \underline{f}(a)\, b \rangle_0 \ = \ \langle a\, \overline{f}(b) \rangle_0 \tag{5.15}$$

for all a and b at which the expressions can be evaluated.

Proof. By linearity of the transformations, it is sufficient to check this when a and b are blades. Further, if a and b have different grades, then all the expressions are 0. Therefore we may assume that $a = a_1 \wedge \cdots \wedge a_k$ and $b = b_1 \wedge \cdots \wedge b_k$ where a_i and b_j are vectors. We then calculate thus:

$$\begin{aligned}
\langle \underline{f}(a)\, b \rangle_0 &= \underline{f}(a) \cdot b \\
&= \big(f(a_1) \wedge \cdots \wedge f(a_k) \big) \cdot (b_1 \wedge \cdots \wedge b_k) \\
&= \det \big(f(a_i) \cdot b_j \big) \\
&= \det \big(a_i \cdot f^t(b_j) \big) \\
&= \langle a\, \overline{f}(b) \rangle_0.
\end{aligned}$$
$\square$

Exercises 5.7.

1. Prove Proposition 5.19. (Hint: How does one calculate $\langle \underline{f}(b \wedge c) \rangle_{p+q}$?)

2. If f_0 and f_1 are linear transformations between vector subspaces such that the appropriate compostions exist, then show that $\overline{f_1 \circ f_0} = \overline{f_0} \circ \overline{f_1}$. (Hint: What can you say about $(f_1 \circ f_0)^t$ where the superscript t indicates the adjoint transformation?)

3. Give an example of a linear transformation $g : \mathbb{G}^n \to \mathbb{G}^n$ which is neither an outermorphism nor an innermorphism.

4. Given outer- and innermorphisms $\underline{f}$ and $\overline{f}$, show that $\underline{f}(a^\dagger) = \big(\underline{f}(a)\big)^\dagger$ and $\overline{f}(a^\dagger) = \big(\overline{f}(a)\big)^\dagger$ for all multivectors a in the domains of the transformations.

5. Let $f : \mathbb{R}^2 \to \mathbb{R}^2$ be the linear transformation that satisfies

$$\begin{aligned}
f(e_1) &= \cos(\theta)\, e_1 + \sin(\theta)\, e_2, \\
f(e_2) &= -\sin(\theta)\, e_1 + \cos(\theta)\, e_2.
\end{aligned}$$

Compute $\underline{f}(x)$ where

$$x \ = \ \lambda_0 + \lambda_1 e_1 + \lambda_2 e_2 + \lambda_3 e_1 e_2.$$

6. Let $f : \mathbb{R}^3 \to \mathbb{R}^3$ be the linear transformation that satisfies

$$f(e_i) = \begin{cases} -e_1 & \text{if } i = 1, \\ e_i & \text{if } i \neq 1. \end{cases}$$

An arbitrary element of $\mathbb{G}^3$ has the form

$$x = \sum_{p=0}^{3} \sum_{I \in \mathcal{O}_p} \lambda_I e_I.$$

Find all $x \in \mathbb{G}^3$ in terms of λ_I and e_I such that $\underline{f}(x) = x$.

7. Let $f : \mathbb{R}^2 \to \mathbb{R}^2$ be the linear transformation that satisfies

$$f(e_i) = \begin{cases} e_1 & \text{if } i = 1, \\ -\sin(\theta)e_1 + \cos(\theta)e_2 & \text{if } i = 2 \end{cases}$$

where θ as a constant such that $0 < \theta < \pi/2$. Compute $\overline{f}(e_1 e_2)$.

8. We give an extended version of the chain rule: Suppose that U, V, and W are open subsets of $\mathbb{R}^p$, $\mathbb{R}^q$, and $\mathbb{R}^r$ respectively. Assume we have $\mathcal{C}^1$ maps

$$U \xrightarrow{g} V \xrightarrow{f} V$$

and set $h = f \circ g$. Choose $x \in U$ and set $y = g(x) \in V$. First show that

$$\wedge^k h'(x) = \left(\wedge^k f'(y) \right) \left(\wedge^k g'(x) \right)$$

where the right-hand side of the equation indicates a composition of linear transformations. Then show that

$$\underline{h'(x)} = \underline{f'(y)} \circ \underline{g'(x)}.$$

5.8   Projections, angles, and orthogonality

If V is a vector subspace of $\mathbb{R}^n$, the reader should understand that its *orthogonal complement*, $V_\perp$, is the set of $x \in \mathbb{R}^n$ such that x is orthogonal to every $y \in V$. It is easily seen that $V_\perp$ is also a vector subspace of $\mathbb{R}^n$.

It is not hard to show that every vector x of $\mathbb{R}^n$ has a unique decomposition $x = x_\| + x_\perp$ such that $x_\| \in V$ and $x_\perp \in V_\perp$. To see this, let $\{u_i\}_{i=1}^n$ be an orthonormal basis for $\mathbb{R}^n$ such that $\{u_i\}_{i=1}^p$ is a basis for V. We break x up into its u_i-components thus: $x = \sum_{i=1}^n (x \cdot u_i) u_i$. Then

$$x_\| = \sum_{i=1}^{p} (x \cdot u_i) u_i \quad \text{and} \quad x_\perp = \sum_{i=p+1}^{n} (x \cdot u_i) u_i.$$

The uniqueness of this decomposition follows from the fact that if we have a second decomposition $x = x_0 + x_1$ where $x_0 \in V$ and $x_1 \in V_\perp$, then we have

$$x \cdot u_i = \begin{cases} (x_0 \cdot u_i) + (x_1 \cdot u_i) = x_0 \cdot u_i & \text{if } 1 \le i \le p, \\ (x_0 \cdot u_i) + (x_1 \cdot u_i) = x_1 \cdot u_i & \text{if } p+1 \le i \le n. \end{cases}$$

Example 5.20. In $\mathbb{R}^3$, let us take

$$V = \text{the } \chi_1\chi_2\text{-plane,}$$
$$V_\perp = \text{the } \chi_3\text{-axis.}$$

Then e_1, e_2 span V and e_3 spans $V_\perp$. Every $x = \lambda_1 e_1 + \lambda_2 e_2 + \lambda_3 e_3$ has the decomposition

$$x_\| = \lambda_1 e_1 + \lambda_2 e_2 \quad \text{and} \quad x_\perp = \lambda_3 e_3.$$

Definition 5.11. Given V, a vector subspace of $\mathbb{R}^n$, we define the *orthogonal projection onto V* to be the linear transformation $P : \mathbb{R}^n \to V$ such that $P(x) = x_\|$. Given that P is a linear transformation, this amounts to the requirement that $P(x) = x$ when $x \in V$ and $P(x) = 0$ when $x \in V_\perp$.

Remark 5.6. The idea of a *projection* is more general than the definition above and goes like this: If by P^2 we mean $P \circ P$, then projections as linear transformations are characterized by the property $P^2 = P$. However, we shall only be interested in orthogonal projections, and when we use the term *projection*, unless we say otherwise, we shall mean an orthogonal one.

Example 5.21. We give an example of a projection that is not orthogonal. Let V be the $\chi_1\chi_2$-plane in $\mathbb{R}^3$ and let $x = (\chi_1, \chi_2, \chi_3) = \chi_1 e_1 + \chi_2 e_2 + \chi_3 e_3$. We take $P : \mathbb{R}^3 \to V$ to be the map

$$P(x) = (\chi_1 - \chi_3)e_1 + (\chi_2 - \chi_3)e_2 = x - \chi_3(e_1 + e_2 + e_3).$$

The map P, given a point $x \in \mathbb{R}^3$, projects it along a line parallel to the vector $e_1 + e_2 + e_3$ till it hits the $\chi_1\chi_2$-plane. This is a linear transformation and satisfies $P^2(x) = P(x)$ so that it is a projection. However it is not orthogonal since if it were, we would have $P(e_3) = 0$ whereas in reality we have $P(e_3) = -e_1 - e_2$.

Now that we have the machinery of geometric algebra, we can do two interesting and useful things: First, we can write explicit formulas for the projection maps onto V and $V_\perp$. Second, we can extend the idea of projection to multivectors by passing from P to the outermorphism $\underline{P}$.

Here is an example:

Example 5.22. We return to the setting of Example 5.20 where V is the $\chi_1\chi_2$-plane and $V_\perp$ is the χ_3-axis. Note that $e_1 e_2$ is a 2-blade parallel to V. Let $a = \alpha_1 e_1 + \alpha_2 e_2 + \alpha_3 e_3$, an arbitrary 1-vector. (See Figure 5.1.) Let $P : \mathbb{R}^3 \to V$ and $Q : \mathbb{R}^3 \to V_\perp$ be the

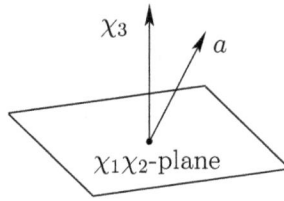

Figure 5.1: V and $V_\perp$ with a 1-vector

(orthogonal) projection maps onto the respective spaces. We show how calculate $P(a)$ and $Q(a)$ using the operations of geometric algebra:

First we left-dot a with $e_1 e_2$ then we divide the result by $e_1 e_2$:

$$\left(a \cdot_L (e_1 e_2)\right)(e_1 e_2)^{-1} = (\alpha_1 e_2 - \alpha_2 e_1)(e_2 e_1)$$
$$= \alpha_1 e_1 + \alpha_2 e_2$$
$$= P(a).$$

Next we repeat this calculation with the wedge product substituted for the left-dot operation:

$$\left(a \wedge (e_1 e_2)\right)(e_1 e_2)^{-1} = \left(\alpha_3 (e_1 e_2 e_3)\right)(e_2 e_1)$$
$$= \alpha_3 e_3 = Q(a).$$

Now we take an arbitrary 2-vector

$$b = \beta_1 e_1 e_2 + \beta_2 e_1 e_3 + \beta_3 e_2 e_3$$

in $\mathbb{R}^3$. If we "project" it onto V, clearly we want the answer to be $\beta_1 e_1 e_2$. If, on the other hand, we "project" it onto $V_\perp$, then it seems reasonable to expect an answer 0. In this second case, b is "too big;" b has dimension 2 while $V_\perp$ is 1-dimensional.

We proceed just as before:

$$\left(b \cdot_L (e_1 e_2)\right)(e_1 e_2)^{-1} = \left[(\beta_1 e_1 e_2 + \beta_2 e_1 e_3 + \beta_3 e_2 e_3) \cdot (e_1 e_2)\right](e_2 e_1)$$
$$= -\beta_1 (e_2 e_1)$$
$$= \beta_1 (e_1 e_2).$$

This is exactly what we wanted and also turns out to be $\underline{P}(b)$. For the projection onto $V_\perp$, we compute, following same pattern used for a,

$$\left(b \wedge (e_1 e_2)\right)(e_1 e_2)^{-1} = \left[(\beta_1 e_1 e_2 + \beta_2 e_1 e_3 + \beta_3 e_2 e_3) \wedge (e_1 e_2)\right](e_2 e_1)$$
$$= 0.$$

Again, this is what we wanted, and we shall show in the next proposition that it is the same as $\underline{Q}(b)$.

Proposition 5.21. *Let V be a nontrivial p-dimensional vector subspace of $\mathbb{R}^n$ and v a p-blade parallel to V. Suppose that $P : \mathbb{R}^n \to V$ is the (orthogonal) projection onto V, and $Q : \mathbb{R}^n \to V_\perp$ is the (orthogonal) projection onto its orthogonal complement. Let $\underline{P}$ and $\underline{Q}$ be the corresponding outermorphisms of $\mathbb{G}^n$ onto $\mathbb{G}(V)$ and $\mathbb{G}(V_\perp)$ respectively. Then for all $a \in \mathbb{G}^n$, we have*

$$\underline{P}(a) = (a \cdot_L v)\, v^{-1} \quad and \quad \underline{Q}(a) = (a \wedge v)\, v^{-1}.$$

A full proof is given in Appendix A.5. From this point on, we shall also feel free to refer to the outermorphism $\underline{P}$ as the *(orthogonal) projection onto V.*

Remark 5.7. One can also write formulas for $\underline{P}$ and $\underline{Q}$ in which v^{-1} appears on the left rather than the right. See Exercise 2 of this section.

We give another formula for the orthogonal projection:

We know that if V is a p-dimensional subspace of $\mathbb{R}^n$ and if $v_1, \ldots, v_n$ is an orthonormal basis for $\mathbb{R}^n$ such that $v_1, \ldots, v_p$ spans V, then the orthogonal projection of $x \in \mathbb{R}^n$ onto V is given by

$$P(x) = (x \cdot v_1) v_1 + \cdots (x \cdot v_p) v_p.$$

If one brings in recipocal bases, this formula can be generalized to bases which are not orthogonal and to k-vectors.

Proposition 5.22. *Suppose V is a p-dimensional vector subspace of $\mathbb{R}^n$ with basis $\{v_i\}_{i=1}^p$ and reciprocal basis $\{m_i\}_{i=1}^p$. Let $P \colon \mathbb{R}^n \to V$ be the orthogonal projection onto V If x is a k-vector in $\mathbb{R}^n$ (where $k \leq p$), then*

$$\underline{P}(x) = \sum_{i_1 < \cdots < i_k} \left(x \cdot \left(v_{i_1} \wedge \cdots \wedge v_{i_k} \right) \right) m_{i_1} \wedge \cdots \wedge m_{i_k}.$$

Proof. Let x be a k-vector in $\mathbb{R}^n$.

We extend $\{v_i\}_{i=1}^p$ from a basis of V to a basis $\{v_i\}_{i=1}^n$ of $\mathbb{R}^n$ in such a way that

1. $v_{p+1}, \ldots, v_n$ are orthogonal to V and

2. $v_{p+1}, \ldots, v_n$ are orthonormal.

If we set $m_i = v_i$ for $i = p+1, \ldots, n$, we see that $\{m_i\}_{i=1}^n$ is the reciprocal basis for $\{v_i\}_{i=1}^n$. (Just compute the dot products.)

We know that $\{m_{i_1} \wedge \cdots \wedge m_{i_k}\}_{i_1 < \cdots < i_k}$ is a basis for the space of k-vectors over $\mathbb{R}^n$, so we can write

$$x = \sum_{i_1 < \cdots < i_k} \lambda_{i_1 \ldots i_k} m_{i_1} \wedge \cdots \wedge m_{i_k}$$

where each $\lambda_{i_1 \ldots i_k}$ is a scalar. If we dot both sides of the last equation by $v_{i_1} \wedge \cdots \wedge v_{i_k}$ and appeal to Proposition 5.16, we find that

$$\lambda_{i_1 \ldots i_k} = x \cdot \left(v_{i_1} \wedge \cdots \wedge v_{i_k} \right).$$

So we now have the equation

$$x = \sum_{i_1 < \cdots < i_k} \left(x \bullet \left(v_{i_1} \wedge \cdots \wedge v_{i_k} \right) \right) m_{i_1} \wedge \cdots \wedge m_{i_k}$$

and apply the projection outermorphism $\underline{P}$ to both sides. Since $x \bullet \left(v_{i_1} \wedge \cdots \wedge v_{i_k} \right)$ is a scalar, this reduces to examining

$$\underline{P}(m_{i_1} \wedge \cdots \wedge m_{i_k}) = P(m_{i_1}) \wedge \cdots \wedge P(m_{i_k}).$$

Since m_{i_j} is a vector of V for $j = 1, \ldots, p$ and orthogonal to V otherwise, the right-hand side of the last equation must be $m_{i_1} \wedge \cdots \wedge m_{i_k}$ if $i_k \leq p$ and 0 otherwise. So we see that

$$\underline{P}(x) = \sum_{i_1 < \cdots < i_k} \left(x \bullet \left(v_{i_1} \wedge \cdots \wedge v_{i_k} \right) \right) m_{i_1} \wedge \cdots \wedge m_{i_k}$$

where it is understood that the sum is over the ordered multi-indices such that $i_k \leq p$.   □

Example 5.23. Let V be the vector subspace $\mathbb{R}^3 \times 0$ of $\mathbb{R}^4$. That is,

$$V = \{(\chi_1, \chi_2, \chi_3, 0) : \chi_i \in \mathbb{R} \text{ for } i = 1, 2, 3\}.$$

We choose a basis

$$v_1 = e_1$$
$$v_2 = e_1 + e_2$$
$$v_3 = e_1 + e_2 + e_3$$

for V. The reciprocal basis is easily checked to be

$$m_1 = e_1 - e_2$$
$$m_2 = e_2 - e_3$$
$$m_3 = e_3.$$

If $P \colon \mathbb{R}^4 \to V$ is the orthogonal projection onto V, then by Proposition 5.22 it must have the form

$$P(x) = (x \bullet v_1) m_1 + (x \bullet v_2) m_2 + (x \bullet v_3) m_3.$$

To check that P is the orthogonal projection, notice that e_1, e_2, e_3 is another basis for V while e_4 is orthogonal to V. If P is the projection we think it is, then it must send each of e_1, e_2, e_3 to itself and e_4 to 0. We see by direct computation that this is indeed the truth:

$$P(e_i) = \begin{cases} e_i & \text{for } i = 1, 2, 3, \\ 0 & \text{for } i = 4. \end{cases}$$

We now move beyond (orthogonal) projections and use geometric algebra to define and compute the angle between hyperplanes and subspaces of $\mathbb{R}^n$ even when the spaces are not of the same dimension. We utilize the ideas of [31] by Macdonald in the following discussion.

First, a fact about orthogonal projections:

Proposition 5.23. *Suppose that a is an element of $\mathbb{G}^n$ and b is a blade in $\mathbb{R}^n$. If $P_b(a)$ is the orthogonal projection of a onto the subspace determined by b, then $|P_b(a)| \leq |a|$.*

Proof. Suppose that b is a p-vector and choose an orthonormal basis $\{u_i\}_{i=1}^n$ for $\mathbb{R}^n$ such that $b = \beta\, u_1 \cdots u_p$ for some nonzero scalar β. We can expand a in terms of this basis: $a = \sum_I \alpha_I u_I$ where the summation is over all possible multi-indices. Now let us introduce the notation $\bar{p} = \{1, 2, \ldots, p\}$. We see that

$$P_b(a) \;=\; \sum_{I \subseteq \bar{p}} \alpha_I\, u_I,$$

that is, it is a sum over only those multi-indices $I = (i_1, \ldots, i_k)$ such that $i_1 < \cdots < i_k \leq p$. As we are dealing with an orthonormal basis, we clearly have

$$|P_b(a)|^2 \;=\; \sum_{I \subseteq \bar{p}} \alpha_I^2 \;\leq\; \sum_I \alpha_I^2 \;=\; |a|^2. \qquad \square$$

Definition 5.12. If a and b are blades in $\mathbb{R}^n$ and $\mathrm{grade}(a) \leq \mathrm{grade}(b)$, then we define the *(undirected) angle between a and b* to be the unique θ in the interval $[0, \pi/2]$ that satisfies

$$\cos(\theta) \;=\; \frac{|P_b(a)|}{|a|}.$$

If A and B are the vector subspaces associated with a and b respectively, then we also say that θ is the *angle between A and B.*

Notice that if we replace a and b by nonzero scalar multiples, then the value of θ does not change. This is because $|P_b(\lambda\, a)| = |\lambda|\,|P_b(a)|$ and $|\lambda\, a| = |\lambda|\,|a|$ and $P_{\lambda b}(a) = P_b(a)$. Thus the angle between subspaces does not depend on which blades we choose to represent them.

Notice also that if we are dealing with pictures, there are two obvious choices for the angle between A and B, the acute angle θ and the obtuse angle $\pi - \theta$. (Figure 5.2) Our definition always chooses the acute angle.

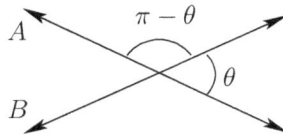

Figure 5.2: The two choices for the angle between A and B

There is, however, a subtle problem in the definition of "angle between" with which we need to deal: Suppose we have $\mathrm{grade}(a) = \mathrm{grade}(b)$. Then we have a choice of definitions for $\cos(\theta)$, either $|P_b(a)|/|a|$ or $|P_a(b)|/|b|$. Can we be sure the two expressions yield the same value?

To see how to deal with this, let a and b be blades with $\mathrm{grade}(a) \leq \mathrm{grade}(b)$. We know that $P_b(a) = (a \cdot_L b)\, b^{-1}$, thus $\big(P_b(a)\big)\, b = a \cdot_L b$. By Proposition 5.12, we know that $|P_b(a)\, b| = |P_b(a)|\,|b|$, so we can write

$$|P_b(a)|\,|b| \;=\; |\big(P_b(a)\big)\, b| \;=\; |a \cdot_L b|.$$

This means that

$$\frac{|P_b(a)|}{|a|} = \frac{|a \cdot_L b|}{|a| \, |b|}.$$

Similarly, when grade$(b) \leq$ grade(a), we have

$$\frac{|P_a(b)|}{|b|} = \frac{|a \cdot_R b|}{|a| \, |b|}.$$

Of course in the case where grade$(a) =$ grade(b), the left-hand and right-hand dot products are the same, so the two expressions for $\cos(\theta)$ will agree.

There is more than one way to think of two subspaces or hyperplanes being "at right angles" with one another. We exhibit two of them below and show their connection with geometric algebra.

Let A and B be two vector subspaces of $\mathbb{R}^n$ and let a and b be blades that are respectively parallel to the subspaces.

Definition 5.13. We say that A and B are *orthogonal* provided $x \cdot y = 0$ for all $x \in A$ and all $y \in B$. If $A = \{0\}$, the space containing only 0, then it is orthogonal to every other vector space. We say that a and b are *orthogonal* if and only if this is true for the associated vector spaces.

Definition 5.14. We say that A and B are *perpendicular* if and only if there exist nonzero vectors $x_0 \in A$ and $y_0 \in B$ such that $x_0 \cdot y = 0$ for all $y \in B$ and $x \cdot y_0 = 0$ for all $x \in A$. We say that a and b are *perpendicular* if and only if this is true for the associated subspaces.

In Figure 5.3 we see on the left A and B that are orthogonal; on the right we see A and B that are perpendicular but not orthogonal.

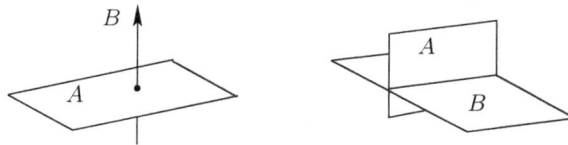

Figure 5.3: Orthogonality and perpendicularity

Suppose that p and q are the dimensions of A and B respectively. If $p, q \geq 1$, then orthogonality clearly implies perpendicularity. However a zero-dimensional subspace is orthogonal to all others and perpendicular to none. And $\mathbb{R}^n$ is orthogonal only to $\{0\}$ and is perpendicular to no subspace.

Here is a characterization of orthogonality; the proof is in Appendix A.5:

Proposition 5.24. *The blades a and b are orthogonal if and only if $ab = a \wedge b$.*

There is similarly a characterization of perpendicularity. It depends on the dot product (or on angles), and the proof is given in Appendix A.5.

Proposition 5.25. *Let a and b be blades in $\mathbb{R}^n$ with respective associated vector subspaces A and B. Then the following statements are equivalent:*

1. *A and B are perpendicular.*

2. $a \cdot_L b = a \cdot_R b = 0.$

3. *The angle between A and B is $\pi/2$.*

Example 5.24. Let $a = \alpha_1 e_1 + \alpha_2 e_2 + \alpha_3 e_3$, an arbitrary vector in $\mathbb{R}^3$, and $b = e_1 e_2$. (This is the setting of Example 5.22 and Figure 5.1.) Assuming $a \neq 0$, we see that a is orthogonal to b if and only if it is perpendicular; and in this case, orthogonality occurs only when $a = \alpha_3 e_3$.

How does this accord with our characterizations of orthogonality and perpendicularity? We note that $a \cdot_R b = 0$ trivially since the grade of a is less than that of b, so we ignore that condition. By Proposition 5.14, we have $ab = a \cdot_L b + a \wedge b$. We see that $ab = a \wedge b$ is equivalent to $a \cdot_L b = 0$, and since $a \cdot_L b = \alpha_1 e_2 - \alpha_2 e_1$, this is equivalent to $\alpha_1 = \alpha_2 = 0$, that is, $a = \alpha_3 e_3$.

Exercises 5.8.

1. Show that if V is a vector subspace of $\mathbb{R}^n$, then $V_\perp$ is also a vector subspace of $\mathbb{R}^n$.

2. Let P, Q, v, and a be as in Proposition 5.21.

   (a) Show that $\underline{P}(a^\dagger) = \left(\underline{P}(a)\right)^\dagger$.

   (b) Show that $\underline{P}(a) = v^{-1}(v \cdot_R a)$. (Hint: Notice that in the formula for $\underline{P}$ given in Proposition 5.21, the blade v can be replaced by $v^\dagger$.)

   (c) Show that $\underline{Q}(a) = v^{-1}(v \wedge a)$.

3. Let V be the subspace $\chi_1 + \chi_2 + \chi_3 = 0$ of $\mathbb{R}^3$. Let P be the projection onto V and Q the projection onto $V_\perp$.

   (a) If $x = \chi_1 e_1 + \chi_2 e_2 + \chi_3 e_3$, then compute $P(x)$ and $Q(x)$.

   (b) If $y = \xi_1 \, e_1 e_2 + \xi_2 \, e_1 e_3 + \xi_3 \, e_2 e_3$, then compute $\underline{P}(y)$.

4. Let V be the subspace of $\mathbb{R}^n$ spanned by $e_{i_1}, \ldots, e_{i_k}$ where $i_1 < \cdots < i_k$. If P is the projection onto V and $I = (i_1, \ldots, i_k)$, show that $\underline{P}(x) = \left(x \cdot_L (e_I^\dagger)\right) e_I$.

5. Find the angle in $\mathbb{R}^4$ between A, the $\chi_1 \chi_2$-plane (a 2-dimensional subspace of $\mathbb{R}^4$) and B, the 3-dimensional subspace $\sum_{i=1}^4 \chi_i = 0$.

6. Let V be a k-dimensional subspace of $\mathbb{R}^n$ and P be the orthogonal projection onto V. Choose an orthonormal basis $\{u_i\}_{i=1}^n$ of $\mathbb{R}^n$ such that $\{u_i\}_{i=1}^k$ spans V. Every $x \in \mathbb{G}^n$ has a unique expansion

$$x = \sum_{p=0}^n \sum_{I \in \mathbb{O}_p} \lambda_I u_I.$$

If $\mathcal{A}_p$ is the set of ordered multi-indices $I = (i_1, \ldots, i_p)$ such that $i_p \le k$, show that

$$\underline{P}(x) = \sum_{p=0}^{n} \sum_{I \in \mathcal{A}_p} \lambda_I u_I.$$

7. Let P be the orthogonal projection onto a subspace V of $\mathbb{R}^n$. Show that $\underline{P}^2 = \underline{P}$.

Chapter 6

The Fundamental Theorem

We now turn our attention back to calculus. Our main aim will be the Fundamental Theorem of geometric calculus. This is a result that contains the fundamental theorem of elementary calculus, the generalized Stokes theorem of differential form theory, and other results.

We first develop a certain amount of machinery.

6.1 The geometric derivative

Recall that if ϕ is a real-valued function on $\mathbb{R}^n$, then

$$\operatorname{grad}\phi(x) = \sum_{i=1}^{n} \frac{\partial \phi}{\partial \chi_i}(x)\, e_i$$

assuming one can perform the differentiations. We are going to generalize this idea of the gradient to multivector fields. Because of the noncommutativity of the geometric product, we shall get two versions of the generalization. This generalization is often referred to as the *vector derivative* or sometimes as the *tangential derivative*. As it is defined not with respect to vectors but to blades or manifolds, we prefer the term *geometric derivative*.

6.1.1 Directional derivatives

We will be primarily interested in functions $f \colon \mathcal{M} \to \mathbb{G}^n$ where $\mathcal{M}$ is a manifold in $\mathbb{R}^n$ and f is $\mathcal{C}^r$ with $r \geq 1$.

Suppose we have a function $f \colon A \to \mathbb{G}^n$ where A is a subset of $\mathbb{R}^n$. If $\{u_i\}_{i=1}^{n}$ is a set of constant vectors that form a basis for $\mathbb{R}^n$, then we can write f as

$$f = \sum_I \phi_I u_I$$

where the summation is over all ordered multi-indices and each ϕ_I is a uniquely determined real-valued function. Because of this, we can readily extend many of the definitions and results of Section 2.2 to this new setting:

1. To say that $f: A \to \mathbb{G}^n$ is $\mathcal{C}^r$ reduces to checking that each ϕ_I is $\mathcal{C}^r$. (Definition 2.5.) This property is independent of our choice of the constant basis $\{u_i\}_{i=1}^n$. (Proposition 2.1.)

2. We say that f is differentiable at x_0 provided there is an open neighborhood V of x_0 in $\mathbb{R}^n$, a linear transformation $f'(x_0): \mathbb{R}^n \to \mathbb{G}^n$, and a map $g: V \to \mathbb{G}^n$ such that

$$f(x_0 + v) - f(x_0) = f'(x_0)(v) + g(v) \quad \text{for } v \in \mathbb{R}^n \tag{6.1}$$

and

$$\lim_{v \to 0} \frac{g(v)}{|v|} = 0.$$

It is understood here that $f'(x_0)(v)$ indicates the action of the linear transformation $f'(x_0)$ on the vector v. (Definition 2.6.)

3. If $f: U \to \mathbb{G}^n$ is $\mathcal{C}^1$ on the open set U, then f is differentiable on U. (Proposition 2.2.)

4. If v is a vector in $\mathbb{R}^n$ and f is at least $\mathcal{C}^1$ at x_0, then the directional derivative of f at x_0 in the direction v is

$$\partial_v f(x_0) = \lim_{\lambda \to 0} \frac{1}{\lambda} \left(f(x_0 + \lambda v) - f(x_0) \right) = f'(x_0) v. \tag{6.2}$$

(Equation (2.3) and Proposition 2.3.)

There is a problem with using (6.2). Suppose f is defined on a manifold, $f: \mathcal{M} \to \mathbb{G}^n$. Notice that $f(x_0 + \lambda v) - f(x_0)$ may be evaluated at points not on $\mathcal{M}$. This expression will still make sense since, if f is at least $\mathcal{C}^1$, then f will have a $\mathcal{C}^1$ extension to an open neighborhood of x_0 in $\mathbb{R}^n$. However the extension is not necessarily unique, so how can we be sure $\partial_v f(x_0)$ is well-defined?

However if v is a tangent vector to $\mathcal{M}$ at x_0 and c is a $\mathcal{C}^1$ curve in $\mathcal{M}$ having the property that $c(\tau_0) = x_0$ and $c'(\tau_0) = v$, then by Proposition 2.7, we have

$$\partial_v f(x_0) = \lim_{\lambda \to 0} \frac{1}{\lambda} \left(f(x_0 + \lambda v) - f(x_0) \right) = \lim_{\lambda \to 0} \frac{(f \circ c)(\tau) - (f \circ c)(\tau_0)}{\tau - \tau_0}.$$

This last equation shows that so long as v is a tangent vector to $\mathcal{M}$, our defining equation for $\partial_v f(x_0)$ depends only on values of f on $\mathcal{M}$, not on which extension of f to $\mathbb{R}^n$ we may use.

Here is a simple corollary to this discussion:

Proposition 6.1. *If v is a tangent vector to the $\mathcal{C}^1$ manifold $\mathcal{M}$ at $x_0 \in \mathcal{M}$ and if $I_{\mathcal{M}}$ is a $\mathcal{C}^1$ function which is the identity on $\mathcal{M}$, then*

$$\partial_v I_{\mathcal{M}}(x_0) = v.$$

Proof. Let c be a $\mathcal{C}^1$ curve in $\mathcal{M}$ that satisfies $c(0) = x_0$ and $c'(0) = v$. Then

$$\partial_v I_\mathcal{M}(x_0) = \lim_{\lambda \to 0} \frac{1}{\lambda} \Big(I_\mathcal{M}(c(\lambda)) - I_\mathcal{M}(c(0)) \Big)$$

$$= \lim_{\lambda \to 0} \frac{1}{\lambda} (c(\lambda)) - c(0)) = v. \qquad \square$$

When we get to the geometric derivative and the Fundamental Theorem of geometric calculus, we shall only need to compute $\partial_v f$ where v is a tangent vector to the manifold, so well-definedness is not necessarily a serious restriction. There is, however, another way to deal with the well-definedness problem, and that is to always construct the extension of f from $\mathcal{M}$ in a well-defined way. We will see how that can be done when we talk later about f being *normally extended*.

Next is a result that is very straightforward, and aside from some minor remarks, we leave its proof to the reader:

Proposition 6.2. *Suppose f and g are $\mathcal{C}^1$ multivector fields and λ is a scalar. If u is a vector, then the following hold:*

1. $\partial_u \langle f \rangle_k = \langle \partial_u f \rangle_k$.

2. $\partial_u(f^\dagger) = (\partial_u f)^\dagger$.

3. $\partial_u(\lambda f) = \lambda \partial_u f$ and $\partial_u(f+g) = \partial_u f + \partial_u g$.

4. *The Leibniz rule holds for the application of ∂_u to the geometric, wedge, left-hand dot, right-hand dot, two-sided dot, and scalar product of multivector fields.*

Recall that by the Leibniz rule, we mean the product rule. Knowing the product rule for derivatives from calculus and the fact that $\partial_u \phi = u \cdot \text{grad} \phi$ for a real-valued function ϕ, it is easily checked that

$$\partial_u(\phi \psi) = (\partial_u \phi) \psi + \phi (\partial_u \psi)$$

for real-valued functions with u a vector. If we say that ∂_u satisfies the Leibniz rule for the geometric product, then we mean

$$\partial_u(f g) = (\partial_u f) g + f (\partial_u g).$$

Example 6.1. Let us check through the steps of establishing the Leibniz rule for the scalar product of multivector fields:

Recall that the scalar product of f and g is $\langle f g^\dagger \rangle_0$. Then the scalar product of $\partial_u f$ and g must be $\langle (\partial_u f) g^\dagger \rangle_0$ while that of f and $\partial_u g$ is $\langle f (\partial_u g)^\dagger \rangle_0$. So the Leibniz rule for the scalar product (if it holds) has the form

$$\partial_u \Big(\langle f g^\dagger \rangle_0 \Big) = \langle (\partial_u f) g^\dagger \rangle_0 + \langle f (\partial_u g)^\dagger \rangle_0. \qquad (6.3)$$

We know from an earlier portion of Proposition 6.2 that

$$\partial_u\left(\langle fg^\dagger\rangle_0\right) = \langle\partial_u(fg^\dagger)\rangle_0.$$

Assuming we have established the Leibniz rule for the geometric product, this becomes

$$\langle(\partial_u f)g^\dagger + f\left(\partial_u(g^\dagger)\right)\rangle_0.$$

Using the earlier fact that $\partial_u(h^\dagger) = (\partial_u h)^\dagger$ and the linearity of $\langle\rangle_0$, this turns into

$$\langle(\partial_u f)g^\dagger\rangle_0 + \langle f\left(\partial_u g\right)^\dagger\rangle_0,$$

and (6.3) is established.

6.1.2   Directional derivatives and coordinates

We want to make sure we understand the notation for differentiation with respect to coordinates; this is, of course, just a special case of the directional derivative.

If we assume f is a multivector-valued function defined on $\mathbb{R}^n$ and we let $x \in \mathbb{R}^n$ have the coordinates $(\chi_1,\ldots,\chi_n)$—that is, $x = (\chi_1,\ldots,\chi_n)$—then slightly generalizing Equation (2.1) and looking back at Equation (6.2), we see that we should set

$$\frac{\partial f}{\partial \chi_i}(x) = \partial_{e_i}f(x) = \lim_{\lambda\to 0}\frac{1}{\lambda}\left(f(x+\lambda e_i)-f(x)\right) = f'(x)e_i. \qquad (6.4)$$

Next we recall that if we have a chart $x\colon U \to \mathbb{R}^n$ and it induces coordinates $(\tau_1,\ldots,\tau_m)$ on $x(U)$, then at the point $x_0 = x(t_0) \in \mathbb{R}^n$, by Equation (2.6) and the discussion that accompanies it, we have

$$\frac{\partial x}{\partial \tau_i}(x_0) = (\partial_{e_i}x)(t_0) = x'(t_0)e_i. \qquad (6.5)$$

Now we want to talk about differentiation with respect to coordinates on a manifold, something we have yet to define.

Suppose that $\mathcal{M}$ is a $\mathcal{C}^1$ m-manifold in $\mathbb{R}^n$ and that $x\colon U \to \mathbb{R}^n$ is a chart (at least $\mathcal{C}^1$) on $\mathcal{M}$. We suppose that the coordinates assigned by x are $(\tau_1,\ldots,\tau_m)$.

Definition 6.1. Let $x_0 = x(t_0) \in \mathcal{M}$, where $t_0 = (\tau_{01},\ldots,\tau_{0m}) \in U$, and suppose the map $f\colon \mathcal{M} \to \mathbb{G}^n$ is at least $\mathcal{C}^1$. Then

$$\frac{\partial f}{\partial \tau_i}(x_0) \overset{\text{def.}}{=} \partial_{e_i}\left(f\circ x\right)(t_0). \qquad (6.6)$$

Notice that Definition 6.1 reduces to (6.4) in the case where $\mathcal{M} = \mathbb{R}^n$ if we take x to be the identity map on $\mathbb{R}^n$.

This notation automatically extends to higher-order derivatives: That is, if f in Definition 6.1 is $\mathcal{C}^k$, then we have

$$\frac{\partial^k f}{\partial \tau_{i_1} \cdots \partial \tau_{i_k}}(x_0) = \left(\partial_{e_{i_1}} \cdots \partial_{e_{i_k}}(f \circ x)\right)(t_0).$$

We check this statement in the case of a mixed, second-order partial derivative:

$$
\begin{aligned}
\frac{\partial^2 f}{\partial \tau_i \partial \tau_j}(x_0) &= \partial_{e_i}\left(\frac{\partial f}{\partial \tau_j} \circ x\right)(t_0) \\
&= \lim_{\lambda \to 0} \frac{1}{\lambda}\left[\frac{\partial f}{\partial \tau_j}(x(t_0 + \lambda e_i)) - \frac{\partial f}{\partial \tau_j}(x(t_0))\right] \\
&= \lim_{\lambda \to 0} \frac{1}{\lambda}\left(\partial_{e_j}(f \circ x)(t_0 + \lambda e_i) - \partial_{e_j}(f \circ x)(t_0)\right) \\
&= \partial_{e_i}\partial_{e_j}(f \circ x)(t_0).
\end{aligned}
$$

On the manifold $\mathcal{M}$, tangent vectors of the form $\partial x/\partial \tau_i$ play a role analogous to the basis vectors e_i of $\mathbb{R}^n$. Each $\partial x/\partial \tau_i$ can be envisioned as tangent to the curve $\tau_i \mapsto (\tau_1, \ldots, \tau_i \ldots, \tau_m)$ in $\mathcal{M}$ along which τ_i increases while all other τ_j are held constant.

The way in which $\partial x/\partial \tau_i$ is analogous to e_i extends to the process of taking partial derivatives of functions:

Proposition 6.3. *Suppose that $(\tau_1, \ldots, \tau_m)$ is a local $\mathcal{C}^1$ coordinatization for the manifold $\mathcal{M}$ and $x_0 = x(t_0) = x(\tau_{01}, \ldots, \tau_{0m})$ lies in the coordinate patch. Let*

$$u_i = \frac{\partial x}{\partial \tau_i}(x_0) = x'(t_0)\,e_i,$$

the tangent vector to $\mathcal{M}$ at x_0. If f is a $\mathcal{C}^1$ multivector field on $\mathcal{M}$ and x_0 lies in the domain of f, then

$$\frac{\partial f}{\partial \tau_i}(x_0) = \partial_{u_i} f(x_0).$$

Proof. We call on Equations (6.6) and (6.2):

$$\frac{\partial f}{\partial \tau_i}(x_0) = \partial_{e_i}(f \circ x)(t_0) = \left[(f \circ x)'(t_0)\right]e_i.$$

Appealing to the chain rule, we see that

$$
\begin{aligned}
\left[(f \circ x)'(t_0)\right]e_i &= f'(x_0)\,x'(t_0)\,e_i \\
&= f'(x_0)\,u_i \\
&= \left(\partial_{u_i} f\right)(x_0)
\end{aligned}
$$

which is the desired result. $\qquad\square$

Example 6.2. The alert reader has likely noticed that the notations $\partial f / \partial \tau_i$ and $\partial x / \partial \tau_i$ seem to be used in contradictory ways.

The coordinate patch $x \colon U \to \mathcal{M}$ maps from $\mathbb{R}^m$ to the manifold and induces coordinates $(\tau_1, \ldots, \tau_m)$ while $f \colon \mathcal{M} \to \mathbb{G}^n$ maps from the manifold into a geometric algebra. But there is a way to replace or reinterpret $\partial x / \partial \tau_i$ in such a way that it becomes a special case of the $\partial f / \partial \tau_i$ notation.

Suppose $I_{\mathcal{M}}$ is the identity map on $\mathcal{M}$, that is, $I_{\mathcal{M}}(x_0) = x_0$ for all $x_0 \in \mathcal{M}$. This clearly extends to a $\mathcal{C}^\infty$ map on $\mathbb{R}^n$, namely the identity map on $\mathbb{R}^n$, so $I_{\mathcal{M}}$ is trivially $\mathcal{C}^r$ on $\mathcal{M}$. (In a later chapter, it turns out to be convenient to consider another way of extending the identity $I_{\mathcal{M}}$ from $\mathcal{M}$ to $\mathbb{R}^n$, what we shall call the *normal identity*.) Suppose $x_0 = x(t_0)$. Then we calculate that

$$
\begin{aligned}
\frac{\partial I_{\mathcal{M}}}{\partial \tau_i}(x_0) &= \partial_{e_i}\big(I_{\mathcal{M}} \circ x\big)(t_0) && \text{(by (6.6))} \\[2mm]
&= \lim_{\lambda \to 0} \frac{1}{\lambda}\left[x(t_0 + \lambda e_i) - x(t_0) \right] \\[2mm]
&= \frac{\partial x}{\partial \tau_i}(x_0) && \text{(by (6.5))}.
\end{aligned}
$$

Because of this, we may sometimes think of x as being the identity map on a manifold (something that is done in [22]) and sometimes as a coordinate patch. To make matters worse, later on we may think of x as representing a typical point on the manifold. It should, however, be clear in any particular instance which usage we have in mind; if not, we shall clarify matters.

In any event, we will use $\partial x / \partial \tau_i$ and $\partial I_{\mathcal{M}} / \partial \tau_i$ interchangably and will keep in mind that $\{(\partial x / \partial \tau_i)(x_0)\}_{i=1}^m$ and $\{(\partial I_{\mathcal{M}} / \partial \tau_i)(x_0)\}_{i=1}^m$ are both the same basis for $T_{x_0}\mathcal{M}$.

Remark 6.1. A useful fact to keep in mind is that if f is a $\mathcal{C}^k$ function on a manifold with coordinates $(\tau_1, \ldots, \tau_m)$, then when we write

$$
\frac{\partial^k f}{\partial \tau_{i_1} \cdots \partial \tau_{i_k}},
$$

the order in which the differentiations are performed does not matter; we can switch $\tau_{i_1}, \ldots, \tau_{i_k}$ around any way we like. This is because when working in $\mathbb{R}^m$ we have equality of mixed partials,

$$
\partial_{e_i}\partial_{e_j}g = \partial_{e_j}\partial_{e_i}g,
$$

and we are simply carrying the partial differentiations over from $\mathbb{R}^m$ to the manifold. (However be careful: This does NOT mean that in general we have $\partial_u \partial_v = \partial_v \partial_u$.)

Two particularly useful instances of this occur when we have a chart x and construct expressions of the form $\partial x / \partial \tau_i$. We know from what was said above that we may consider $\partial x / \partial \tau_i$ to actually be a derivative of the identity function on $\mathcal{M}$:

$$
\frac{\partial x}{\partial \tau_i} = \frac{\partial I_{\mathcal{M}}}{\partial \tau_i}.
$$

Clearly, mixed partials of $I_{\mathcal{M}}$ are equal, so we feel free to write down relations such as

$$\frac{\partial^2 x}{\partial \tau_i \partial \tau_j} = \frac{\partial^2 x}{\partial \tau_j \partial \tau_i}. \tag{6.7}$$

Furthermore, if we set $u_i = \partial x / \partial \tau_i$, then by Proposition 6.3, we have

$$\partial_{u_j} u_i = \frac{\partial^2 x}{\partial \tau_j \partial \tau_i}. \tag{6.8}$$

An immediate consequence of this is that if u_i is as defined here, then

$$\partial_{u_j} u_i = \partial_{u_i} u_j. \tag{6.9}$$

We now think about coordinates in a different way.

Suppose we have a $\mathcal{C}^r$ p-manifold $\mathcal{M}$ ($r \geq 1$) in $\mathbb{R}^n$. Assume the coordinates $(\tau_1, \ldots, \tau_p)$ are induced on $\mathcal{M}$ by a $\mathcal{C}^r$ chart $x \colon U \to \mathbb{R}^n$. We know that $(\tau_1, \ldots, \tau_p)$ is a p-tuple of real numbers, the coordinates of a point of $\mathcal{M}$ induced by the chart x.

However we can also consider each τ_i to be a map which assigns to a point of $\mathcal{M}$ the ith component of its induced coordinates. To be more precise, suppose that $x_0 \in \mathcal{M}$, that $x_0 = x(t_0)$, and that $t_0 = (\tau_{01}, \ldots, \tau_{0p})$. Notice that we have a well-defined function

$$x_0 = x(\tau_{01}, \ldots, \tau_{0p}) \mapsto (\tau_{01}, \ldots, \tau_{0p}) \mapsto \tau_{0i}.$$

We denote this map as τ_i and have $\tau_i(x_0) = \tau_{0i}$. Obviously $\tau_i = \pi_i \circ x^{-1}$ where π_i is the projection of an ordered p-tuple to its ith component. Since, by Definition 2.14, x^{-1} is $\mathcal{C}^r$, the same must be true of the map τ_i. We immediately deduce the following:

Proposition 6.4. *Assuming τ_i is $\mathcal{C}^1$,*

$$\frac{\partial \tau_i}{\partial \tau_j} = \begin{cases} 1 & \text{if } i = j, \\ 0 & \text{if } i \neq j. \end{cases}$$

Proof. Suppose that $x_0 = x(t_0)$. By Equation (6.6), we have

$$\frac{\partial \tau_i}{\partial \tau_j}(x_0) = \partial_{e_j}(\tau_i \circ x)(t_0).$$

Since $\tau_i = \pi_i \circ x^{-1}$, we see that $\tau_i \circ x = \pi_i$ and compute that

$$\partial_{e_j}(\tau_i \circ x)(t_0) = (\partial_{e_j} \pi_i)(t_0) = \delta_{ij}$$

where δ_{ij} is Kronecker's delta. $\qquad\square$

Suppose ϕ is a real-valued function and we form the composition $\phi(\tau_1, \ldots, \tau_p)$ where $\tau_1, \ldots, \tau_p$ are coordinate functions. Then computing $\partial(\phi(\tau_1, \ldots, \tau_p))/\partial \tau_i$ reduces to an ordinary computation in calculus.

Suppose, for instance, we want to compute $\partial\big(\phi(\tau_1,\tau_2)\big)/\partial\tau_1$. Choose $x_0 \in \mathcal{M}$ and suppose that $x_0 = x(\tau_{01},\tau_{02})$. Then appealing carefully to our definitions and the fact that $\tau_i = \pi_i \circ x^{-1}$, we have

$$
\begin{aligned}
\frac{\partial\big(\phi(\tau_1,\tau_2)\big)}{\partial\tau_1}(x_0) &= \partial_{e_1}\big(\phi(\tau_1,\tau_2)\circ x\big)(t_0) \\
&= \partial_{e_1}\big(\phi(\pi_1,\pi_2)\big)(t_0) \\
&= \lim_{\lambda\to 0}\frac{1}{\lambda}\big(\phi(\tau_{01}+\lambda,\,\tau_{02}) - \phi(\tau_{01},\tau_{02})\big).
\end{aligned}
$$

This last expression is exactly the one for computing $\partial\phi(\tau_1,\tau_2)/\partial\tau_1$ that is found in introductory calculus.

Example 6.3. Suppose $\phi(\tau_1,\tau_2) = \tau_1^3\sin(4\tau_2)$, where τ_1,τ_2 are the coordinate functions on a manifold. Then

$$
\begin{aligned}
\frac{\partial\big(\phi(\tau_1,\tau_2)\big)}{\partial\tau_1} &= 3\tau_1^2\sin(4\tau_2), \\
\frac{\partial\big(\phi(\tau_1,\tau_2)\big)}{\partial\tau_2} &= 4\tau_1^3\cos(4\tau_2).
\end{aligned}
$$

Here are two formulas of the type that one sees in an introductory calculus class. Recall that $\{d\sigma_i\}_{i=1}^m$ is the reciprocal frame for the tangent vectors $\{\partial x/\partial\sigma_i\}_{i=1}^m$ (Definition 5.7).

Proposition 6.5. *Suppose that $(\sigma_1,\ldots,\sigma_m)$ and $(\tau_1,\ldots,\tau_m)$ are two sets of $\mathcal{C}^1$ local coordinates on a manifold $\mathcal{M}$. If f is a $\mathcal{C}^1$ multivector field on $\mathcal{M}$, then at all points of $\mathcal{M}$ where the functions are defined, we have*

$$
\frac{\partial f}{\partial\sigma_i} = \sum_{j=1}^m \frac{\partial f}{\partial\tau_j}\frac{\partial\tau_j}{\partial\sigma_i} \quad\text{and}\quad d\sigma_i = \sum_{j=1}^m \frac{\partial\sigma_i}{\partial\tau_j}d\tau_j.
$$

We shall not prove this result here, partly because we shall not need it for the Fundamental Theorem of geometric calculus and partly because we shall prove it later (Proposition 7.7) as part of a more general result and in a more general setting.

6.1.3   A reciprocal frames result

Next we need to know something about the relation between different reciprocal bases of a given vector space:

Proposition 6.6. *Let $\{u_1,\ldots,u_p\}$ and $\{v_1,\ldots,v_p\}$ be bases for a p-dimensional subspace W of $\mathbb{R}^n$. Let $\{m_1,\ldots,m_p\}$ and $\{n_1,\ldots,n_p\}$ be the reciprocal bases for $\{u_i\}_{i=1}^p$ and $\{v_i\}_{i=1}^p$ respectively. Then the following hold:*

1. $u_i = \sum_{j=1}^p (u_i\cdot n_j)\,v_j$ and $m_i = \sum_{j=1}^p (m_i\cdot v_j)\,n_j$.

2. $\sum_{j=1}^{p} (v_i \cdot m_j)(u_j \cdot n_k) = \delta_{ik}$, *Kronecker's delta, where*

$$\delta_{ik} = \begin{cases} 1 & \text{if } i = k, \\ 0 & \text{if } i \neq k. \end{cases}$$

This is, again, an easy proposition to establish. One has only, for example, to note that one can expand each u_i in terms of v_j thus, $u_i = \sum_j \alpha_{ij} v_j$, and then find a way to determine α_{ij}. We leave the proof to the reader.

6.1.4 Geometric derivatives; definitions and properties

We now show that certain expressions that are suggestive of gradients are also independent of the choice of basis.

Proposition 6.7. *Let W be a p-dimensional linear subspace of $\mathbb{R}^n$. Suppose that $\{u_1, \ldots, u_p\}$ and $\{v_1, \ldots, v_p\}$ are two bases of W and suppose that $\{m_1, \ldots, m_p\}$ and $\{n_1, \ldots, n_p\}$ are their respective reciprocal bases. If f is a $\mathcal{C}^1$ multivector field on W, then*

$$\sum_{i=1}^{p} (\partial_{u_i} f) m_i = \sum_{i=1}^{p} (\partial_{v_i} f) n_i \quad \text{and} \quad \sum_{i=1}^{p} m_i (\partial_{u_i} f) = \sum_{i=1}^{p} n_i (\partial_{v_i} f).$$

Proof. We prove only the first of the formulas.

It is sufficient to consider the case where $f = \phi w$ where $\phi : W \to \mathbb{R}$ is $\mathcal{C}^1$ and w is a fixed multivector. We appeal to Proposition 6.6 and calculate thus:

$$\sum_{i=1}^{p} \left(\partial_{u_i}(\phi w) \right) m_i = \sum_{i=1}^{p} (\text{grad}(\phi) \cdot u_i) w m_i$$

$$= \sum_{i=1}^{p} \left(\text{grad}(\phi) \cdot \left(\sum_{j=1}^{p} (u_i \cdot n_j) v_j \right) \right) w \left(\sum_{k=1}^{p} (m_i \cdot v_k) n_k \right)$$

$$= \sum_{j,k=1}^{p} \sum_{i=1}^{p} (u_i \cdot n_j)(m_i \cdot v_k) (\text{grad}(\phi) \cdot v_j) w n_k$$

$$= \sum_{j,k=1}^{p} \delta_{jk} (\text{grad}(\phi) \cdot v_j) w n_k$$

$$= \sum_{j=1}^{p} \left(\partial_{v_j}(\phi w) \right) n_j. \qquad \square$$

A $\mathcal{C}^k$ multivector field is a map $f : U \to \mathbb{G}^n$ where U is an open subset of $\mathbb{R}^n$. If f is given, then notice that $\sum_{i=1}^{p} (\partial_{u_i} f) m_i$ in Proposition 6.7 is completely determined by two things: One is the vector subspace W and the other is the point x at which it is evaluated. Let us ignore x and give our attention to the role of W.

We know that the vector subspace W is completely determined by a p-blade w which is parallel to it. So W is also completely determined by λw where λ is any nonzero scalar; in particular, it is determined by $-w$. Thus the interesting expression $\sum_{i=1}^{p} (\partial_{u_i} f) m_i$ is independent of any choice of orientation of W.

There is one mathematical construction to which there is a naturally associated unit p-blade field: a p-manifold which is at least $\mathcal{C}^1$. This always has a choice of two orientations at each point. These considerations lead us to the following definition:

Definition 6.2. Suppose that $\mathcal{M}$ is a $\mathcal{C}^1$ p-manifold and f is a $\mathcal{C}^1$ multivector field whose domain is a subset of $\mathcal{M}$. Given a point x_0 lying in $\mathcal{M}$ and the domain of f, let $\{u_1, \ldots, u_p\}$ be a continuous frame of tangent vectors to $\mathcal{M}$ defined in an open neighborhood of x_0 and $\{m_1, \ldots, m_p\}$ the reciprocal frame. We then set

$$\left(\vec{\nabla}_{\mathcal{M}} f\right)(x_0) \overset{\text{def.}}{=} \sum_{i=1}^{p} \left(\partial_{u_i} f\right)(x_0)\, m_i(x_0),$$

$$\left(\overleftarrow{\nabla}_{\mathcal{M}} f\right)(x_0) \overset{\text{def.}}{=} \sum_{i=1}^{p} m_i(x_0)\, \left(\partial_{u_i} f\right)(x_0).$$

By the product of $\partial_{u_i} f$ and m_i we mean the geometric product, and we call $\vec{\nabla}_{\mathcal{M}} f$ and $\overleftarrow{\nabla}_{\mathcal{M}} f$ respectively the *right and left geometric derivatives of f on $\mathcal{M}$*. If we are given not an arbitrary manifold but a vector subspace W determined by a p-blade w, we shall also feel free to designate the geometric derivatives on W by symbols such as

$$\vec{\nabla}_w f, \quad \vec{\nabla}_W f, \quad \overleftarrow{\nabla}_w f, \quad \text{and} \quad \overleftarrow{\nabla}_W f.$$

In the case where we compute geometric derivatives on $\mathbb{R}^n$, we permit ourselves to write $\vec{\nabla} f$ and $\overleftarrow{\nabla} f$ without any subscript $\mathcal{M}$ or w or W. If $\vec{\nabla} = \overleftarrow{\nabla}$, we shall simply write ∇.

In the literature of geometric calculus (see [22, 32, 42]), the symbol ∂ or ∂_x is often used for the geometric derivative. We have reserved it for directional derivatives as in $\partial_v f$; in [22] this is designated $v \cdot \partial f$. Furthermore, what we have called the geometric derivative tends to be called the *vector derivative* and refers only to $\overleftarrow{\nabla}_{\mathcal{M}} f$, not to $\vec{\nabla}_{\mathcal{M}} f$. It does not really matter whether one uses $\vec{\nabla}$ or $\overleftarrow{\nabla}$; there is, as we shall shortly explain, an easy way to switch between them, and there are times when it is useful to be able to make that switch.

We note also that the term *derivative* as used in introductory calculus is synonomous with the idea of rate of change. However that is not true here. We discuss the connection between the geometric derivative and the rate of change of a multivector field in the next section.

Example 6.4. Suppose that ϕ is a real-valued $\mathcal{C}^1$ function defined on an open subset of $\mathbb{R}^n$. The directional derivatives of ϕ will be scalars so that $\vec{\nabla} \phi = \overleftarrow{\nabla} \phi = \nabla \phi$. Since $\{e_1, \ldots, e_n\}$ is its own reciprocal basis, we calculate

$$\nabla \phi = \sum_{i=1}^{n} \frac{\partial \phi}{\partial \chi_i}\, e_i = \text{grad}(\phi).$$

However suppose we consider the one-dimensional manifold $\mathcal{M}_i$ consisting of $x = (\chi_1, \ldots, \chi_n)$ such that $\chi_j = 0$ for all $j \neq i$. Then

$$\nabla_{\mathcal{M}_i} \phi = \frac{\partial \phi}{\partial \chi_i} e_i.$$

Now let us turn our attention to the two-dimensional manifold $\mathcal{M}$ such that $\chi_j = 0$ for all $j \geq 3$. In that case,

$$\nabla_{\mathcal{M}} \phi = \frac{\partial \phi}{\partial \chi_1} e_1 + \frac{\partial \phi}{\partial \chi_2} e_2.$$

Example 6.5. Let S^1 be the unit circle $\chi_1^2 + \chi_2^2 = 1$ in $\mathbb{R}^2$. Define $g : \mathbb{R} \to S^1$ by

$$x = g(\theta) = \cos(\theta) e_1 + \sin(\theta) e_2.$$

The restriction of g to every interval J of length less than 2π is a $\mathcal{C}^\infty$ chart on S^1. We assign S^1 the orientation

$$u(x) = g'(\theta) 1 = \partial_1 g(\theta) = -\sin(\theta) e_1 + \cos(\theta) e_2$$

where $x = g(\theta)$. Then for $x \in S^1$, we have

$$(\partial_u u)(x) = \frac{d}{d\theta} (u \circ g)(\theta) = -\cos(\theta) e_1 - \sin(\theta) e_2 = -x$$

which is the inward directed unit normal to x on the unit circle. Notice that $\{x, u(x)\}$ is a pair of orthonormal vectors and $x u(x) = e_1 e_2$ and that the reciprocal frame of $\{u\}$ is, trivially, $\{u\}$. It follows that

$$\vec{\nabla}_{S^1} u = (\partial_u u) u = -x u = -e_1 e_2 \quad \text{and}$$
$$\overleftarrow{\nabla}_{S^1} u = u (\partial_u u) = -u x = e_1 e_2.$$

Proposition 6.8. *Let f and g be $\mathcal{C}^1$ multivector fields on a $\mathcal{C}^1$ manifold $\mathcal{M}$ and let λ be a scalar. Then the following hold:*

1. *Geometric derivatives are linear over the multivector fields on which they operate. That is, for the right geometric derivative for example, we have*

$$\vec{\nabla}_{\mathcal{M}}(f + g) = \vec{\nabla}_{\mathcal{M}} f + \vec{\nabla}_{\mathcal{M}} g \quad \text{and} \quad \vec{\nabla}_{\mathcal{M}}(\lambda f) = \lambda \vec{\nabla}_{\mathcal{M}} f.$$

2. $\left(\vec{\nabla}_{\mathcal{M}} f\right)^\dagger = \overleftarrow{\nabla}_{\mathcal{M}}(f^\dagger).$

3. *If c is a constant multivector, then*

$$\vec{\nabla}_{\mathcal{M}}(cf) = c\left(\vec{\nabla}_{\mathcal{M}} f\right) \quad \text{and} \quad \vec{\nabla}_{\mathcal{M}}(fc) = \left(\vec{\nabla}_{\mathcal{M}} f\right)c$$

where we are taking the geometric product of c and the multivector-valued functions.

Proof. We prove only the second part.

Suppose that $\{u_i\}_{i=1}^m$ and $\{m_i\}_{i=1}^m$ are a frame of vectors and the associated reciprocal frame for tangent spaces to $\mathcal{M}$. Then

$$\left(\vec{\nabla}_{\mathcal{M}} f\right)^\dagger = \left(\sum_{i=1}^m \left(\partial_{u_i} f\right) m_i\right)^\dagger$$

$$= \sum_{i=1}^m m_i \left(\partial_{u_i} f\right)^\dagger$$

$$= \sum_{i=1}^m m_i \partial_{u_i} \left(f^\dagger\right)$$

$$= \overleftarrow{\nabla}_{\mathcal{M}} \left(f^\dagger\right). \qquad \square$$

There is, of course, a version of Proposition 6.8 for $\overleftarrow{\nabla}_{\mathcal{M}}$.

However the Leibniz rule (product rule) does not carry over to our various products in geometric algebra, at least not in the obvious form. To be sure, it *almost* carries over.

To see what we mean, we calculate a geometric derivative for the geometric product of multivector fields f and g:

$$\vec{\nabla}_{\mathcal{M}}(fg) = \sum_{i=1}^p \partial_{u_i}(fg)\, m_i$$

$$= \sum_{i=1}^p \left(\partial_{u_i} f\right) g\, m_i + \sum_{i=1}^p f\left(\partial_{u_i} g\right) m_i.$$

That is, we get

$$\vec{\nabla}_{\mathcal{M}}(fg) = \sum_{i=1}^p \left(\partial_{u_i} f\right) g\, m_i + f\left(\vec{\nabla}_{\mathcal{M}} g\right).$$

We do *not* get $\vec{\nabla}_{\mathcal{M}}(fg) = (\vec{\nabla}_{\mathcal{M}} f) g + f(\vec{\nabla}_{\mathcal{M}} g)$.

If we consider $\overleftarrow{\nabla}_{\mathcal{M}}$, we find that

$$\overleftarrow{\nabla}_{\mathcal{M}}(fg) = (\overleftarrow{\nabla}_{\mathcal{M}} f) g + \sum_{i=1}^p m_i f\left(\partial_{u_i} g\right). \qquad (6.10)$$

In [22] and [42], the notation ∂ is used for $\overleftarrow{\nabla}$, and dots are introduced over functions and operators to show just where ∂ operates. In these works, (6.10) would be written in approximately the form

$$\partial(fg) = \overset{\bullet}{\partial}\overset{\bullet}{f} g + \overset{\bullet}{\partial} f \overset{\bullet}{g},$$

which yields something closer to the standard form of the Leibniz rule.

If we consider other products, that is the wedge product or the various forms of the dot product, then we see in all of them that we do not get quite the Leibniz rule but something that looks close to it.

We now combine coordinates and geometric derivatives.

Suppose a $\mathcal{C}^1$ manifold $\mathcal{M}$ has local coordinates $(\tau_1, \ldots, \tau_m)$ that are induced by a chart x. Recall that the associated basis vectors for the tangent spaces are denoted $\partial x / \partial \tau_i$, and the corresponding reciprocal vectors are $d\tau_1, \ldots, d\tau_m$. That is,

$$\frac{\partial x}{\partial \tau_i} \cdot d\tau_j = \delta_{ij}. \tag{6.11}$$

(In a later chapter, we will show that the d in $d\tau_i$ may be regarded as the *exterior derivative*, an important operation in differential form theory.)

We then obtain the following immediate result with the second conclusion following from the first:

Corollary 6.1. *Let $\mathcal{M}$ be a $\mathcal{C}^1$ manifold with local coordinates $(\tau_1, \ldots, \tau_m)$ and with the reciprocal tangent vector frame $\{d\tau_i\}_{i=1}^m$ defined by (6.11). If f is a $\mathcal{C}^1$ multivector field on $\mathcal{M}$, then the following hold:*

1. $\overrightarrow{\nabla}_{\mathcal{M}} f = \sum_{i=1}^m \frac{\partial f}{\partial \tau_i} d\tau_i$ *and* $\overleftarrow{\nabla}_{\mathcal{M}} f = \sum_{i=1}^m d\tau_i \frac{\partial f}{\partial \tau_i}.$

2. $d\tau_i = \overrightarrow{\nabla}_{\mathcal{M}} \tau_i = \overleftarrow{\nabla}_{\mathcal{M}} \tau_i.$

Example 6.6. Let (χ_1, χ_2) be the cartesian coordinates and (ρ, θ) be the polar coordinates of $\mathbb{R}^2$. We know we may regard the cartesian coordinates as functions of the polar coordinates: $\chi_1 = \rho \cos(\theta)$ and $\chi_2 = \rho \sin(\theta)$. Then it makes sense to compute $\partial \chi_i / \partial \rho$ and $\partial \chi_i / \partial \theta$. Appealing to Corollary 6.1, we see that

$$d\chi_1 = \overrightarrow{\nabla}_{\mathbb{R}^2} \chi_1 = \cos(\theta) d\rho - \rho \sin(\theta) d\theta,$$
$$d\chi_2 = \overrightarrow{\nabla}_{\mathbb{R}^2} \chi_2 = \sin(\theta) d\rho + \rho \cos(\theta) d\theta.$$

Example 6.7. Let $x = \sum_{i=1}^n \chi_i e_i$. We may regard this as an arbitrary point in $\mathbb{R}^n$, as a parametrization $(\chi_1, \ldots, \chi_n) \mapsto x$ of $\mathbb{R}^n$, or as a vector-valued function $f(x) = x$ on $\mathbb{R}^n$. We calculate

$$\overrightarrow{\nabla} x = \sum_{j=1}^n \frac{\partial x}{\partial \chi_j} e_j = \sum_{i,j=1}^n \frac{\partial \chi_i}{\partial \chi_j} e_i e_j = n.$$

(Remember that $\overrightarrow{\nabla} = \overrightarrow{\nabla}_{\mathbb{R}^n}$.)

Exercises 6.1.

1. Prove Proposition 6.2.

2. Prove Proposition 6.6.

3. Let W be the subspace $\chi_1 + \chi_2 + \chi_3 = 0$ of $\mathbb{R}^3$. We see that $\{u_1, u_2\}$ is a basis where

$$u_1 = e_1 - e_3 \quad \text{and} \quad u_2 = e_2 - e_3.$$

If ϕ is a real-valued $\mathcal{C}^1$ function on W, then show that

$$\nabla_W \phi = \text{grad}(\phi) - \frac{1}{3} \left(\frac{\partial \phi}{\partial \chi_1} + \frac{\partial \phi}{\partial \chi_2} + \frac{\partial \phi}{\partial \chi_3} \right) (e_1 + e_2 + e_3).$$

4. Given a $\mathcal{C}^1$ vector field $f(x) = \sum_{i=1}^n \phi_i(x) e_i$ on $\mathbb{R}^n$, show that

$$(\vec{\nabla} f)(x) = \operatorname{div}(f)(x) + \sum_{i<j} \left(\frac{\partial \phi_i}{\partial \chi_j}(x) - \frac{\partial \phi_j}{\partial \chi_i}(x) \right) e_i e_j$$

where

$$\operatorname{div}(f)(x) = \sum_{i=1}^n \frac{\partial \phi_i}{\partial \chi_i}(x).$$

5. We use the notation of Example 6.5. Let $f : S^1 \to \mathbb{R}^2$ be a $\mathcal{C}^1$ vector field on S^1. Show that

   (a) $\vec{\nabla}_{S^1} f = (\partial_u f) \cdot u + +[(\partial_u f) \cdot x] e_1 e_2$.

   (b) If we write f in terms of the standard basis, that is, $f = \psi_1 e_1 + \psi_2 e_2$ where ψ_1 and ψ_2 are $\mathcal{C}^1$ real-valued functions, then

   $$\vec{\nabla}_{S^1} f = -\sin(\theta) \, \partial_u \psi_1 + \cos(\theta) \, \partial_u \psi_2$$
   $$+ \left[\cos(\theta) \, \partial_u \psi_1 + \sin(\theta) \, \partial_u \psi_2 \right] e_1 e_2.$$

   (c) If we write f in terms of the basis $\{x, u(x)\}$, that is, $f = \phi_1 u + \phi_2 x$ where ϕ_1 and ϕ_2 are $\mathcal{C}^1$ real-valued functions, then

   $$\vec{\nabla}_{S^1} f = \partial_u \phi_1 - \phi_2 - (\phi_1 + \partial_u \phi_2) e_1 e_2.$$

6. Prove Proposition 6.8. Hint for the third part: The function f can be written in the form $\sum_I \phi_I e_I$ where each ϕ is a real-valued function and c can be written in the form $\sum_J \gamma_J e_J$ where each γ_J is a real constant. Of course I and J are multi-indices.

7. Prove Corollary 6.1.

8. Suppose that $(\sigma_1, \ldots, \sigma_p)$ and $(\tau_1, \ldots, \tau_p)$ are two sets of $\mathcal{C}^1$ local coordinates on a manifold $\mathcal{M}$. Show that

$$d\tau_i = \sum_{j=1}^p \frac{\partial \tau_i}{\partial \sigma_j} d\sigma_j.$$

9. Let f be the multivector field $f(x) = x^k$, $k = 1, 2, \ldots$, on $\mathbb{R}^n$. (By x^k, we understand the geometric product $xx \cdots x$ with k factors.) Compute $\vec{\nabla} f$.

10. If $\phi : \mathbb{R}^n \to \mathbb{R}$ is $\phi(x) = |x|^2$ for $x \neq 0$, compute $\vec{\nabla}\phi$.

11. Let $\mathcal{M}$ be the upper hemisphere of the unit sphere in $\mathbb{R}^3$ with the "north pole" removed. That is, $\mathcal{M}$ is

$$\chi_1^2 + \chi_2^2 + \chi_3^2 = 1 \quad \text{where } 0 < \chi_3 < 1.$$

Here are two different local parametrizations:

$$x = x(\chi_1, \chi_2) = \chi_1 e_1 + \chi_2 e_2 + \chi_3 e_3,$$
$$x = x(\phi, \theta) = \cos(\phi)\cos(\theta) e_1 - \sin(\phi)\cos(\theta) e_2 + \sin(\theta) e_3$$

where for the first parametrization it is understood that $\chi_3 = \sqrt{1 - \chi_1^2 - \chi_2^2}$ and for the second one we have $0 < \theta < \pi/2$ and ϕ restricted to some interval sufficiently small so that the parametrization will be a one-to-one map. In what follows, keep in mind that $\partial x/\partial \chi_1$, $\partial x/\partial \chi_2$, $\partial x/\partial \phi$, and $\partial x/\partial \theta$ are tangent vectors to $\mathcal{M}$ and $d\chi_1, d\chi_2, d\phi, d\theta$ are the associated reciprocal vectors.

(a) Prove that $\chi_1 \vec{\nabla}_{\mathcal{M}}(\chi_1) + \chi_2 \vec{\nabla}_{\mathcal{M}}(\chi_2) + \chi_3 \vec{\nabla}_{\mathcal{M}}(\chi_3) = 0$.

(b) Compute $d\chi_i$ in terms of ϕ and θ for $i = 1, 2$.

(c) If $f : \mathcal{M} \to \mathbb{G}^3$ is the multivector field

$$f(x) = f(\chi_1, \chi_2, \chi_3) = \chi_1 d\chi_1 + \chi_2 d\chi_2,$$

then rewrite $f(x)$ in terms of ϕ and θ.

12. In what follows, we write functions in terms of the coordinates (χ_1, χ_2) of $x \in \mathbb{R}^2$. Remember that this forces

$$\frac{\partial x}{\partial \chi_i} = d\chi_i = e_i.$$

Compute $\vec{\nabla} f$ in terms of χ_i and $d\chi_i$:

(a) $f(\chi_1, \chi_2) = \cos(\chi_2) d\chi_1 - \chi_1 \sin(\chi_2) d\chi_2$.

(b) $f(\chi_1, \chi_2) = \sin(\chi_2) d\chi_1 + \chi_1 \cos(\chi_2) d\chi_2$.

(c) $f(\chi_1, \chi_2) = e^{\chi_2} d\chi_1 + e^{\chi_1} d\chi_2$.

(d) $f(\chi_1, \chi_2) = |x|^r (-\chi_2 d\chi_1 + \chi_1 d\chi_2)$ where r is a fixed real number.

(e) $f(\chi_1, \chi_2) = |x|^r (\chi_1 d\chi_1 + \chi_2 d\chi_2)$ where r is a fixed real number.

13. Suppose that $\mathcal{M}$ is a $\mathcal{C}^1$ manifold in $\mathbb{R}^n$ with coordinates $(\tau_1, \ldots, \tau_p)$ and $f : \mathcal{M} \to \mathbb{G}^n$ is a $\mathcal{C}^1$ multivector field. Show that

$$\frac{\partial}{\partial \tau_i} \langle f \rangle_k = \left\langle \frac{\partial f}{\partial \tau_i} \right\rangle_k.$$

14. Given a product in $\mathbb{G}^n$ and a "differential operator," the Leibniz rule is not always satisfied. Find a specific example in some $\mathbb{R}^n$ such that

$$\vec{\nabla}(fg) \neq (\vec{\nabla}f) g + f(\vec{\nabla}g).$$

15. We know that for the right-hand geometric derivative of a geometric product, instead of the Leibniz rule, we obtain

$$\vec{\nabla}_{\mathcal{M}}(fg) = \sum_{i=1}^{p} (\partial_{u_i} f) g \, m_i + \sum_{i=1}^{p} f(\partial_{u_i} g) \, m_i.$$

Write the corresponding equations for the wedge product, the right-hand, left-hand, and two-sided dot product, and the scalar product of multivector fields.

6.2   Change of a multivector field over a cell

The ideas of change and rate of change are some of the foundation stones of calculus. And the fact that we can interpret certain mathematical expressions in terms of these ideas is often key to applying calculus to real world problems.

However these ideas only make sense in certain carefully delimited settings. We understand what is meant by the change of f over an interval or an arc and, in certain cases, over a 2- or 3-dimensional region. We also understand what is meant by the rate of change of f at a point in a given direction; this is the *directional derivative* of calculus, and it makes sense in $\mathbb{R}^n$ and even more general spaces.

We want to extend these notions: We will try to form a conception of what we mean by *change of f over a cell* and *rate of change of f over a manifold*.

NOTICE: From this point on, we assume all functions are differentiable as many times as required to write down any given expression.

6.2.1   Change; the familiar story

Change over an arc

If $\mathcal{A}$ is an arc in $\mathbb{R}^n$ that runs from P to Q and ϕ is defined on $\mathcal{A}$, then

$$\text{change of } \phi \text{ over } \mathcal{A} \; = \; \phi(Q) - \phi(P).$$

It can also be useful to talk about *mean change*. On the arc $\mathcal{A}$,

$$\text{mean change of } \phi \; = \; \frac{\phi(Q) - \phi(P)}{\text{length of } \mathcal{A}} \; = \; \frac{\text{change of } \phi}{\text{length of } \mathcal{A}}.$$

The rate of change of ϕ is connected to mean change by the fact that if one lets the interval or arc shrink down to a point x_0, then the mean change converges to the rate of change:

$$\lim_{P,Q \to x_0} \frac{\phi(Q) - \phi(P)}{\text{length of } \mathcal{A}} \; = \; \partial_v \phi(x_0) \; = \; \text{grad}\phi(x_0) \cdot v$$

where in this case (Figure 6.1) we take v to be the unit tangent vector to $\mathcal{A}$ in the direction of the orientation of $\mathcal{A}$, and

$$\text{grad}\, \phi(x_0) \; \overset{\text{def.}}{=} \; \sum_{i=1}^{n} \frac{\partial \phi}{\partial \chi_i}(x_0)\, e_i.$$

There are two things we should notice here:

The first is that the change of ϕ over $\mathcal{A}$ is computed by evaluating ϕ only on the boundary of $\mathcal{A}$, $\partial \mathcal{A} = \{P, Q\}$.

The second is that we are making use of an "orientation" of $\partial \mathcal{A}$ and it is related to the orientation of $\mathcal{A}$.

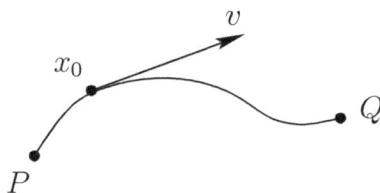

Figure 6.1: Imagine P and Q shrinking toward x_0

What do we mean by an orientation on P and Q? They are only points, not cells, unless we wish to confer an honorary title and call them 0-cells. If we do think of them that way, the obvious thing to do is to assign them the numbers $+1$ and -1 as their orientations. Then

$$\text{change of } \phi \text{ over } \mathcal{A} \ = \ \phi(P)(-1) + \phi(Q)(+1).$$

To see what the relation of the orientation of $\partial \mathcal{A}$ is to that of $\mathcal{A}$, notice that if the orientation of $\mathcal{A}$ is from Q to P, then we want

$$\text{change of } \phi \text{ over } \mathcal{A} \ = \ \phi(P)(+1) + \phi(Q)(-1) \ = \ \phi(P) - \phi(Q).$$

We describe what happens here by saying that the orientation of $\mathcal{A}$ *induces* an orientation of $\partial \mathcal{A}$.

Change of flux over a 2- or 3-dimensional region

Think of a vector field f in 2- or 3-dimensional space and an arc or membrane $\mathcal{M}$ on which the vector field is impinging. Imagine that f represents something like a fluid flow in two dimensions passing through an arc or an electric or magnetic field impinging on a membrane in three dimensions. (See Figure 6.2.)

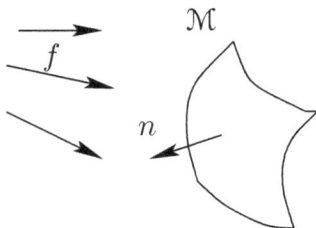

Figure 6.2: Vector field impinging on membrane in $\mathbb{R}^3$

To calculate the associated flux through $\mathcal{M}$ means to find out how much "stuff" passes through it per unit of time. A standard way to do this is to first find a unit normal vector n to $\mathcal{M}$ and then to evaluate the integral $\int_{\mathcal{M}} f \cdot n$.

Suppose that $\mathcal{M}$ is a 2-cell in $\mathbb{R}^2$ or a 3-cell in $\mathbb{R}^3$ and f is a 1-vector field defined over $\mathcal{M}$. We are now concerned to calculate the *change of flux* across the cell $\mathcal{M}$.

Let n be the outward unit normal vector at every point of $\partial\mathcal{M}$. (See Figure 6.3) Notice that if, at a given $x \in \partial\mathcal{M}$ we have $f(x)\cdot n(x) > 0$, then we naturally think of this

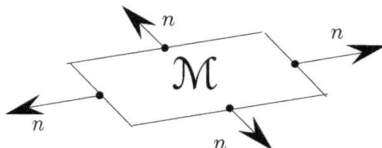

Figure 6.3: Outward normal unit vector n on $\partial\mathcal{M}$

as giving us *flux out* at x. If, on the other hand, we have $f(x)\cdot n(x) < 0$, then we have *flux in* at x. If we ask ourselves what we might mean by the *change of flux across* $\mathcal{M}$, it seems natural to set

$$\text{change in flux across } \mathcal{M} \;=\; \text{flux out} - \text{flux in}$$

(or possibly flux in $-$ flux out depending on what is convenient at any given moment). But this amounts to

$$\text{change in flux across } \mathcal{M} \;=\; \int_{\partial\mathcal{M}} f\cdot n.$$

Notice that just as in the case of the arc, we compute change by evaluating over the boundary of the cell.

We can also replace the outward normal unit vector n by an orientation ∂w of $\partial\mathcal{M}$ if we simultaneously replace f by an appropriate k-vector field. Here is how we do that:

Recall that $e = e_1 e_2$ is the standard orientation of $\mathbb{R}^2$ while $e = e_1 e_2 e_3$ is that of $\mathbb{R}^3$. Let e also be the orientation of our 2- or 3-cell $\mathcal{M}$. We *induce* an orientation at all points of $\partial\mathcal{M}$ by setting $\partial w = ne$. (For $m = 2$, this will be the counterclockwise orientation on the boundary of the 2-cell. For $m = 3$, it will be a tangent 2-vector field on $\partial\mathcal{M}$.) We can write $ne = (-1)^r e^\dagger n$ where r is determined by the value of m. We see that we have

$$f\cdot n \;=\; \langle fn\rangle_0 \;=\; \langle (-1)^r fe(-1)^r e^\dagger n\rangle_0 \;=\; \langle F\,\partial w\rangle_0$$

where $F = (-1)^r fe$. Thus

$$\text{change in flux across } \mathcal{M} \;=\; \int_{\partial\mathcal{M}} \langle F\,\partial w\rangle_0.$$

We do not say anything at this point about rate of change of f. This is because we know from vector analysis that we can get very different looking expressions for whatever ought to correspond to rate of change; the basic ones occur in Green's theorem, Gauss's divergence theorem, and Stokes' theorem. However we do know that we should have

$$\text{change of } f \text{ (or } F\text{) over } \mathcal{M} \;=\; \int_{\mathcal{M}} \text{rate of change of } f \text{ (or } F\text{)}.$$

6.2.2 Change of f over a cell; the general case

We now give a definition which does not have the precision, generality, and finality that we expect of a mathematical definition. It is, nevertheless, a useful step to pinning down the idea of *change of f over a cell*.

Definition 6.3. Suppose that $\mathcal{M}$ is a cell and that we assign to its boundary, $\partial \mathcal{M}$, an orientation, ∂w. Let f be a multivector field defined on $\mathcal{M}$, and let us set

$$\text{change of } f \text{ over } \mathcal{M} \stackrel{\text{def.}}{=} \int_{\partial \mathcal{M}} f(\partial w) \tag{6.12}$$

where $f(\partial w)$ is the geometric product.

So what are the shortcomings of this definition?

For one thing, if our cell is an arc with endpoints P and Q, we know that we want to have $\partial w(P) + \partial w(Q) = 0$; this is an extra condition. However this is a minor blemish since for a cell $\mathcal{M}$ of dimension 2 or greater, we will be able to establish as a theorem the equivalent condition $\int_{\partial \mathcal{M}} \partial w = 0$.

For another, it is unclear how we go about assigning this orientation ∂w to the boundary. In two and three dimensions, it looks as though this ought to have something to do with the outward unit normal vector. However if we have, for example, a 3-cell in $\mathbb{R}^7$, the concept of the outward unit normal vector evaporates and it is not at all clear what is a good way to find ∂w. We shall go into the idea of *induced orientation* carefully and in detail in the next section. We shall see that there are at least two desirable ways to construct an induced orientation, $\overleftarrow{\partial} w$ and $\overrightarrow{\partial} w$.

For a third thing, given the noncommutative nature of the geometric product, there is a very obvious and plausible variant on Definition 6.3, namely

$$\text{change of } f \text{ over } \mathcal{M} = \int_{\partial \mathcal{M}} (\partial w) f. \tag{6.13}$$

How do we choose between these two possibilities? We do not. Let us instead adopt the point of view that *both* (6.12) and (6.13) are reasonable ways to define change over a cell.

Later on, when we consider how ∂w is *induced* by an orientation w on $\mathcal{M}$ and that the process of inducing orientation has a noncommutative aspect, then we may want to expand our list of reasonable definitions of change over a cell, (6.12) and (6.13), to include even more exotic expressions such as

$$\int_{\partial \mathcal{M}} (\overleftarrow{\partial} w) f, \quad \int_{\partial \mathcal{M}} f(\overrightarrow{\partial} (w^{\dagger})), \quad \text{etc.}$$

Of course these will not, in general, all give the same answer, but they will often only differ by a sign, and which of these is the *right* choice will vary depending on which problem in geometry or physics we happen to be dealing with.

6.2.3   Rate of change

Here we find a wonderful use for the concept of the geometric derivative. We look ahead at a particular form of what will turn out to be the Fundamental Theorem of Geometric Calculus:

$$\int_{\partial \mathcal{M}} f(\overrightarrow{\partial} w) = \int_{\mathcal{M}} (\overrightarrow{\nabla}_{\mathcal{M}} f) \, w \qquad (6.14)$$

where it is understood that $\mathcal{M}$ is a cell, that w is a given orientation of $\mathcal{M}$, and that $\overrightarrow{\partial} w$ is an orientation that is "induced" on $\partial \mathcal{M}$ by w. The left-hand side of (6.14) is the change of f over $\mathcal{M}$. We can see from this that what we want to be the rate of change of f is something like this:

Definition 6.4. Let f be a multivector field defined on some open subset of $\mathbb{R}^n$, let x be a point in the domain of f where x lies in the manifold $\mathcal{M}$. Then

$$\text{rate of change of } f \text{ over } \mathcal{M} \text{ at } x \stackrel{\text{def.}}{=} \left(\overrightarrow{\nabla}_{\mathcal{M}} f(x) \right) w$$

where w is an orientation of $\mathcal{M}$.

Of course this is not uniquely defined since there are two choices for the orientation of $\mathcal{M}$, namely $\pm w$. Actually the situation is even more complicated: We have several reasonable candidates for the rate of change of f over $\mathcal{M}$. For example,

$$w \left(\overrightarrow{\nabla}_{\mathcal{M}} f \right), \quad \left(\overleftarrow{\nabla}_{\mathcal{M}} f \right) w, \quad \left(\overrightarrow{\nabla}_{\mathcal{M}} f \right) w^{\dagger}, \quad \text{etc.}$$

Which one is the "right one" depends on which problem one is dealing with.

Notice two things about this approach to the idea of rate of change: First, if we let $\mathcal{M}$ be an interval J in $\mathbb{R}$ and f be a real-valued function ϕ on J, then w must be ± 1 and $\left(\overrightarrow{\nabla}_J \phi \right) w = \pm \phi'$, plus-or-minus the derivative of ϕ. We are used to the idea this is the rate of change of ϕ where the sign tells us which way we are moving when we calculate the rate of change. Second, Equation (6.14) amounts to

$$\text{change of } f \text{ over } \mathcal{M} = \int_{\mathcal{M}} (\text{rate of change of } f),$$

one of the important properties that we insist the idea of change over a cell must satisfy.

Exercises 6.2.

1. If $e = e_1 \cdots e_m$ and n is a vector in $\mathbb{R}^m$, find r such that $ne = (-1)^r e^{\dagger} n$.

2. Check the validity of Equation 6.14 in the case where $\mathcal{M}$ is the unit square $\mathcal{I}^2$ and f is a real-valued $\mathcal{C}^1$ function ϕ. Assume that $w = e_1 e_2$ and that the orientation $\overrightarrow{\partial} w$ of the boundary of the square is given by

$$\overrightarrow{\partial} w = \begin{cases} e_1 & \text{on } 0 \leq \chi_1 \leq 1, \; \chi_2 = 0, \\ e_2 & \text{on } \chi_1 = 1, \; 0 \leq \chi_2 \leq 1, \\ -e_1 & \text{on } 0 \leq \chi_1 \leq 1, \; \chi_2 = 1, \\ -e_2 & \text{on } \chi_1 = 0, \; 0 \leq \chi_2 \leq 1. \end{cases}$$

6.3 Induced orientation

6.3.1 A classical example; the right-hand rule

A particularly clear instance of induced orientation occurs when $\mathcal{M}$ is a 2-cell or membrane in $\mathbb{R}^3$ and one considers the classical Stokes' theorem. (We do not attempt any proof of Stokes' theorem here. We simply summon its statement from the realm of vector analysis.)

For every $x \in \mathcal{M}$, let $n(x)$ be a unit normal vector to $\mathcal{M}$ at the point x; we assume the assignment $x \mapsto n(x)$ is some sort of "nice" continuously differentiable assignment. (Keep in mind here that "normal" means orthogonal or perpendicular to the manifold.) We know from our study of geometric algebra that assigning a unit normal vector field to $\mathcal{M}$ can be thought of as a disguised instance of assigning an orientation w to $\mathcal{M}$. (It will turn out, though it may not be clear right now, that the exact w we want is the one defined by the geometric algebra equation $nw = e_1 e_2 e_3$.)

We may think of either n or w as providing an orientation for $\mathcal{M}$. Associated with this orientation is an *induced orientation* ∂w on $\partial \mathcal{M}$. See Figure 6.4. The induced orientation and n are related by the requirement that a screw pointing in the direction n and rotating in the direction of ∂w must be a *right-hand* screw. This is sometimes referred to as the *right-hand rule*.

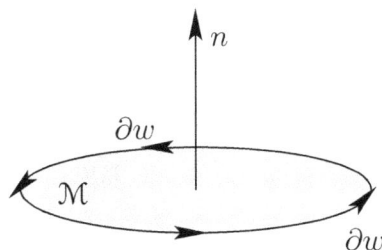

Figure 6.4: Induced orientation on 2-dimensional membrane

Stokes' theorem shows us why this might be a useful idea: To state the theorem, we first suppose that we have a vector field

$$f(x) = \phi_1(x) e_1 + \phi_2(x) e_2 + \phi_3(x) e_3$$

defined on a neighborhood of $\mathcal{M}$. We then set

$$\text{curl}(f) \stackrel{\text{def.}}{=} \left(\frac{\partial \phi_3}{\partial \chi_2} - \frac{\partial \phi_2}{\partial \chi_3} \right) e_1 + \left(\frac{\partial \phi_1}{\partial \chi_3} - \frac{\partial \phi_3}{\partial \chi_1} \right) e_1 + \left(\frac{\partial \phi_2}{\partial \chi_1} - \frac{\partial \phi_1}{\partial \chi_2} \right) e_3.$$

Then Stokes' theorem says

$$\int_{\partial \mathcal{M}} f \cdot \partial w = \int_{\mathcal{M}} \text{curl}(f) \cdot n.$$

It will turn out later that Stokes' theorem is only one case of the Fundamental Theorem of geometric calculus.

6.3.2   Induced orientation on the unit cube

If we have an arbitrary $\mathcal{C}^1$ cell $\mathcal{M}$ with orientation w, what should we mean by the induced orientation ∂w on $\partial \mathcal{M}$? We begin with the straightforward problem of deciding what we mean by induced orientation on the boundary of the unit cube.

Recall that the p-dimensional unit cube is $\mathcal{I}^p = \mathcal{I} \times \cdots \times \mathcal{I}$ p-times where $\mathcal{I} = [0,1]$. Let us take $e = e_1 \cdots e_p$ as the orientation of $\mathcal{I}^p$. By $\partial \mathcal{I}^p$, the boundary of $\mathcal{I}^p$, we mean the set of points $x = (\chi_1, \ldots, \chi_p)$ such that at least one χ_i is 0 or 1. Then what should we mean by ∂e, the induced orientation on $\partial \mathcal{I}^p$?

Assuming $p \geq 1$, by the $(p-1)$-face $\mathcal{I}_{ij}^{p-1}$ of $\mathcal{I}^p$ we mean the set of $x = (\chi_1, \ldots, \chi_p)$ belonging to $\mathcal{I}^p$ such that $\chi_i = j$. We understand here that $i = 1, \ldots, p$ and $j = 0, 1$.

Thus for $\mathcal{I}^2$, by $\mathcal{I}_{20}^1$ we mean the line segment consisting of the points (χ_1, χ_2) such that $0 \leq \chi_1 \leq 1$ and $\chi_2 = 0$. (See Figure 6.5.) For $\mathcal{I}^3$, by $\mathcal{I}_{31}^2$ we mean the square consisting of the points (χ_1, χ_2, χ_3) such that $0 \leq \chi_1, \chi_2 \leq 1$ and $\chi_3 = 1$.

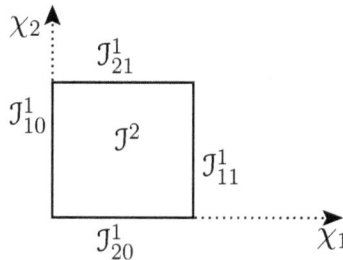

Figure 6.5: The 1-faces of $\mathcal{I}^2$

We will be mostly concerned about $(p-1)$- faces, but more generally, by a *k-dimensional face of $\mathcal{I}^p$* (or *k-face*), we mean a set of points $(\chi_1, \ldots, \chi_p)$ such that $p-k$ of the coordinates have been assigned the values 0 or 1.

For instance, in $\mathcal{I}^2$, both $(0,1)$ and $(1,1)$ are 0-faces. The set of points $(\chi_1, 0)$, where χ_1 is permitted to range through all values between 0 and 1, is a 1-face of $\mathcal{I}^2$.

On the other hand, in $\mathcal{I}^3$, the set of points $(\chi_1, 1, \chi_3)$ is a 2-face.

Notice that 0-faces are just the *vertices* of $\mathcal{I}^p$ while the 1-faces are the *edges*.

If we think of $\mathcal{I}^p$ in the cases $p = 2$ and $p = 3$, then the $(p-1)$-dimensional faces $\mathcal{I}_{ij}^{p-1}$ have easily determined outward unit normal vectors. Considering these cases, it is not hard to see that for general p we ought to define the outward normal unit vector n by

$$n = -e_i \text{ on } \mathcal{I}_{i0}^{p-1} \quad \text{and} \quad n = e_i \text{ on } \mathcal{I}_{i1}^{p-1}.$$

That is, $n = (-1)^{j-1} e_i$ on $\mathcal{I}_{ij}^{p-1}$. See Figure 6.6.

We know that ∂e on $\mathcal{I}_{ij}^{p-1}$ ought to be a unit tangent $(p-1)$-blade to $\mathcal{I}_{ij}^{p-1}$ and n is orthogonal to $\mathcal{I}_{ij}^{p-1}$. We consider several ways we might construct ∂e.

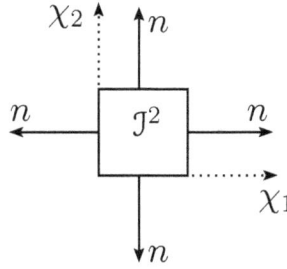

Figure 6.6: Unit normal vectors on $\partial \mathfrak{J}^2$

We might, for example, define $\partial_0 e$ by $\partial_0 e = n e$. We see in this case that on $\mathfrak{J}^{p-1}_{ij}$ we have

$$\partial_0 e \;=\; (-1)^{j-1} e_i e \;=\; (-1)^{i+j} (e_1 \wedge \cdots \wedge \widehat{e_i} \wedge \cdots \wedge e_p).$$

It is easy to think of variations we might play on this game and write a short but not exhaustive, list of some of the ways we might define ∂e:

$$\partial_0 e = n e, \quad \partial_1 e = e n, \quad \partial_2 e = n e^{\dagger}$$
$$\partial_3 e = e^{\dagger} n, \quad \partial_4 e = -n e, \quad \partial_5 e = -n e^{\dagger}.$$

And of course we can also ask how we want to define $\partial(-e)$. We single out two possibilities and take care of e and $-e$ in one fell swoop:

Definition 6.5. Suppose n is the outward unit normal vector to $\partial \mathfrak{J}^p$ and w is an orientation of $\mathfrak{J}^p$. This means that either $w = e_1 \cdots e_p$ or $w = -e_1 \cdots e_p$. Then we set

$$\overleftarrow{\partial} w \stackrel{\text{def.}}{=} w n \quad \text{and} \quad \overrightarrow{\partial} w \stackrel{\text{def.}}{=} n w.$$

Remark 6.2. We have

$$\overrightarrow{\partial} e \;=\; (-1)^{j-1} e_i e \;=\; (-1)^{i+j} (e_1 \wedge \cdots \wedge \widehat{e_i} \wedge \cdots \wedge e_p) \quad \text{on } \mathfrak{J}^{p-1}_{ij}. \tag{6.15}$$

Notice also that by the definition of $\overrightarrow{\partial}$, if $p = 1$, then $\overrightarrow{\partial} e = \pm 1$ on the two endpoints of the interval $\mathfrak{J}$.

Remark 6.3. There is a logical problem with this discussion in that a given point of $\partial \mathfrak{J}^p$ can belong to more than one $(p-1)$-face $\mathfrak{J}^{p-1}_{ij}$ of $\mathfrak{J}^p$. For example, the point $x = (\chi_1, 1, 0)$ of $\mathfrak{J}^3$ belongs to both $\mathfrak{J}^2_{21}$ and $\mathfrak{J}^2_{30}$. Our definition of ∂e then assigns more than one value to such a point, a most unsatisfactory state of affairs.

Yet we propose to ignore this problem. This is not quite cheating as the reader might be inclined to suppose. The reason we can do this is that ultimately we are interested in the boundary of a cell only in so far as we can evaluate an integral over it, and the set of points at which the induced orientation fails to be well-defined is also a set over which the integral is zero.

If, for instance, we are working on $\mathcal{I}^3$, the set of points at which the induced orientation is not well-defined is the edges of the unit cube, a one-dimensional set. If we integrate over $\partial \mathcal{I}^3$, we will use an an area integral, a two-dimensional integral, and a one-dimensional set makes a zero contribution to such an integral.

Example 6.8. Assuming $\mathcal{I}^2$ has the orientation $e = e_1 e_2$, then $\overrightarrow{\partial} e$ has the following values on $\partial \mathcal{I}^2$:

$$
\overrightarrow{\partial} e = \begin{cases}
-e_2 & \text{on } \mathcal{I}_{10}^1 \\
e_1 & \text{on } \mathcal{I}_{20}^1 \\
e_2 & \text{on } \mathcal{I}_{11}^1 \\
-e_1 & \text{on } \mathcal{I}_{21}^1.
\end{cases}
$$

See Figure 6.7.

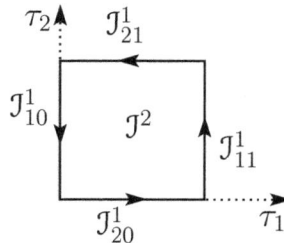

Figure 6.7: Induced orientation on $\mathcal{I}^2$

Here are some simple properties of induced orientation on $\partial \mathcal{I}^p$. Proofs are left to the readers.

Proposition 6.9. *Assume that w is an orientation of $\mathcal{I}^p$, that is, $w = \pm e_1 \cdots e_p$. Then the following hold:*

1. $\overrightarrow{\partial}(-w) = -\overrightarrow{\partial} w$ and $\overleftarrow{\partial}(-w) = -\overleftarrow{\partial} w$.

2. $\overleftarrow{\partial} w = (-1)^{p-1} \overrightarrow{\partial} w$.

3. $\overleftarrow{\partial}(w^\dagger) = (\overrightarrow{\partial} w)^\dagger$

6.3.3   Induced orientation on the boundary of a cell

We are now ready to extend the concept of induced orientation to the boundary of a $\mathcal{C}^1$ p-cell $\mathcal{M}$ with a given orientation w.

$\mathcal{M}$ has a $\mathcal{C}^1$ parametrization $x : \mathcal{I}^p \to \mathcal{M}$. We define the $(p-1)$-*dimensional faces* $\mathcal{M}_{ij}^{p-1}$ of $\mathcal{M}$ to be the images under x of the $(p-1)$-faces of $\mathcal{I}^p$:

$$\mathcal{M}_{ij}^{p-1} \overset{\text{def.}}{=} x\left(\mathcal{I}_{ij}^{p-1}\right) \quad \text{where } i = 1, \dots, p \text{ and } j = 0, 1.$$

See Figure 6.8.

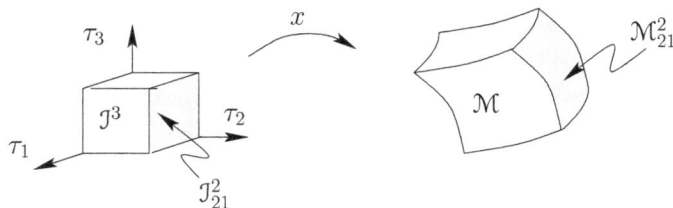

Figure 6.8: Defining $(p-1)$-faces of a p-cell

Though we are really mostly concerned with $(p-1)$-faces of p-cells, more generally, we say that A, a subset of $\mathcal{M}$, is a k-face of $\mathcal{M}$ provided there is a k-face B of $\mathcal{I}^p$ such that $A = x(B)$.

This makes it look as though being a k-face is a property of the parametrization x, but this is wrong. Though we do not prove it (this would take us too far afield), the following *cell face assumption* can be justified by topological and analytical considerations:
Cell face assumption. If $\mathcal{M}$ is a $\mathcal{C}^1$ p-cell and $x_1, x_2 : \mathcal{I}^p \to \mathcal{M}$ are $\mathcal{C}^1$ parametrizations of $\mathcal{M}$, then $x_2^{-1} \circ x_1$ takes k-faces of $\mathcal{I}^p$ onto k-faces of $\mathcal{I}^p$ for $k = 0, 1, \dots, p$.

This shows that being a k-face of a cell is independent of choice of parametrization.

Choose a point x in the face $\mathcal{M}_{ij}^{p-1}$. It is the image under the map x of some $t \in \mathcal{I}_{ij}^{p-1}$; that is, $x = x(t)$. Now if we apply the linear transformation $x'(t)$ to a vector u that is tangent to $\mathcal{I}_{ij}^{p-1}$ at the the point t, then $x'(t)u$ will be a vector that is tangent to $\mathcal{M}_{ij}^{p-1}$ at the point x. (You may want to consult Exercise 5 of this section in connection with this statement.)

This is true in particular for the case $u = e_k$ where $k \neq i$, and we know that

$$e_k \xrightarrow{x'(t)} \frac{\partial x}{\partial \tau_k}(t).$$

If we consult Equation (6.15), this means that

$$\frac{\partial x}{\partial \tau_1}(t) \wedge \dots \wedge \widehat{\frac{\partial x}{\partial \tau_i}(t)} \wedge \dots \wedge \frac{\partial x}{\partial \tau_p}(t)$$

is a tangent blade to $\mathcal{M}_{ij}^{p-1}$ at x. (Recall that a hat over a vector means that it is omitted.) This suggests the following definition:

Definition 6.6. Suppose $\mathcal{M}$ is a $\mathcal{C}^1$ p-cell with orientation w. If $x = x(\tau_1, \ldots, x_p)$ is a $\mathcal{C}^1$ parametrization $x : \mathcal{I}^p \to \mathcal{M}$ of the cell that agrees with the orientation w, then we define the *(right-hand) induced orientation* $\overrightarrow{\partial} w$ on $\partial \mathcal{M}$ by

$$\overrightarrow{\partial} w(x) \stackrel{\text{def.}}{=} \lambda\,(-1)^{i+j} \frac{\partial x}{\partial \tau_1}(t) \wedge \cdots \wedge \widehat{\frac{\partial x}{\partial \tau_i}(t)} \wedge \cdots \wedge \frac{\partial x}{\partial \tau_p}(t)$$

whenever $x = x(t) \in \mathcal{M}_{ij}^{p-1}$ where λ is the unique positive constant that insures $|\overrightarrow{\partial} w| = 1$. This amounts to

$$\overrightarrow{\partial} w(x) = \frac{x'(t)\,\overrightarrow{\partial} e}{|x'(t)\,\overrightarrow{\partial} e|}$$

where $x'(t)$ is the outermorphism induced by the linear transformation $x'(t)$, and $x'(t)$ is operating on the $(p-1)$-blade $\overrightarrow{\partial} e$. We define the *(left-hand) induced orientation* on $\partial \mathcal{M}$ by

$$\overleftarrow{\partial} w(x) \stackrel{\text{def.}}{=} \frac{x'(t)\,\overleftarrow{\partial} e}{|x'(t)\,\overleftarrow{\partial} e|}.$$

This is equivalent to

$$\overleftarrow{\partial} w = (-1)^{p-1} \overrightarrow{\partial} w. \tag{6.16}$$

Example 6.9. We endow $\mathcal{I}^2$ with the orientation $e = e_1 e_2$ and construct a 2-cell $\mathcal{M}$ with the parametrization $x : \mathcal{I}^2 \to \mathcal{M}$ where

$$x(\tau_1, \tau_2) = (\tau_1 + \tau_2^2 + 2)\,e_1 + \tau_2\,e_2.$$

See Figure 6.9. We then calculate that

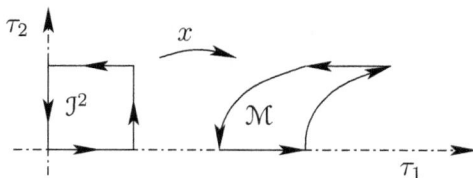

Figure 6.9: Induced orientation on a 2-cell

$$x'(t)e_1 = \frac{\partial x}{\partial \tau_1} = e_1,$$

$$x'(t)e_2 = \frac{\partial x}{\partial \tau_2} = 2\tau_2 e_1 + e_2.$$

From this we finally compute that at $x = (\chi_1, \chi_2) = x(\tau_1, \tau_2)$ on $\partial \mathcal{M}$, we have

$$
\overrightarrow{\partial} w(x) = \begin{cases}
\frac{2\chi_2 e_1 + e_2}{\sqrt{4\chi_2^2 + 1}} & \text{on } \mathcal{M}_{10}^1 \\[2mm]
-\frac{2\chi_2 e_1 + e_2}{\sqrt{4\chi_2^2 + 1}} & \text{on } \mathcal{M}_{11}^1 \\[2mm]
e_1 & \text{on } \mathcal{M}_{20}^1 \\[2mm]
-e_1 & \text{on } \mathcal{M}_{21}^1.
\end{cases}
$$

Remark 6.4. When defining induced orientation ($\overrightarrow{\partial} w$ or $\overleftarrow{\partial} w$) on $\partial \mathcal{M}$, we have the same problem as was discussed in Remark 6.3, namely that the induced orientation may not be well-defined at certain points. We treat this problem the same way as before: If $\mathcal{M}$ is a p-cell, the set of problem points will be a $(p-2)$-dimensional set, and if we are using a standard theory of integration—say, Riemann or Lebesgue integration—then an integral over $\partial \mathcal{M}$ will vanish on the troublesome set.

Though Definition 6.6 makes it appear that induced orientation depends on the parametrization of the cell, this is not true.

Proposition 6.10. *If we are given the orientation w of a cell, then the induced orientations $\overrightarrow{\partial} w$ and $\overleftarrow{\partial} w$ are independent of the choice of parametrization of the cell.*

A proof of this is given in Appendix A.6.

Induced orientation on the boundary of a $\mathcal{C}^1$ cell satisfies the same properties as in the case of $\mathcal{I}^p$ in Proposition 6.9:

Proposition 6.11. *If $\mathcal{M}$ is a $\mathcal{C}^1$ cell with orientation w, then $\overrightarrow{\partial}(-w) = -\overrightarrow{\partial} w$ and $\overleftarrow{\partial}(-w) = -\overleftarrow{\partial} w$.*

A partial proof. We consider only the simple case where $\mathcal{M} = \mathcal{I}^2$ and only $\overrightarrow{\partial}$, but this exhibits all the important features.

In Definition 6.6 of induced orientation, we need to know which face $\mathcal{M}_{ij}^{p-1}$ of a cell a point x on the boundary belongs to in order to assign the correct value of $\overrightarrow{\partial} w$ to x. Although $\overrightarrow{\partial} w(x)$ is independent of the parametrization of the cell, it turns out that the notation $\mathcal{M}_{ij}^{p-1}$ *does* depend on the parametrization, and this can turn out to be important.

Our orientations of $\mathcal{I}^2$ are $w = e_1 e_2$ and $-w = -e_1 e_2$. If we want to apply Definition 6.6, we need parametrizations which give these orientations. For w and $-w$ respectively, let us adopt the parametrizations

$$
x(\tau_1, \tau_2) = \tau_1 e_1 + \tau_2 e_2 \quad \text{and} \quad y(\tau_1, \tau_2) = (1 - \tau_1) e_1 + \tau_2 e_2 \tag{6.17}
$$

where $0 \le \tau_1, \tau_2 \le 1$. We then have

$$
w = \frac{\partial x}{\partial \tau_1} \wedge \frac{\partial x}{\partial \tau_2} = e_1 \wedge e_2 \quad \text{and} \quad -w = \frac{\partial y}{\partial \tau_1} \wedge \frac{\partial y}{\partial \tau_2} = (-e_1) \wedge e_2.
$$

Recall that in Definition 6.6 when we used the notation $\mathcal{M}_{ij}^{p-1}$ for a $(p-1)$-face of the p-cell $\mathcal{M}$, then j took on the values 0 and 1 while i ranged over $1, \ldots, p$ and told us that we were considering the face on which τ_i took on the value j.

Let us, in a similar fashion, use the notations $\mathcal{A}_{ij}$ and $\mathcal{B}_{ij}$ for the 1-faces of $\mathcal{J}^2$ where these notations correspond to the parametrizations x and y respectively. As before, $j = 0, 1$ while $i = 1, 2$. If we think of x and y as mapping from (τ_1, τ_2)-space to (χ_1, χ_2)-space, we see that, for example, when using the y parametrization, $\mathcal{B}_{10}$ is the face corresponding to $\tau_1 = 0$ and lies in the line $\chi_1 = 1$. In Figure 6.10 we see that

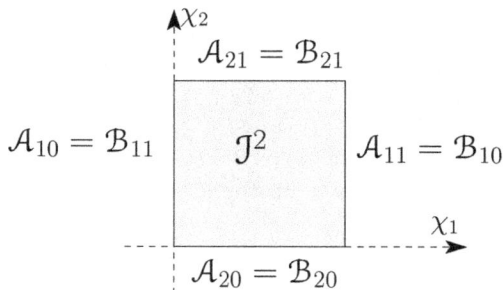

Figure 6.10: Labels on $\partial \mathcal{J}^2$

$$\mathcal{A}_{10} = \mathcal{B}_{11}, \quad \mathcal{A}_{11} = \mathcal{B}_{10}, \quad \mathcal{A}_{20} = \mathcal{B}_{20}, \quad \mathcal{A}_{21} = \mathcal{B}_{21}.$$

We can now justify our desired result by checking the value of $\overrightarrow{\partial} w$ and $\overrightarrow{\partial}(-w)$ on each face of $\mathcal{J}^2$. By Definition 6.6,

$$\overrightarrow{\partial} w = (-1)^{1+1} \widehat{e_1} \wedge e_2 = e_2 \quad \text{on } \mathcal{A}_{11}$$

and

$$\overrightarrow{\partial}(-w) = (-1)^{1+0} \widehat{(-e_1)} \wedge e_2 = -e_2 \quad \text{on } \mathcal{B}_{10}.$$

The other faces yield similar results. □

Proposition 6.12. *If $\mathcal{M}$ is a $\mathcal{C}^1$ p-cell with orientation w, then $\overleftarrow{\partial} w = (-1)^{p-1} \overrightarrow{\partial} w$.*

Proof. Let $x \colon \mathcal{J}^p \to \mathcal{M}$ be a $\mathcal{C}^1$ parametrization that agrees with w. Suppose that $x_0 \in \partial \mathcal{M}$ and $x_0 = x(t_0)$. Then appealing to Definition 6.6 and Proposition 6.9, we have

$$\begin{aligned}
\overleftarrow{\partial} w(x_0) &= \lambda \, x'(t_0) \, \overleftarrow{\partial} e(t_0) \\
&= (-1)^{p-1} \lambda \, x'(t_0) \, \overrightarrow{\partial} e(t_0) \\
&= (-1)^{p-1} \overrightarrow{\partial} w(x_0)
\end{aligned}$$

where λ is a normalizing constant. □

Proposition 6.13. *If $\mathcal{M}$ is a $\mathcal{C}^1$ cell with orientation w, then $\overleftarrow{\partial}(w^\dagger) = \left(\overrightarrow{\partial} w\right)^\dagger$.*

Proof. Let us suppose $\mathcal{M}$ is a p-cell. We make use of Propositions 6.11 and 6.12, and Definition 5.1. Of course w and $w^\dagger$ are grade p while the associated induced orientations are grade $p-1$. Then

$$\overleftarrow{\partial}(w^\dagger) = (-1)^{\frac{p(p-1)}{2}} \overleftarrow{\partial} w,$$

$$\overleftarrow{\partial} w = (-1)^{p-1} \overrightarrow{\partial} w,$$

$$\overrightarrow{\partial} w = (-1)^{\frac{(p-1)(p-2)}{2}} \left(\overrightarrow{\partial} w\right)^\dagger.$$

It follows that $\overleftarrow{\partial}(w^\dagger) = (-1)^r \left(\overrightarrow{\partial} w\right)^\dagger$ where

$$r = \frac{p(p-1)}{2} + p - 1 + \frac{(p-1)(p-2)}{2} = p(p-1).$$

This last number must be even, so $\overleftarrow{\partial}(w^\dagger) = \left(\overrightarrow{\partial} w\right)^\dagger$. $\qquad\square$

We have talked about induced orientation on the boundary of a cell. However it is often easy to extend this idea to other orientable manifolds with boundaries.

Suppose we have, for example, a disk $\mathcal{M}$ which is "nicely" decomposed into cells as in Figure 6.11. That is, $\mathcal{M} = \bigcup_{i=0}^4 \mathcal{M}_i$. (Strictly speaking, $\mathcal{M}$ itself is not a cell

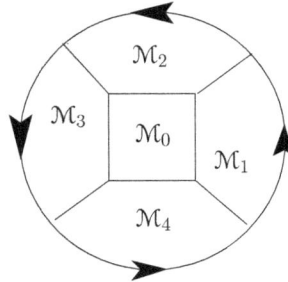

Figure 6.11: A disk decomposed into cells

according to our Definition 2.7; it has no "corners.") Let the orientation of $\mathcal{M}$ and each cell $\mathcal{M}_i$ be $w_{\mathcal{M}} = w_{\mathcal{M}_i} = e_1 e_2$. Now we know how to compute the induced orientation $\overrightarrow{\partial} w_{\mathcal{M}_i}$ on each cell $\mathcal{M}_i$, so we take the induced orientation on $\mathcal{M}$ to be

$$\overrightarrow{\partial} w_{\mathcal{M}} = \sum_{i=0}^4 \overrightarrow{\partial} w_{\mathcal{M}_i}.$$

Now suppose x is an interior point of $\mathcal{M}$ that lies on a 1-dimensional face $\mathcal{A}$ of the smaller cells which is common to $\partial \mathcal{M}_i$ and $\partial \mathcal{M}_j$ where $i \neq j$. Both $\partial \mathcal{M}_i$ and $\partial \mathcal{M}_j$ induce an orientation of $\mathcal{A}$ at the point x, and we see they do it in such a way that $\overrightarrow{\partial} w_{\mathcal{M}_i}(x) + \overrightarrow{\partial} w_{\mathcal{M}_j}(x) = 0.$

Of course, things can be more complicated than having x lie on the boundaries of only two of the small cells. We see in Figure 6.11 four points that lie on the boundaries of three different small cells. If x is such a point, it will lie on three 1-cells $\mathcal{A}_1, \mathcal{A}_2, \mathcal{A}_3$ which are each common to two different small, 2-dimensional cells, and by considering each $\mathcal{A}_k$ separately, we can again show that the sum of the induced orientations at x is zero.

The conclusion here is that the induced orientation shows up only on the boundary of the disk $\mathcal{M}$ in Figure 6.11 and is indicated by the arrows on the boundary.

There is another way of getting at the induced orientation of the boundary of a manifold: We introduce the idea of the *outward normal unit vector n* to the boundary.

If $x \in \partial \mathcal{M}$ and $\mathcal{M}$ has orientation w while $\partial \mathcal{M}$ has orientation $\overrightarrow{\partial} w$, then we define n at x by the equation

$$\overrightarrow{\partial} w(x) \;=\; n(x)\, w(x). \tag{6.18}$$

For x in the boundary of $\mathcal{M}$, we know by the way we have introduced the idea of manifold that $\mathcal{M}$ has a slight extension beyond $\partial \mathcal{M}$; thus both the tangent spaces $T_x \mathcal{M}$ and $T_x \partial \mathcal{M}$ exist, and $T_x \partial \mathcal{M}$ is a subspace of $T_x \mathcal{M}$. Since $n = (\overrightarrow{\partial} w) w^\dagger$, we see that n must be a 1-vector in $T_x \mathcal{M}$. We leave it as an exercise for the reader to show that (6.18) implies n must be orthogonal to $\overrightarrow{\partial} w$ and a unit vector.

Keep in mind that n is normal to (that is, orthogonal to) the *boundary* of $\mathcal{M}$ and do not confuse it with a vector N that is normal to $\mathcal{M}$. We picture these concepts for a flat 2-cell in $\mathbb{R}^3$ in Figure 6.12.

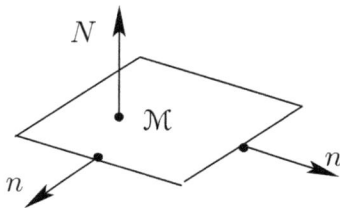

Figure 6.12: The outward normal unit vector n to $\partial \mathcal{M}$

We do not try to justify the idea that n points *out of* $\mathcal{M}$ rather than in to its interior; for one thing, we lack a precise definition of *outward*. However we note that earlier, when we discussed these ideas for the unit p-cube with orientation e, we had the equation $\overrightarrow{\partial} e = n e$ and it was clear that n pointed outward. Beyond this, one can construct simple examples and verify for them that n does indeed point outward.

In the case where one is working with a cell and wishes to use $\overleftarrow{\partial} w$ rather than $\overrightarrow{\partial} w$, we define the outward normal unit vector by

$$\overleftarrow{\partial} w \;=\; w n. \tag{6.19}$$

Example 6.10. Suppose that $\mathcal{B}$ is the unit ball in $\mathbb{R}^3$ with center at the origin, that is, the set of points $x = (\chi_1, \chi_2, \chi_3)$ such that $|x| \leq 1$, and we endow it with orientation $w = e_1 e_2 e_3$. Can we find a formula in terms of the coordinates of x for $\overrightarrow{\partial} w$ on $\partial \mathcal{B}$?

It is intuitively clear that the unit outward normal to $\partial \mathcal{B}$ at x ought to be

$$n(x) = x = \chi_1 e_1 + \chi_2 e_2 + \chi_3 e_3.$$

Then

$$\overrightarrow{\partial} w(x) = nw = \chi_1 (e_2 e_3) - \chi_2 (e_1 e_3) + \chi_3 (e_1 e_2).$$

Exercises 6.3.

1. Make a drawing of the unit cube $\mathcal{I}^3$ superimposed on the χ_1-, χ_2-, and χ_3-axes. In your drawing indicate and label the 2-dimensional faces $\mathcal{I}^2_{30}$ and $\mathcal{I}^2_{31}$.

2. Recall our list of possible definitions of induced orientation on the unit cube $\mathcal{I}^p$ with $e = e_1 \cdots e_p$:

$$\partial_0 e = ne, \quad \partial_1 e = en, \quad \partial_2 e = ne^\dagger$$
$$\partial_3 e = e^\dagger n, \quad \partial_4 e = -ne, \quad \partial_5 e = -ne^\dagger.$$

For each of $i = 1, \ldots, 5$, we have $\partial_i e = (-1)^r \partial_0 e$. Find r for all of these.

3. Show that $\overleftarrow{\partial} e$ and $\overrightarrow{\partial} e$ assign the clockwise and counterclockwise orientation respectively to $\partial \mathcal{I}^2$.

4. Prove Proposition 6.9.

5. Suppose that $\mathcal{M}$ and $\mathcal{N}$ are $\mathcal{C}^1$ p-manifolds lying in $\mathbb{R}^m$ and $\mathbb{R}^n$ respectively and that $f : \mathcal{M} \to \mathcal{N}$ is a one-to-one $\mathcal{C}^1$ map. Let x_0 be a point in $\mathcal{M}$ and $y_0 = f(x_0) \in \mathcal{N}$. Show that if u is a tangent vector to $\mathcal{M}$ at x_0, then $v = f'(x_0) u$ is a tangent vector to $\mathcal{N}$ at y_0. (Hint: Use the characterization of tangent vectors in terms of curves in the manifold that is found in Section 4.5.)

6. If $\mathcal{M}$ is a $\mathcal{C}^1$ cell with orientation w and n is defined by $\overrightarrow{\partial} w = nw$, show that n is a unit vector and is orthogonal to the vector subspace defined by $\overrightarrow{\partial} w$.

7. Show that if we have a cell with orientation w and we define 1-vectors n_0 and n_1 on the boundary of the cell by

$$\overrightarrow{\partial} w = n_0 w \quad \text{and} \quad \overleftarrow{\partial} w = w n_1,$$

then $n_0 = n_1$.

8. Give a general proof of Proposition 6.11 for a $\mathcal{C}^1$ cell.

9. (a) Sometimes a manifold $\mathcal{M}$ in $\mathbb{R}^3$ will be specified by an equation of the form $\phi(x) = \phi(\chi_1, \chi_2, \chi_3) = 0$. (For example, if we take $\phi(x) = \chi_1^2 + \chi_2^2 + \chi_3^2 - 1$, then $\mathcal{M}$ is the unit 2-sphere centered at the origin.) Show that if ϕ is $\mathcal{C}^1$ and x_0 is a point on $\mathcal{M}$, then $\vec{\nabla}\phi(x_0)$ is an orthogonal vector to $\mathcal{M}$. (That is, $\vec{\nabla}\phi(x_0)$ must be orthogonal to the tangent space $T_{x_0}\mathcal{M}$.)

   (b) Let $\mathcal{M}$ be the manifold $\chi_1^2 + \chi_2^2 - \chi_3^2 = 1$ in $\mathbb{R}^3$. Find an orientation w for $\mathcal{M}$.

6.4   Oriented integrals

We know from Section 4.6 how to integrate k-vector fields over a manifold $\mathcal{M}$. If f is a multivector field on $\mathcal{M}$, we set

$$\int_{\mathcal{M}} f \overset{\text{def.}}{=} \sum_k \int_{\mathcal{M}} \langle f \rangle_k.$$

Thus, for example, if $f(x) = \phi_0(x) + \phi_1(x)e_1 + \phi_2(x)e_2 + \phi_3(x)e_1e_2$ is a multivector field on $\mathcal{M}$, we have

$$\int_{\mathcal{M}} f = \int_{\mathcal{M}} \phi_0 + \left(\int_{\mathcal{M}} \phi_1 \right) e_1 + \left(\int_{\mathcal{M}} \phi_2 \right) e_2 + \left(\int_{\mathcal{M}} \phi_3 \right) e_1e_2.$$

Notice our extended definition implies

$$\int_{\mathcal{M}} \langle f \rangle_k = \left\langle \int_{\mathcal{M}} f \right\rangle_k.$$

So now we can integrate multivector fields.

It is clear that this extended definition of the integral of f over a manifold must satisfy linearity properties:

$$\int_{\mathcal{M}} (f + g) = \int_{\mathcal{M}} f + \int_{\mathcal{M}} g \quad \text{and} \quad \int_{\mathcal{M}} \lambda f = \lambda \int_{\mathcal{M}} f$$

where λ is a real number constant. This property of factoring a constant out from under the integral can be extended to cases in which the constant is a multivector and it is "multiplied" by f in different ways:

Proposition 6.14. *Suppose that f is a multivector field on a manifold $\mathcal{M}$ and a is a constant multivector. Assuming that the products are all defined and the functions are integrable, the following hold:*

1. $\int_{\mathcal{M}} af = a \int_{\mathcal{M}} f$ *where the multiplication is the geometric product.*

2. $\int_{\mathcal{M}} \langle af \rangle_0 = \left\langle a \int_{\mathcal{M}} f \right\rangle_0.$

3. $\displaystyle\int_{\mathcal{M}} a \bullet_L f = a \bullet_L \left(\int_{\mathcal{M}} f \right).$

4. $\displaystyle\int_{\mathcal{M}} a \bullet_R f = a \bullet_R \left(\int_{\mathcal{M}} f \right).$

5. $\displaystyle\int_{\mathcal{M}} a \wedge f = a \wedge \left(\int_{\mathcal{M}} f \right).$

Of course each of these formulas has a version with the opposite "handedness" involving fa, $\langle fa \rangle_0$, etc. We leave the proof of the proposition to the exercises.

We remarked in Section 4.6 that integrals of the form $\int_{\mathcal{M}} f$ are *unoriented* integrals. This lack of orientation is indicated by the fact that if we use a parametrization $x : U \to \mathcal{M}$ to evaluate the integral and our coordinates are $\tau_1, \ldots, \tau_m$, then we have absolute values appearing in this formula:

$$\int_{\mathcal{M}} f = \int_U (f \circ x)(t) \left| \frac{\partial x}{\partial \tau_1}(t) \wedge \cdots \wedge \frac{\partial x}{\partial \tau_m}(t) \right| d\tau_1 \cdots d\tau_m. \tag{6.20}$$

Remark 6.5. Notice that each $d\tau_i$ on the right-hand side of the last equation indicates the variable with respect to which one intends to integrate; it should not be confused with the reciprocal vector $d\tau_i$. This use of the notation $d\tau_i$ occurs at a number of places from here on; it should be clear from context whether we mean a reciprocal vector or a tag for integration.

Yet though the integral of (6.20) is unoriented, we know from elementary calculus that integrals of real-valued, single-variable function are oriented, a fact manifested by the formula

$$\int_\alpha^\beta \phi(\tau) d\tau = -\int_\beta^\alpha \phi(\tau) d\tau.$$

If we wanted to think of such an integral as unoriented, we might change the symbolism thus:

$$\int_\alpha^\beta \phi(\tau) d\tau = \int_{[\alpha,\beta]} \phi(\tau) d\tau,$$

assuming $\alpha < \beta$. This indicates that we integrate over the set $[\alpha, \beta]$ but that we do not specify a direction in which to integrate.

How might orientation—in the sense of geometric calculus—enter into this situation? Orientations of $[\alpha, \beta]$ must be unit vectors in $\mathbb{R}$; so the only two possibilities are $+1$ and -1. Then our familiar calculus symbolism and our unoriented integrals can be connected thus:

$$\int_\alpha^\beta \phi(\tau) d\tau = \int_{[\alpha,\beta]} \phi(\tau)(+1) d\tau,$$

$$\int_\beta^\alpha \phi(\tau) d\tau = \int_{[\alpha,\beta]} \phi(\tau)(-1) d\tau.$$

Notice that the pattern we are using is

$$\int_{\text{set}} (\text{function}) \times (\text{orientation}).$$

This leads us to the following idea:

If $\mathcal{M}$ is a manifold with orientation w and f is a multivector field on $\mathcal{M}$, then we shall understand that

$$\text{the integral of } f \text{ over the oriented manifold } \mathcal{M} \; = \; \int_{\mathcal{M}} f\, w.$$

The product $f w$ is the geometric product, so we are forced to admit some possible variations on this concept of *oriented integral*. For example,

$$\int_{\mathcal{M}} w f, \quad \int_{\mathcal{M}} f w^\dagger, \quad \text{etc.}$$

We shall regard them all as oriented integrals over $\mathcal{M}$; the particular one we want will depend on the problem at hand. As we shall show in this section, one can sometimes make a strong argument in favor of the forms $\int_{\mathcal{M}} f w^\dagger$ and $\int_{\mathcal{M}} w^\dagger f$.

However a little thought shows us that we can do even better than the forms indicated above. A far more general oriented integral would involve two multivector fields f and g as in $\int_{\mathcal{M}} f w g$. We consider how to use a coordinatization of the manifold to evaluate such a creature:

Proposition 6.15. *Let $x : U \to \mathcal{M}$ be a $\mathcal{C}^1$ parametrization of the manifold $\mathcal{M}$ with associated coordinates $\tau_1, \ldots, \tau_m$. Assume that x agrees with the orientation w of $\mathcal{M}$ and that f and g are integrable multivector fields on $\mathcal{M}$. Then*

$$\int_{\mathcal{M}} f w g = \int_U \left[(f \circ x)(t) \left(\frac{\partial x}{\partial \tau_1}(t) \wedge \cdots \wedge \frac{\partial x}{\partial \tau_m}(t) \right) (g \circ x)(t) \right] d\tau_1 \cdots d\tau_m$$

(where $d\tau_1, \ldots, d\tau_m$ in the last integral refer to the variables with respect to which we integrate, not to reciprocal vectors).

Proof. Since x agrees with w, we have

$$w = \frac{\frac{\partial x}{\partial \tau_1} \wedge \cdots \wedge \frac{\partial x}{\partial \tau_m}}{\left| \frac{\partial x}{\partial \tau_1} \wedge \cdots \wedge \frac{\partial x}{\partial \tau_m} \right|}.$$

If we replace f by $f w g$ in Equation (6.20), the result is immediate. $\qquad \square$

Example 6.11. Let $\mathcal{A}$ be an arc from P to Q with orientation (unit tangent vector) w. Let us integrate the constant function 1 over the oriented arc. We choose an parametrization $x : \mathcal{I} \to \mathcal{A}$ that agrees with the orientation of $\mathcal{A}$. This means that

$$\frac{dx}{d\tau} = w, \quad x(0) = P, \quad \text{and } x(1) = Q.$$

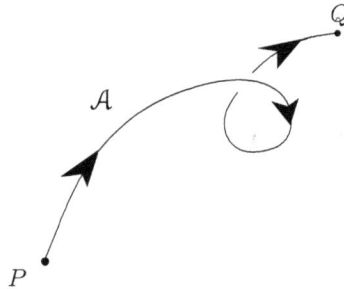

Figure 6.13: Oriented arc from P to Q

Then the desired integral is

$$\int_{\mathcal{A}} 1\,w \;=\; \int_{[0,1]} \frac{dx}{d\tau}\,d\tau$$
$$= x(1) - x(0) \;=\; Q - P.$$

Example 6.12. Let $\mathcal{H}$ be the half-circle $\chi_2^2 + \chi_3^2 = 1$, $\chi_3 \geq 0$, in the $\chi_2\chi_3$-plane and $\mathcal{M}$ be the 2-surface

$$\mathcal{M} \;=\; [0,1] \times \mathcal{H} \subseteq \mathbb{R}^3.$$

See Figure 6.14. Here is a parametrization $x : [0,1] \times [-\frac{\pi}{2}, \frac{\pi}{2}] \to \mathcal{M}$ of $\mathcal{M}$:

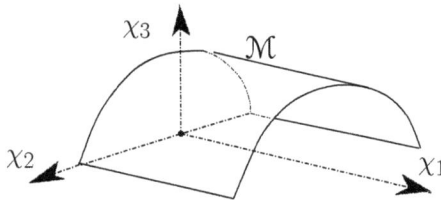

Figure 6.14: Half-cylinder

$$x(\tau_1, \tau_2) \;=\; \tau_1 e_1 + \cos(\tau_2)\,e_2 + \sin(\tau_2)\,e_3.$$

Let us assume the orientation w of $\mathcal{M}$ agrees with this parametrization. How would we evaluate the integral of a real-valued function $\phi(x) = \phi(\chi_1, \chi_2, \chi_3)$ over this oriented half-cylinder?

First step: We calculate

$$\frac{\partial x}{\partial \tau_1} \;=\; e_1,$$
$$\frac{\partial x}{\partial \tau_2} \;=\; -\sin(\tau_2)\,e_2 + \cos(\tau_2)\,e_3.$$

Next, we note that

$$\frac{\partial x}{\partial \tau_1} \wedge \frac{\partial x}{\partial \tau_2} = -\sin(\tau_2)\, e_1 e_2 + \cos(\tau_2)\, e_1 e_3.$$

Finally, we plug this information into the equation of Proposition 6.15 and obtain the desired formula:

$$\int_{\mathcal{M}} \phi\, w =$$

$$= \int_{-\frac{\pi}{2}}^{\frac{\pi}{2}} \int_0^1 \phi\big(\tau_1, \cos(\tau_2), \sin(\tau_2)\big) \big[-\sin(\tau_2)\, e_1 e_2 + \cos(\tau_2)\, e_1 e_3 \big]\, d\tau_1\, d\tau_2.$$

An important instance of an oriented integral is the integral over the boundary of a manifold using the induced orientation.

The basic case is that of a cell. Suppose $\mathcal{M}$ is an m-cell with orientation w and $x : \mathcal{I}^m \to \mathcal{M}$ is a parametrization that agrees with w. The integral of f over the oriented $\partial \mathcal{M}$ is

$$\int_{\partial \mathcal{M}} f\, \vec{\partial} w \overset{\text{def.}}{=} \sum_{i=1}^m \sum_{j=0}^1 \int_{\mathcal{M}_{ij}^{m-1}} f\, \vec{\partial} w$$

where $\mathcal{M}_{ij}^{m-1}$ is an $(m-1)$-face of $\mathcal{M}$. Recall that each $(m-1)$-face of $\mathcal{M}$ is the image of the corresponding $(m-1)$-face of the unit cube $\mathcal{I}^m$: That is $\mathcal{M}_{ij}^{m-1} = x\big(\mathcal{I}_{ij}^{m-1}\big)$. It follows from Proposition 6.15 and Definition 6.6 that the integral of f over a single oriented face of the cell can be evaluated by the formula

$$\int_{\mathcal{M}_{ij}^{m-1}} f\, \vec{\partial} w =$$

$$= (-1)^{i+j} \int_{\mathcal{I}_{ij}^{m-1}} (f \circ x)(t) \left(\frac{\partial x}{\partial \tau_1}(t) \wedge \cdots \wedge \widehat{\frac{\partial x}{\partial \tau_i}}(t) \wedge \cdots \wedge \frac{\partial x}{\partial \tau_m}(t) \right) dt$$

where $t = (\tau_1, \ldots, \tau_m) \in \mathcal{I}^m$ (with τ_i set equal to the constant j which must be either 0 or 1) and $dt = d\tau_1 \cdots \widehat{d\tau_i} \cdots d\tau_m$.

Example 6.13. Let us see what this last formula looks like in the case where $\mathcal{M}$ is a 3-cell and we wish to integrate over the oriented face $\mathcal{M}_{2,1}^2$. Since $(-1)^{2+1} = -1$, we have

$$\int_{\mathcal{M}_{2,1}^2} f\, \vec{\partial} w =$$

$$= -\int_0^1 \int_0^1 (f \circ x)(\tau_1, 1, \tau_3) \left[\frac{\partial x}{\partial \tau_1}(\tau_1, 1, \tau_3) \wedge \frac{\partial x}{\partial \tau_3}(\tau_1, 1, \tau_3) \right] d\tau_1\, d\tau_3.$$

Example 6.14. Suppose we write out the whole integral over $\partial \mathcal{M}$ in the case where $\mathcal{M}$ is a 2-cell:

$$\int_{\partial \mathcal{M}} f\, \vec{\partial} w = \int_{\mathcal{M}_{1,0}^1} + \int_{\mathcal{M}_{1,1}^1} + \int_{\mathcal{M}_{2,0}^1} + \int_{\mathcal{M}_{2,1}^1}$$

$$= -\int_0^1 (f \circ x)(0, \tau_2) \left[\frac{\partial x}{\partial \tau_2}(0, \tau_2) \right] d\tau_2$$

$$+ \int_0^1 (f \circ x)(1, \tau_2) \left[\frac{\partial x}{\partial \tau_2}(1, \tau_2) \right] d\tau_2$$

$$+ \int_0^1 (f \circ x)(\tau_1, 0) \left[\frac{\partial x}{\partial \tau_1}(\tau_1, 0) \right] d\tau_1$$

$$- \int_0^1 (f \circ x)(\tau_1, 1) \left[\frac{\partial x}{\partial \tau_1}(\tau_1, 1) \right] d\tau_1.$$

Here is a result that can be useful whether an integral is oriented or not; it can also be used to change the order and "handedness" of induced orientations.

Proposition 6.16. *If f is integrable over the manifold $\mathcal{M}$, then*

$$\left(\int_{\mathcal{M}} f \right)^{\dagger} = \int_{\mathcal{M}} f^{\dagger}.$$

Next is an important example of an oriented integral that is particularly easy to evaluate. The basic trick is to write f in terms of the manifold coordinates.

Proposition 6.17. *Let $\mathcal{M}$ be a $\mathcal{C}^1$ p-cell with coordinates $(\tau_1, \ldots, \tau_p)$ where $\alpha_{i0} \leq \tau_i \leq \alpha_{i1}$ for each i and the coordinatization agrees with the orientation w of $\mathcal{M}$. Suppose f is a $\mathcal{C}^1$ p-vector field on $\mathcal{M}$ such that f can be written in the form*

$$f(x) = \phi(t) d\tau_1 \wedge \cdots \wedge d\tau_p \tag{6.21}$$

where $x = x(t) = x(\tau_1, \ldots, \tau_p)$. Then

$$\int_{\mathcal{M}} f w^{\dagger} = \int_{\mathcal{M}} w^{\dagger} f = \int_{\alpha_{p0}}^{\alpha_{p1}} \cdots \int_{\alpha_{10}}^{\alpha_{11}} \phi(\tau_1, \ldots, \tau_p) d\tau_1 \ldots d\tau_p. \tag{6.22}$$

Remark 6.6. The reader should understand that in Equation (6.21), $d\tau_1, \ldots, d\tau_p$ are reciprocal vectors while in (6.22), they indicate the variables with respect to which the integrations should be carried out.

Example 6.15. Let $\mathcal{M}$ be the rectangle in $\mathbb{R}^3$ with base the origin and edges $e_1 + e_3$ and e_2. (See Figure 6.15.) It is not hard to check that $(e_1 e_2 - e_2 e_3)/\sqrt{2}$ is a unit blade parallel to $\mathcal{M}$, and we take this to be the orientation w of the rectangle. We wish to integrate $f w^{\dagger}$ over $\mathcal{M}$ using (6.22) where

$$f(x) = f(\chi_1, \chi_2, \chi_3) = 4(\chi_1^2 + \chi_2^2 + \chi_3^2)(e_1 e_2 - e_2 e_3).$$

We introduce the parametrization $\chi_1 = \chi_3 = \tau_1$, $\chi_2 = \tau_2$, where $0 \leq \tau_1, \tau_2 \leq 1$. We next calculate

$$\frac{\partial x}{\partial \tau_1} = e_1 + e_3, \qquad\qquad \frac{\partial x}{\partial \tau_2} = e_2,$$

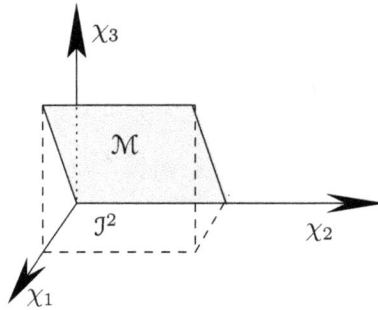

Figure 6.15: A rectangle in $\mathbb{R}^3$

$$d\tau_1 \;=\; \frac{1}{2}(e_1 + e_3), \qquad\qquad d\tau_2 \;=\; e_2.$$

It is straightforward to see that

$$f(x) \;=\; (16\tau_1^2 + 8\tau_2^2)\,(d\tau_1 \wedge d\tau_2).$$

Applying (6.22), we obtain

$$\int_{\mathcal{M}} f\, w^\dagger \;=\; \int_0^1 \int_0^1 (16\tau_1^2 + 8\tau_2^2)\, d\tau_1\, d\tau_2 \;=\; 8.$$

Proof of Proposition 6.17. We consider only $\int_{\mathcal{M}} f\, w^\dagger$. The other case is proved similarly.

Let us set

$$R \;=\; [\alpha_{10}, \alpha_{11}] \times \cdots [\alpha_{p0}, \alpha_{p1}],$$

a rectangle in $\mathbb{R}^p$. Using the fact that our parametrization agrees with the orientation of $\mathcal{M}$ and appealing to Proposition 6.15, we see that

$$\int_{\mathcal{M}} f\, w^\dagger \;=\; \int_R (f \circ x) \left(\frac{\partial x}{\partial \tau_1} \wedge \cdots \wedge \frac{\partial x}{\partial \tau_p} \right)^\dagger$$

$$=\; \int_R \phi\, (d\tau_1 \wedge \cdots \wedge d\tau_p) \left(\frac{\partial x}{\partial \tau_1} \wedge \cdots \wedge \frac{\partial x}{\partial \tau_p} \right)^\dagger.$$

By Proposition 5.16,

$$(d\tau_1 \wedge \cdots \wedge d\tau_p) \left(\frac{\partial x}{\partial \tau_1} \wedge \cdots \wedge \frac{\partial x}{\partial \tau_p} \right)^\dagger \;=\; 1.$$

Therefore the integral of $f\, w^\dagger$ over $\mathcal{M}$ reduces to the integral of ϕ over R, and

$$\int_R \phi \;=\; \int_{\alpha_{p0}}^{\alpha_{p1}} \cdots \int_{\alpha_{10}}^{\alpha_{11}} \phi(\tau_1, \ldots, \tau_p)\, d\tau_1 \ldots d\tau_p. \qquad\qquad \square$$

Exercises 6.4.

1. Prove Proposition 6.16. (Hint: f has an expansion of the form $\sum_I \phi_I e_I$ where the sum is over all ordered multi-indices.)

2. Assuming that $\mathcal{M}$ is a cell with orientation w and f is integrable over $\partial \mathcal{M}$, show that

$$\left(\int_{\partial \mathcal{M}} f \, \overrightarrow{\partial}(w^\dagger) \right)^\dagger = \int_{\partial \mathcal{M}} (\overleftarrow{\partial} w) f^\dagger.$$

3. If $\mathcal{M}$ is the surface of Example 6.12, show that $\int_{\mathcal{M}} w = \pi e_1 e_2$.

4. If $e = e_1 \cdots e_p$, show that

$$\int_{\partial \mathfrak{I}^p} \overleftarrow{\partial} e = \int_{\partial \mathfrak{I}^p} \overrightarrow{\partial} e = 0.$$

5. Recall that by the change of a multivector field f over a cell $\mathcal{M}$ we mean $\int_{\partial \mathcal{M}} f \, \overrightarrow{\partial} w$. Let

$$\mathcal{M} = [\alpha_0, \alpha_1] \times [\beta_0, \beta_1] \times [\gamma_0, \gamma_1],$$

a 3-dimensional rectangle in $\mathbb{R}^3$. We take the orientation of $\mathcal{M}$ to be $w = e_1 e_2 e_3$. Compute the change of $\phi(x) = \phi(\chi_1, \chi_2, \chi_3) = \chi_i$ over $\mathcal{M}$. Now replace ϕ by $f(x) = \chi_i e_i$ and compute the change of f over $\mathcal{M}$.

6. Let $\mathcal{M}$ be the set of points $(\chi_1, \chi_2) \in \mathbb{R}^2$ satisfying $1 \leq \chi_1^2 + \chi_2^2 \leq 4$ and $\chi_2 \geq 0$. Let $\mathcal{M}$ have orientation $w = e_1 e_2$ and suppose f, a multivector field on $\mathcal{M}$, is defined by

$$f(\chi_1, \chi_2) = \left(\sqrt{\chi_1^2 + \chi_2^2} \right) e_1 e_2.$$

Notice that if we use (χ_1, χ_2) as our coordinatization of $\mathcal{M}$, then we have

$$\frac{\partial x}{\partial \chi_1} = d\chi_1 = e_1 \quad \text{and} \quad \frac{\partial x}{\partial \chi_2} = d\chi_2 = e_2.$$

Suppose we have a second coordinatization of $\mathcal{M}$ by (ρ, θ) where

$$\chi_1 = \rho \cos(\theta) \quad \text{and} \quad \chi_2 = \rho \sin(\theta)$$

with the understanding that $1 \leq \rho \leq 2$ and $-\pi/2 \leq \theta \leq \pi/2$.

   (a) Sketch $\mathcal{M}$.

   (b) Compute $d\rho \wedge d\theta$.

   (c) Use Proposition 6.17 to compute $\int_{\mathcal{M}} f w^\dagger$.

7. Prove Proposition 6.14.

6.5   The Fundamental Theorem of geometric calculus

We will tend to refer to the formula

$$\int_{\partial \mathcal{M}} f \, \vec{\partial} w \;=\; \int_{\mathcal{M}} \left(\vec{\nabla}_{\mathcal{M}} f \right) w \tag{6.23}$$

as the *Fundamental Theorem of geometric calculus* where it is understood that $\mathcal{M}$ is a $\mathcal{C}^2$ cell with orientation w and f is a $\mathcal{C}^1$ multivector field on the cell.

This reference to *the* Fundamental Theorem is misleading; there are variations on (6.23) that are just as important, perhaps more so. However (6.23) is closest in appearance to the formulas one encounters in calculus and vector analysis: The integrals in them take the form $\int f(x)\, dx$ with the differential on the right.

In (6.23), one sees that the orientations w and $\vec{\partial} w$ are explicitly separated from the function being integrated (that is f) and the differential (that is, the symbol dx or dt that tells us the variable with respect to which the integration is carried out). Indeed, we make use of the dx symbol only when getting down to the details of a numerical calculation or as reciprocal vectors to tangent vectors associated with a coordinate system. However in works such as [8] and [22], the orientation is incorporated into the differential and is not explicitly displayed.

The reader should also note that our presentation differs from that of others is that we tend to work with (6.23) which can be thought of as the *right-hand version* of the Fundamental Theorem, while the form most often encountered in works on geometric algebra is the *left-hand version*. However, as we exhibit, it is very easy to switch back and forth between right-hand and left-hand versions of integrals and other concepts; it does not greatly matter which version one uses. The form of the Fundamental Theorem that we shall prove will be more general than the right-hand and left-hand versions.

We shall not immediately prove the Fundamental Theorem but shall first show some things we can do with the right-hand version:

6.5.1   Some applications

Example 6.16. Let $\mathcal{M}$ be the interval $[\alpha, \beta]$ in $\mathbb{R}$, and let it have orientation $w = 1$, the number (vector) that "runs" from left to right. The outward unit normal to $\partial \mathcal{M}$ is $n = -1$ at α and $n = 1$ at β. So if we use Equation (6.18), we see that $\vec{\partial} w = -1$ at α and $+1$ at β.

Next suppose we have a $\mathcal{C}^1$ real-valued function ϕ defined over $\mathcal{M} = [\alpha, \beta]$. We see that $\vec{\nabla}_{\mathcal{M}} \phi$ reduces to ϕ', the derivative of elementary calculus. If we integrate $\phi \, \vec{\partial} w$ over the boundary of $\mathcal{M}$, since this is a two-point set, $\partial \mathcal{M} = \{\alpha, \beta\}$, we interpret the integral as

$$\int_{\partial \mathcal{M}} \phi \, \vec{\partial} w \;=\; \phi(\alpha) \, \vec{\partial} w(\alpha) + \phi(\beta) \, \vec{\partial} w(\beta) \;=\; \phi(\beta) - \phi(\alpha).$$

Therefore the equation $\int_{\mathcal{M}} \left(\vec{\nabla}_{\mathcal{M}} \phi \right) w = \int_{\partial \mathcal{M}} \phi \, \vec{\partial} w$ becomes

$$\int_{\alpha}^{\beta} \phi' \;=\; \phi(\beta) - \phi(\alpha).$$

That is, the fundamental theorem of elementary, single variable calculus is tucked inside the Fundamental Theorem of geometric calculus, though this may not be obvious at first glance.

Example 6.17. Suppose $\mathcal{M}$ is a $\mathcal{C}^2$ cell with orientation w. Since 1 is a constant, we have $\vec{\nabla}_{\mathcal{M}} 1 = 0$, and the Fundamental Theorem tells us that

$$\int_{\partial \mathcal{M}} \vec{\partial} w = 0.$$

Example 6.18 (Green's theorem and a little more). Green's theorem in the plane says something like this:

$$\int_{\partial \mathcal{M}} \phi_1 \, d\chi_1 + \phi_2 \, d\chi_2 = \int_{\mathcal{M}} \left(\frac{\partial \phi_2}{\partial \chi_1} - \frac{\partial \phi_1}{\partial \chi_2} \right) d\chi_1 \, d\chi_2 \tag{6.24}$$

where $\mathcal{M}$ is some appropriate "region" in $\mathbb{R}^2$, the boundary of $\mathcal{M}$ has "counterclockwise" orientation, and ϕ_1, ϕ_2 are sufficiently differentiable. Let us see that this result falls out of the Fundamental Theorem.

Let $\mathcal{M}$ be a $\mathcal{C}^2$ 2-cell in $\mathbb{R}^2$ with orietation $e = e_1 e_2$. Let $f = \phi_1 e_1 + \phi_2 e_2$ be a $\mathcal{C}^1$ 1-vector field defined on an open neighborhood of $\mathcal{M}$. The Fundamental Theorem states

$$\int_{\partial \mathcal{M}} f \, \vec{\partial} w = \int_{\mathcal{M}} (\vec{\nabla} f) \, w$$

or, in this case,

$$\int_{\partial \mathcal{M}} f \, \vec{\partial} e = \int_{\mathcal{M}} (\vec{\nabla} f) \, e \tag{6.25}$$

(where, we remind ourselves, $\vec{\nabla} f$ refers to the geometric derivative with respect to the manifold $\mathbb{R}^2$). Now f and $\vec{\partial} e$ are 1-vector fields on $\partial \mathcal{M}$, so we can write

$$\int_{\partial \mathcal{M}} f \, \vec{\partial} e = \int_{\partial \mathcal{M}} (f \cdot \vec{\partial} e + f \wedge \vec{\partial} e).$$

Let us assume a $\mathcal{C}^2$ parametrization $x : \mathcal{I} \to \mathcal{M}_{ij}^1$ of the 1-face $\mathcal{M}_{ij}^1$ of $\mathcal{M}$ and write $x(\tau) = \chi_1(\tau) e_1 + \chi_2(\tau) e_2$. Then

$$\frac{dx}{d\tau}(\tau) = \chi_1'(\tau) e_1 + \chi_2'(\tau) e_2.$$

If we now appeal to Proposition 6.15 and calculate a little bit, we see that

$$\int_{\mathcal{M}_{ij}^1} f \, \vec{\partial} e = \int_{\mathcal{I}} (\phi_1 e_1 + \phi_2 e_2)(\chi_1' e_1 + \chi_2' e_2)$$

$$= \int_0^1 \left[\phi_1(\tau) \chi_1'(\tau) + \phi_2(\tau) \chi_2'(\tau) \right] d\tau$$

$$+ \left(\int_0^1 \left[\phi_1(\tau) \chi_2'(\tau) - \phi_2(\tau) \chi_1'(\tau) \right] d\tau \right) e_1 e_2.$$

Using calculus-style notation, we can write

$$\int_0^1 \left[\phi_1(\tau)\,\chi_1'(\tau) + \phi_2(\tau)\,\chi_2'(\tau) \right] d\tau \;=\; \int_{\mathcal{M}_{ij}^1} \phi_1\,d\chi_1 + \phi_2\,d\chi_2. \qquad (6.26)$$

(Warning: Keep in mind that (6.26) defines a shorthand notation for a scalar-valued line integral; we are not necessarily thinking here of $d\chi_i$ as a reciprocal vector.) Next notice that along $\mathcal{M}_{ij}^1$, the outward unit normal vector to $\partial\mathcal{M}$ is given by

$$n \;=\; \frac{\chi_2'(\tau)\,e_1 - \chi_1'(\tau)\,e_2}{\sqrt{\chi_1'(\tau)^2 + \chi_2'(\tau)^2}}.$$

To see that n is orthogonal to $\mathcal{M}_{ij}^1$, one need only dot it with the tangent vector $dx/d\tau$; it is clear that this is a unit vector; and we leave it to the reader to convince him or herself that n is directed "outward." Calling on Equation (6.20), we find that

$$\int_{\mathcal{M}_{ij}^1} f \cdot n \;=\; \int_0^1 \left[\phi_1(\tau)\,\chi_2'(\tau) - \phi_2(\tau)\,\chi_1'(\tau) \right] d\tau.$$

Since these calculations work on every face of $\partial\mathcal{M}$, we have shown that

$$\int_{\partial\mathcal{M}} f\,\vec{\partial}e \;=\; \int_{\partial\mathcal{M}} \phi_1\,d\chi_1 + \phi_2\,d\chi_2 + \left(\int_{\partial\mathcal{M}} f \cdot n \right) e, \qquad (6.27)$$

a sum of terms of grade 0 and 2.

We now turn to the integral over the 2-cell $\mathcal{M}$. We calculate

$$\vec{\nabla}f \;=\; (\partial_{e_1} f)\,e_1 + (\partial_{e_2} f)\,e_2$$

$$= \left(\frac{\partial\phi_1}{\partial\chi_1}\,e_1 + \frac{\partial\phi_2}{\partial\chi_1}\,e_2 \right) e_1 + \left(\frac{\partial\phi_1}{\partial\chi_2}\,e_1 + \frac{\partial\phi_2}{\partial\chi_2}\,e_2 \right) e_2$$

$$= \frac{\partial\phi_1}{\partial\chi_1} + \frac{\partial\phi_2}{\partial\chi_2} + \left(\frac{\partial\phi_1}{\partial\chi_2} - \frac{\partial\phi_2}{\partial\chi_1} \right) e.$$

Therefore,

$$\int_{\mathcal{M}} (\vec{\nabla}f)\,e \;=\; \int_{\mathcal{M}} \left(\frac{\partial\phi_1}{\partial\chi_1} + \frac{\partial\phi_2}{\partial\chi_2} \right) e + \int_{\mathcal{M}} \left(\frac{\partial\phi_1}{\partial\chi_2} - \frac{\partial\phi_2}{\partial\chi_1} \right) e^2$$

$$= \left(\int_{\mathcal{M}} (\operatorname{div}f) \right) e + \int_{\mathcal{M}} \left(\frac{\partial\phi_2}{\partial\chi_1} - \frac{\partial\phi_1}{\partial\chi_2} \right)$$

where

$$\operatorname{div}f \;=\; \frac{\partial\phi_1}{\partial\chi_1} + \frac{\partial\phi_2}{\partial\chi_2}.$$

If we now look back at Equation (6.27) and equate the grade zero parts on each side, we obtain

$$\int_{\partial\mathcal{M}} \phi_1\,d\chi_1 + \phi_2\,d\chi_2 \;=\; \int_{\mathcal{M}} \left(\frac{\partial\phi_2}{\partial\chi_1} - \frac{\partial\phi_1}{\partial\chi_2} \right)$$

which is Green's theorem. If we equate the grade two parts, we obtain

$$\int_{\partial \mathcal{M}} (f \cdot n) = \int_{\mathcal{M}} (\operatorname{div} f)$$

which is a two-dimensional version of Gauss's divergence theorem.

Example 6.19 (Gauss's divergence theorem in n-dimensions). Let $\mathcal{M}$ be a $\mathcal{C}^2$ n-cell in $\mathbb{R}^n$ with orientation $e = e_1 \cdots e_n$. Let $f : \mathcal{M} \to \mathbb{R}^n$ be a $\mathcal{C}^1$ 1-vector field. If we consider the Fundamental Theorem, parts of the same grade on each side of the equation must be equal. In particular,

$$\int_{\partial \mathcal{M}} \langle f \, \vec{\partial} e \rangle_n = \int_{\mathcal{M}} \langle (\vec{\nabla} f) \, e \rangle_n = \int_{\mathcal{M}} \langle (\vec{\nabla} f) \rangle_0 \, e \tag{6.28}$$

which we claim is the same thing as Gauss's divergence theorem. Let m be the outward unit normal vector to $\partial \mathcal{M}$. We know that $\vec{\partial} e = me$. Appealing to the definition of dot product, we calculate

$$\langle f \, \vec{\partial} e \rangle_n = \langle f m e \rangle_n = \langle f m \rangle_0 \, e = (f \cdot m) \, e.$$

This gives us information about one side of (6.28). Next we write f in the form $f = \sum_{i=1}^{n} \phi_i \, e_i$, and

$$\operatorname{div} f \stackrel{\text{def.}}{=} \sum_{i=1}^{n} \frac{\partial \phi_i}{\partial \chi_i}.$$

Then

$$\vec{\nabla} f = \sum_{i=1}^{n} (\partial_{e_i} f) \, e_i = \sum_{i,j=1}^{n} \frac{\partial \phi_i}{\partial \chi_j} \, e_j e_i.$$

Notice that the grade 0 terms of this expression are those for which $i = j$, so $\langle \vec{\nabla} f \rangle_0 = \operatorname{div} f$. Then

$$\langle \vec{\nabla} f e \rangle_n = (\langle \vec{\nabla} f \rangle_0) \, e = (\operatorname{div} f) \, e.$$

Substituting back into (6.28) and dividing out e, we obtain

$$\int_{\partial \mathcal{M}} (f \cdot m) = \int_{\mathcal{M}} \operatorname{div} f$$

which is Gauss's divergence theorem.

Example 6.20. Recall that we called Equation (6.23) the *right-hand* Fundamental Theorem. It is easy to manufacture from that a *left-hand* version. We remind ourselves that

$$f^{\dagger\dagger} = f,$$
$$\vec{\partial}(w^\dagger) = (\overleftarrow{\partial} w)^\dagger,$$

$$\vec{\nabla}_M(f^\dagger) = (\overleftarrow{\nabla}_M f)^\dagger$$

$$\left(\int g\right)^\dagger = \int g^\dagger.$$

We now replace f by $f^\dagger$ and w by $w^\dagger$ in (6.23) to obtain

$$\int_M \left(\vec{\nabla}_M(f^\dagger)\right) w^\dagger = \int_{\partial M} f^\dagger \, \vec{\partial}(w^\dagger).$$

Next we apply reversion to both sides of the last equation and obtain

$$\int_M w(\overleftarrow{\nabla}_M f) = \int_{\partial M} (\overleftarrow{\partial} w) f. \tag{6.29}$$

This, it turns out, is the version of the Fundamental Theorem which is most customary in geometric calculus. We see that it is quite easy to switch between the right-hand and left-hand versions of the Fundamental Theorem.

6.5.2 Statement and proof of the Fundamental Theorem

Instead of the right-hand or left-hand version of the Fundamental Theorem, we shall prove a *two-sided version*:

Theorem 6.1. *Let $\mathcal{M}$ be a $\mathcal{C}^2$ m-cell in $\mathbb{R}^n$ with orientation w, and let f and g be $\mathcal{C}^1$ multivector fields on $\mathcal{M}$. Then*

$$\int_{\partial \mathcal{M}} f(\vec{\partial} w) g = \int_{\mathcal{M}} (\vec{\nabla}_{\mathcal{M}} f) w g + (-1)^{m-1} \int_{\mathcal{M}} f w(\overleftarrow{\nabla}_{\mathcal{M}} g) \tag{6.30}$$

and

$$\int_{\partial \mathcal{M}} f(\overleftarrow{\partial} w) g = (-1)^{m-1} \int_{\mathcal{M}} (\vec{\nabla}_{\mathcal{M}} f) w g + \int_{\mathcal{M}} f w(\overleftarrow{\nabla}_{\mathcal{M}} g). \tag{6.31}$$

Notice that Equation (6.30) implies (6.31); since $\overleftarrow{\partial} w = (-1)^{m-1} \vec{\partial} w$, all we need to do is multiply by $(-1)^{m-1}$.

If $\mathcal{M}$ is an arc running from P to Q, then we interpret the integral over the boundary by

$$\int_{\partial \mathcal{M}} f(\vec{\partial} w) g \stackrel{\text{def.}}{=} f(Q) g(Q) - f(P) g(P).$$

(Notice also that since $\overleftarrow{\partial} w = (-1)^{m-1} \vec{\partial} w$, when $m = 1$, we have $\overleftarrow{\partial} w = \vec{\partial} w$.) Because of this, it is convenient to give a separate proof in the case $m = 1$.

The one-dimensional case

Suppose that $\mathcal{M}$ is a 1-cell (arc) running from P to Q. Let $x : \mathcal{I} \to \mathcal{M}$ be a $\mathcal{C}^2$ parametrization that agrees with w. Then

$$\frac{dx}{d\tau} = \left|\frac{dx}{d\tau}\right| w \quad \text{and} \quad d\tau = \frac{1}{\left|\frac{dx}{d\tau}\right|} w$$

where $d\tau$ is the reciprocal vector for the tangent vector $dx/d\tau$. Then

$$\vec{\nabla}_{\mathcal{M}} f(x) = \frac{df}{d\tau}(x)\, d\tau = \frac{d(f \circ x)}{d\tau}(t)\, \frac{1}{\left|\frac{dx}{d\tau}\right|}\, w$$

where $x = x(\tau)$. A similar formula holds for $\overleftarrow{\nabla}_{\mathcal{M}} g(x)$ with w now on the left instead of the right. In the calculation that follows, notice that at the second line we switch from an integral over $\mathcal{M}$ to one over $\mathcal{I} = [0, 1]$, which necessitates the introduction of an extra $|dx/d\tau|$ term; and in the second and third lines, $d\tau$ is not a reciprocal vector but merely a signal that we integrate with respect to τ:

$$
\begin{aligned}
\int_{\mathcal{M}} \left(\vec{\nabla}_{\mathcal{M}} f\right) w g + (-1)^{1-1} \int_{\mathcal{M}} f w \left(\overleftarrow{\nabla}_{\mathcal{M}} g\right) &= \int_{\mathcal{M}} \frac{df}{d\tau} \frac{1}{\left|\frac{dx}{d\tau}\right|} g + \int_{\mathcal{M}} f \frac{1}{\left|\frac{dx}{d\tau}\right|} \frac{dg}{d\tau} \\
&= \int_0^1 \frac{d(f \circ x)}{d\tau}(\tau)\, (g \circ x)(\tau)\, d\tau + \int_0^1 (f \circ x)(\tau)\, \frac{d(g \circ x)}{d\tau}(\tau)\, d\tau \\
&= \int_0^1 \frac{d}{d\tau}\big((f \circ x)(g \circ x)\big)(\tau)\, d\tau \\
&= f\big(x(1)\big)\, g\big(x(1)\big) - f\big(x(0)\big)\, g\big(x(0)\big) \\
&= f(Q)\, g(Q) - f(P)\, g(P).
\end{aligned}
$$

This gives us the Fundamental Theorem for $m = 1$. From this point on, we shall assume $m \geq 2$.

Lemmas

If we take the difference of integrals of g on opposite faces of a unit cube, then we can "fill in" the integral on the whole cube by integrating the appropriate partial derivative of g between those opposite faces.

Example 6.21. Consider the 2-faces $\mathcal{I}_{0,3}^2$ and $\mathcal{I}_{1,3}^2$ of $\mathcal{I}^3$. We convert a difference of integrals on those faces to an integral on $\mathcal{I}^3$ thus:

$$
\begin{aligned}
\int_{\mathcal{I}_{1,3}^2} g - \int_{\mathcal{I}_{0,3}^2} g &= \int_0^1 \int_0^1 g(\tau_1, \tau_2, 1)\, d\tau_1\, d\tau_2 - \int_0^1 \int_0^1 g(\tau_1, \tau_2, 0)\, d\tau_1\, d\tau_2 \\
&= \int_0^1 \int_0^1 \int_0^1 \frac{\partial g}{\partial \tau_3}(\tau_1, \tau_2, \tau_3)\, d\tau_1\, d\tau_2\, d\tau_3.
\end{aligned}
$$

An appropriate general statement of this idea is this:

Lemma 6.1. *If g is a $\mathcal{C}^1$ multivector field on $\mathcal{I}^m$, then*

$$\sum_{j=0}^1 (-1)^{j-1} \int_{\mathcal{I}_{ij}^{m-1}} g = \int_{\mathcal{I}^m} \frac{\partial g}{\partial \tau_i}.$$

Continuing with the proof of the Fundamental Theorem, let $x : \mathcal{I}^m \to \mathcal{M}$ be a $\mathcal{C}^2$ parametrization of $\mathcal{M}$ that agrees with w and suppose that $\tau_1, \ldots, \tau_m$ are the coordinates assigned to the cell by x. Let us set

$$n_i = (-1)^{i-1} \frac{\partial x}{\partial \tau_1} \wedge \cdots \wedge \widehat{\frac{\partial x}{\partial \tau_i}} \wedge \cdots \wedge \frac{\partial x}{\partial \tau_m}.$$

Notice that n_i is a $\mathcal{C}^1$ $(m-1)$-vector field that is defined on all of $\mathcal{M}$ and that on $\mathcal{M}_{ij}^{m-1}$ we have $\vec{\partial} w \overset{\text{ray}}{=} (-1)^{j-1} n_i$.

Lemma 6.2. *Everywhere on the cell $\mathcal{M}$ we have*

$$\sum_{i=1}^{m} \frac{\partial n_i}{\partial \tau_i} = 0.$$

Proof. We introduce the notation

$$u_i(x) = \frac{\partial x}{\partial \tau_i}(t) \quad \text{when } x = x(t).$$

Thus $u_i \circ x = \partial x / \partial \tau_i$. It follows from our definition of the partial derivative with respect to τ_k of a function defined on $\mathcal{M}$ that

$$\frac{\partial u_i}{\partial \tau_k}(x) = \frac{\partial (u_i \circ x)}{\tau_k}(t) = \frac{\partial^2 x}{\partial \tau_k \partial \tau_i}(t).$$

Since x is $\mathcal{C}^2$, the order in which one computes a mixed second order partial derivative is irrelevant; or, if one wishes, one can say that $\partial u_i / \partial \tau_k = \partial u_k / \partial \tau_i$. We now establish the validity of the lemma:

$$\sum_{i=1}^{m} \frac{\partial n_i}{\partial \tau_i} = \sum_{i=1}^{m} (-1)^{i-1} \frac{\partial}{\partial \tau_i} \left(u_1 \wedge \cdots \wedge \widehat{u_i} \wedge \cdots \wedge u_m \right)$$

$$= \sum_{i=1}^{m} \sum_{k<i} (-1)^{i+k} \frac{\partial^2 x}{\partial \tau_i \partial \tau_k} \left(u_1 \wedge \cdots \wedge \widehat{u_k} \wedge \cdots \wedge \widehat{u_i} \wedge \cdots \wedge u_m \right)$$

$$+ \sum_{i=1}^{m} \sum_{k>i} (-1)^{i+k-1} \frac{\partial^2 x}{\partial \tau_i \partial \tau_k} \left(u_1 \wedge \cdots \wedge \widehat{u_i} \wedge \cdots \wedge \widehat{u_k} \wedge \cdots \wedge u_m \right)$$

$$= 0.$$

The last step follows from the fact that for any i and k that are distinct, a term will appear twice, once with one sign, once with the oppposite one.   □

The case $m \geq 2$ for the Fundamental Theorem

Remainder of the proof for Theorem 6.1. We sum the oriented integral of f and g over the boundary of $\mathcal{M}$ and then invoke Proposition 6.15 thus:

$$\int_{\partial \mathcal{M}} f(\vec{\partial} w) g = \sum_{i=1}^{m} \sum_{j=0}^{1} \int_{\mathcal{M}_{ij}^{m-1}} f(\vec{\partial} w) g$$

$$= \sum_{i=1}^{m} \sum_{j=0}^{1} (-1)^{j-1} \int_{\mathcal{I}_{ij}^{m-1}} (f \circ x)(n_i \circ x)(g \circ x).$$

By Lemma 6.1 and the product rule, this becomes

$$\sum_{i=1}^{m} \int_{\mathcal{I}^m} \frac{\partial}{\partial \tau_i} \big[(f \circ x)(n_i \circ x)(g \circ x) \big] =$$

$$= A + B + C$$

where

$$A = \int_{\mathcal{I}^m} \sum_{i=1}^{m} \left(\frac{\partial (f \circ x)}{\partial \tau_i} \right) (n_i \circ x)(g \circ x)$$

$$B = \int_{\mathcal{I}^m} (f \circ x) \left(\sum_{i=1}^{m} \frac{\partial (n_i \circ x)}{\partial \tau_i} \right) (g \circ x)$$

$$C = \int_{\mathcal{I}^m} (f \circ x) \sum_{i=1}^{m} (n_i \circ x) \left(\frac{\partial (g \circ x)}{\partial \tau_i} \right).$$

By Lemma 6.2, $B = 0$.

Next recall that $d\tau_i$ is the reciprocal vector for $\partial x / \partial \tau_i$. (There is a subtle point to consider here in that we may treat $d\tau_i$ either as a function of the point $x = x(t)$ in $\mathcal{M}$ or as a function of $t \in \mathcal{I}^m$. In what we are about to do, it does not matter which way we think of $d\tau_i$.) Let $u = u_1 \wedge \cdots \wedge u_m$. By Proposition 5.5,

$$(d\tau_i) u = \sum_{k=1}^{m} (-1)^{k-1} (d\tau_i \cdot u_k)(u_1 \wedge \cdots \wedge \widehat{u_k} \wedge \cdots \wedge u_m) + d\tau_i \wedge u.$$

Since $d\tau_i(x)$ is a tangent vector to $\mathcal{M}$ at x and $T_x^1 \mathcal{M}$ is generated by $\{u_k(x)\}_{k=1}^{m}$, we see that $d\tau_i \wedge u = 0$. We also know that $d\tau_i \cdot u_k = \delta_{ik}$, so

$$(d\tau_i) u = (-1)^{i-1} u_1 \wedge \cdots \wedge \widehat{u_i} \wedge \cdots \wedge u_m = n_i.$$

One can show similarly that

$$u(d\tau_i) = (-1)^{m-1} n_i.$$

We therefore have

$$A = \int_{\mathcal{I}^m} \sum_{i=1}^{m} \frac{\partial (f \circ x)}{\partial \tau_i} d\tau_i (u \circ x)(g \circ x) = \int_{\mathcal{M}} (\overrightarrow{\nabla}_{\mathcal{M}} f) w g.$$

The last step was justified by Proposition 6.15 and the definition of the geometric derivative. Similarly,

$$C = (-1)^{m-1} \int_{\mathcal{I}^m} (f \circ x)(u \circ x) \sum_{i=1}^{m} d\tau_i \frac{\partial (g \circ x)}{\partial \tau_i} = (-1)^{m-1} \int_{\mathcal{M}} f w (\overleftarrow{\nabla}_{\mathcal{M}} g).$$

We have thus established the Fundamental Theorem. $\qquad \square$

Exercises 6.5.

1. Let

$$f(x) = \frac{1}{n} \sum_{i=1}^{n} (-1)^{n-i} \chi_i (e_1 \cdots \hat{e}_i \cdots e_n)$$

on $\mathbb{R}^n$ where $x = (\chi_1, \ldots, \chi_n)$. Show that if $\mathcal{M}$ is a $\mathcal{C}^2$ n-cell in $\mathbb{R}^n$ with orientation $w = e_1 \cdots e_n$, then

$$\int_{\partial \mathcal{M}} f \, \vec{\partial} w^\dagger = \text{vol}(\mathcal{M}).$$

2. Recall that by the change of a multivector field f over a cell $\mathcal{M}$ we mean $\int_{\mathcal{M}} f \, \vec{\partial} w$. Consider the n-dimensional rectangle

$$\mathcal{M} = [\alpha_{10}, \alpha_{11}] \times \cdots \times [\alpha_{n0}, \alpha_{n1}]$$

in $\mathbb{R}^n$ with orientation $w = e_1 \cdots e_n$. Compute the change over $\mathcal{M}$ of the multivector field $f : \mathbb{R}^n \to \mathbb{G}^n$ defined by

$$f(x) = f(\chi_1, \ldots, \chi_n) = \sum_{i=1}^{n} \chi_i v_i$$

where each v_i is a fixed multivector.

3. Suppose that $\rho_0 > 0$ and $0 < \theta_0 < \pi/2$. Let $\mathcal{A}_{\theta_0}$ be the portion of the circle $\chi_1^2 + \chi_2^2 = \rho_0^2$ in the first quadrant of the $\chi_1\chi_2$-plane which has one end at the point $(\rho_0, 0)$ and subtends an angle of θ_0. Suppose that ϕ is a real-valued $\mathcal{C}^\infty$ function defined on $\mathcal{A}_{\theta_0}$ such that

$$\vec{\nabla}\phi(x) = \frac{1}{|x|^2} (-\chi_2 \, d\chi_1 + \chi_1 \, d\chi_2)$$

where $x = (\chi_1, \chi_2)$. (It should also be remembered that if we are using the cartesian coordinates of $\mathbb{R}^2$ for our parametrization, then $d\chi_i = e_i$.)

   (a) Find the change of ϕ over $\mathcal{A}_{\theta_0}$.

   (b) Find a ϕ that satisfies these conditions.

4. Suppose $0 < \rho_0 < \rho_1$ and let $\mathcal{A}$ be the annulus

$$\rho_0^2 \leq \chi_1^2 + \chi_2^2 \leq \rho_1^2$$

in $\mathbb{R}^2$ where (χ_1, χ_2) are the cartesian coordinates of $x \in \mathcal{A}$. Let $\mathcal{A}$ have orientation $w = e_1 e_2$. For the cases below, find the change of f over $\mathcal{A}$.

   (a)
$$f(\chi_1, \chi_2) = \frac{1}{|x|^2} (\chi_1 \, d\chi_1 + \chi_2 \, d\chi_2).$$

   (b)
$$f(\chi_1, \chi_2) = |x|^2 (-\chi_2 \, d\chi_1 + \chi_1 \, d\chi_2).$$

   (c)
$$f(\chi_1, \chi_2) = \frac{1}{|x|^2} (-\chi_2 \, d\chi_1 + \chi_1 \, d\chi_2).$$

Chapter 7

Miscellaneous topics

This last chapter is an attempt to show that geometric algebra and geometric calculus are tools that can be used to address a wide variety of mathematical topics. It is a limited attempt is several ways.

First, the topics on which we touch are developed only superficially; each would deserve a book on its own. For example, we say very little about the theory of differential forms other than to point out how it arises naturally as a part of geometric calculus. And we do no more with complex analysis than show how it fits into geometric calculus, prove the Clifford-Cauchy integral formula and the fundamental theorem of algebra, and hint that this is part of what is known as Clifford analysis.

Second, there are many topics, accessible using geometric algebra and geometric calculus, on which we touch not at all. For instance tensor analysis (see the discussion of extensors in Chapter 3 of [22]) or projective geometry (see [19] or [23]) or the theory of spinors (see [9] and [14]) or the very nice way in which geometric algebra can be used to treat such geometric transformations as rotations and reflections (see [9]).

Third, a discussion of this nature amounts to a somewhat arbitrary choice of topics of interest to the author and is a reflection of the limits of his knowledge and capacity.

7.1   Linear systems and simplexes

Let us begin with some simple topics of an algebraic and geometric nature.

7.1.1   Systems of linear equations

Everyone is familiar from college algebra with solving systems of linear equations of the form

$$\alpha_{11}\chi_1 + \cdots + \alpha_{1q}\chi_q = \beta_1$$
$$\vdots \qquad\qquad\qquad (7.1)$$
$$\alpha_{p1}\chi_1 + \cdots + \alpha_{pq}\chi_q = \beta_p$$

for $\chi_1, \ldots, \chi_q$. Let us approach this familiar problem and its related concepts from the point of view of geometric algebra.

Minors and rank

Let A be a $p \times q$ matrix with real entries:

$$A = \begin{pmatrix} \alpha_{11} & \alpha_{12} & \cdots & \alpha_{1q} \\ \alpha_{21} & \alpha_{22} & \cdots & \alpha_{2q} \\ \vdots & & & \\ \alpha_{p1} & \alpha_{p2} & \cdots & \alpha_{pq} \end{pmatrix}. \tag{7.2}$$

We want a notation for submatrices of A. Let $I = (i_1, \ldots, i_r)$ be an ordered multi-index from $\{1, \ldots, p\}$ (so that $1 \le i_1 < \cdots < i_r \le p$) and $J = (j_1, \ldots, j_s)$, an ordered multi-index from $\{1, \ldots, q\}$. We then set

$$A_{I \times J} \overset{\text{def.}}{=} \begin{pmatrix} \alpha_{i_1 j_1} & \cdots & \alpha_{i_1, j_s} \\ \vdots & & \\ \alpha_{i_r j_1} & \cdots & \alpha_{i_r j_s} \end{pmatrix} \tag{7.3}$$

which is the form of an arbitrary $r \times s$ submatrix of A.

By an $r \times r$ *minor of* A we mean a number of the form $\det\left(A_{I \times J}\right)$ where $A_{I \times J}$ is a (square) $r \times r$ submatrix of A.

It turns out that minors arise naturally in the setting of geometric algebra. In the next result, where we show this, recall that $\mathcal{O}_r$ is the set of ordered multi-indices of length r and that if $I = (i_1, \ldots, i_r)$, then $a_I = a_{i_1} \wedge \cdots \wedge a_{i_r}$.

Proposition 7.1. *Let A be the $p \times q$ matrix of (7.2). Let U be a p-dimensional vector subspace of some $\mathbb{R}^n$ and suppose that $\{u_i\}_{i=1}^p$ is a basis for U. Set $a_j = \sum_{i=1}^p \alpha_{ij} u_i$ for $j = 1, \ldots, q$. We see that we may consider each a_j to be the jth column vector of A. Then if $1 \le r \le \min(p, q)$ and J is the ordered multi-index $(j_1, \ldots, j_r)$ from $\{1, \ldots, q\}$, it follows that*

$$a_J = a_{j_1} \wedge \cdots \wedge a_{j_r} = \sum_{I \in \mathcal{O}_r} \det\left(A_{I \times J}\right) u_I.$$

Proof. We see that a_J is an r-vector in U and $\{u_K\}_{K \in \mathcal{O}_r}$ is a basis for $\wedge^r U$, so we must be able to write a_J in the form

$$a_J = \sum_{K \in \mathcal{O}_r} \lambda_{KJ} u_K \tag{7.4}$$

where each λ_{KJ} is a real scalar. Now let $\{m_i\}_{i=1}^p$ be the reciprocal basis to $\{u_j\}_{j=1}^p$. Let $I = (i_1, \ldots, i_r)$ be an ordered multi-index from the index set $\{1, \ldots, p\}$, form $m_I = m_{i_1} \wedge \cdots \wedge m_{i_r}$, and dot both sides of (7.4) by $m_I^\dagger$. By Proposition 5.16, we obtain $\lambda_{IJ} = a_J \cdot m_I^\dagger$. Definition 3.6 tells us that

$$a_J \cdot m_I^\dagger = \det \begin{pmatrix} a_{j_1} \cdot m_{i_1} & \cdots & a_{j_1} \cdot m_{i_r} \\ \vdots & & \\ a_{j_r} \cdot m_{i_1} & \cdots & a_{j_r} \cdot m_{i_r} \end{pmatrix}.$$

We compute

$$a_{j_k} \cdot m_{i_s} = \sum_{i=1}^{p} \alpha_{ij_k}(u_i \cdot m_{i_s}) = \alpha_{i_s j_k}.$$

Then

$$a_J \cdot m_I^\dagger = \det \begin{pmatrix} \alpha_{i_1 j_1} & \cdots & \alpha_{i_r,j_1} \\ \vdots & & \\ \alpha_{i_1 j_r} & \cdots & \alpha_{i_r j_r} \end{pmatrix}$$
$$= \det\left((A_{I \times J})^T\right)$$
$$= \det(A_{I \times J})$$

(where we recall B^T denotes the transpose of a matrix B). This is the desired result. $\square$

Earlier, when discussing charts on manifolds, we assumed the reader understood what we meant by the rank of a linear transformation $f'(x_0)$. The following discussion is independent of our previous remarks and can be read without reference to them. Here we define rank for matrices and linear transformations, and the earlier discussion had the second use of rank in mind.

Definition 7.1. By the *rank* of the matrix A, we mean the largest r such that A has a nonzero $r \times r$ minor.

Example 7.1. Let

$$A = \begin{pmatrix} 3 \\ -2 \\ 5 \end{pmatrix},$$

$$B = \begin{pmatrix} 1 & 1 & 0 \\ 0 & 0 & 0 \\ 0 & 1 & 1 \end{pmatrix},$$

$$C = \begin{pmatrix} 1 & 0 & 0 & 1 \\ 0 & 1 & 0 & 1 \\ 1 & 1 & 0 & 2 \end{pmatrix}.$$

Then A, B, C have, respectively, rank 1, 2, and 2.

Corollary 7.1. *If A is a $p \times q$ matrix with real entries, then*

$$rank(A) = rank(A^T)$$
$= $ *the maximal number of linearly independent column vectors of A*
$= $ *the maximal number of linearly independent row vectors of A.*

Proof. Assume A has the form of (7.2). We can take the column vectors to be $a_j = \sum_{i=1}^{p} \alpha_{ij} e_i$ for $j = 1, \ldots, q$. By Propositon 7.1, every $r \times r$ minor $\det(A_{I \times J})$ will occur as a coefficient in the expansion of a_J in terms of the r-blades e_I. Note that for a given $J = (j_1, \ldots, j_r)$, the following are equivalent:

$$a_J = a_{j_1} \wedge \cdots \wedge a_{j_r} = 0.$$

$\{a_{j_1}, \ldots, a_{j_r}\}$ is a linearly dependent set.   (7.5)

Every minor $\det(A_{I \times J})$ is zero where $I \in \mathcal{O}_r$.

The following are also equivalent for a given J:

$$a_J = a_{j_1} \wedge \cdots \wedge a_{j_r} \neq 0.$$

$\{a_{j_1}, \ldots, a_{j_r}\}$ is a linearly independent set.   (7.6)

Some minor $\det(A_{I \times J})$ is nonzero where $I \in \mathcal{O}_r$.

If (7.5) holds, then $\mathrm{rank}(A) < r$. If (7.6) holds, then $\mathrm{rank}(A) \geq r$. It follows from this that the rank of A must be the maximal number of linearly independent column vectors of A.

To see the connection between the rank of A and A^T, first it is easily checked that for ordered multi-indices I and J from index sets $\{1, \ldots, p\}$ and $\{1, \ldots, q\}$ respectively, we have $(A^T)_{J \times I} = (A_{I \times J})^T$. Then

$$\det \left((A^T)_{J \times I} \right) = \det \left((A_{I \times J})^T \right) = \det(A_{I \times J}).$$

It follows that A and A^T have exactly the same $r \times r$ minors and must therefore have the same rank. Since the column vectors of A^T are the row vectors of A, we see the relation between rank and the number of independent column and row vectors.   □

We now introduce the idea of *rank* for a linear transformation:

Definition 7.2. Suppose that U and V are finite-dimensional vector spaces over $\mathbb{R}$ and $f: U \to V$ is a linear transformation. Then

$$\mathrm{rank}(f) \stackrel{\mathrm{def.}}{=} \dim f(U).$$

Example 7.2. The linear transformation $f_1: \mathbb{R}^n \to \mathbb{R}^n$ given by

$$f_1(\chi_1, \chi_2, \ldots, \chi_n) = (\chi_1, 0, \ldots, 0)$$

is the orthogonal projection of $\mathbb{R}^n$ onto the subspace $\mathbb{R} \times 0 \times \cdots \times 0$. It has rank 1. The linear transformation $f_2: \mathbb{R}^3 \to \mathbb{R}^3$ given by

$$\begin{aligned} f_2(\chi_1, \chi_2, \chi_3) = {} & \left(\chi_1 \cos(\theta) - \chi_2 \sin(\theta) \right) e_1 \\ & + \left(\chi_1 \sin(\theta) + \chi_2 \cos(\theta) \right) e_2 + \chi_3 e_3 \end{aligned}$$

amounts to a rotation of the $e_1 e_2$-plane through the angle θ while the e_3-axis is held fixed. It has rank 3.

The next result shows that the concepts of rank for a matrix and a linear transformation, when viewed properly, are really the same thing and thus that the rank of a matrix must be the same if we rewrite it in terms of a different set of basis vectors.

Proposition 7.2. *Suppose we have a linear transformation $f \colon U \to V$ between finite-dimensional vector spaces over $\mathbb{R}$. Let $\{u_i\}_{i=1}^q$ and $\{v_j\}_{j=1}^p$ be bases for U and V respectively and suppose that f has the form*

$$f(u_i) = \sum_{j=1}^p \alpha_{ji} v_j$$

so that the $p \times q$ matrix A of (7.2) is the matrix of f with respect to these bases. Then

$$rank(f) = rank(A) = rank(f^t)$$

where $f^t \colon V \to U$ is the adjoint transformation of f.

Proof. The formula given for $f(u_i)$ amounts to the matrix equation

$$\begin{pmatrix} \alpha_{11} & \alpha_{12} & \cdots & \alpha_{1q} \\ \alpha_{21} & \alpha_{22} & \cdots & \alpha_{2q} \\ \vdots & & & \\ \alpha_{p1} & \alpha_{p2} & \cdots & \alpha_{pq} \end{pmatrix} \begin{pmatrix} 0 \\ \vdots \\ 1 \\ \vdots \\ 0 \end{pmatrix} = \begin{pmatrix} \alpha_{1i} \\ \alpha_{2i} \\ \vdots \\ \alpha_{pi} \end{pmatrix}. \tag{7.7}$$

The rank of f must be the largest r such that $\{f(u_{i_1}), \ldots, f(u_{i_r})\}$ is a linearly independent set where $i_1 < \cdots < i_r$. But we see from (7.7) that this is the same thing as asking for the maximal number of linearly independent column vectors of A, and by Corollary 7.1, this is the rank of A.

Next, let $\{m_i\}_{i=1}^q$ and $\{n_j\}_{j=1}^p$ be the reciprocal bases for $\{u_i\}_{i=1}^q$ and $\{v_j\}_{j=1}^p$ respectively. We know from Proposition 5.17 that the matrix of f^t, expressed in terms of $\{m_i\}_{i=1}^q$ and $\{n_j\}_{j=1}^p$, is A^T. Thus $rank(f^t) = rank(A^T) = rank(A)$. □

Solving a system of linear equations

Suppose we write down a system of linear equations:

$$\begin{aligned} \alpha_{11}\chi_1 + & \quad \cdots \quad + \alpha_{1m}\chi_m = & \beta_1 \\ \vdots \quad & \qquad \qquad \vdots \qquad \quad \vdots \\ \alpha_{n1}\chi_1 + & \quad \cdots \quad + \alpha_{nm}\chi_m = & \beta_n. \end{aligned} \tag{7.8}$$

Typically we assume α_{ij} and β_j are given and we want to know if there exist $\chi_1, \ldots, \chi_m$ that satisfy (7.8). Assuming the answer is yes, we want to know how to solve for $\chi_1, \ldots, \chi_m$.

Let us set

$$a_i = \sum_{j=1}^n \alpha_{ji} e_j \quad \text{and} \quad b = \sum_{j=1}^n \beta_j e_j.$$

Then (7.8) becomes the vector equation

$$a_1\chi_1 + \cdots + a_m\chi_m = b \tag{7.9}$$

where $a_i, b \in \mathbb{R}^n$ and $\chi_j \in \mathbb{R}$.

To expedite discussion of the relation between the standard presentation of this material and our presentation here, we first introduce the matrix

$$A = \begin{pmatrix} \alpha_{11} & \cdots & \alpha_{1m} \\ \vdots & & \\ \alpha_{n1} & \cdots & \alpha_{nm} \end{pmatrix}.$$

obtained by writing down the coefficients associated with the variables χ_j, and then the *augmented matrix*

$$A^+ = \begin{pmatrix} \alpha_{11} & \cdots & \alpha_{1m} & \beta_1 \\ \vdots & & & \\ \alpha_{n1} & \cdots & \alpha_{nm} & \beta_n \end{pmatrix}.$$

Nonexistence of a solution

Equation 7.9 requires b to be a linear combination of $a_1, \ldots, a_m$, but can we be sure of this? After all, (7.8) is a *given* system of equations; we do not know if the source of the system is trustworthy or not. If b is not a linear combination of $a_1, \ldots, a_m$, this means (7.9) does not have a solution. This is also equivalent to A^+ having one more linearly independent vector than A, and in terms of matrices, this amounts to

$$\text{rank}(A) < \text{rank}(A^+).$$

The solution of the system

From now on, we assume b is a linear combination of $a_1, \ldots, a_m$, that is, a solution of (7.9) exists.

However $a_1, \ldots, a_m$ are not necessarily linearly independent. Let us find a maximal linearly independent subset of $\{a_1, \ldots, a_m\}$; without loss of generality, we may assume this is $\{a_1, \ldots, a_k\}$. That is $a_1 \wedge \cdots \wedge a_k \neq 0$, but for any $i > k$, we have $a_1 \wedge \cdots \wedge a_k \wedge a_i = 0$. We can then go back to Equation 7.9 and rewrite it in terms of $a_1, \ldots, a_k$. Because of this, without loss of generality, assume that $\{a_1, \ldots, a_m\}$ is already a linearly independent set.

In terms of matrices, this amounts to

$$\text{rank}(A) = \text{rank}(A^+) = m.$$

The equivalent geometric algebra conditions are

$$a_1 \wedge \cdots \wedge a_m \neq 0 \quad \text{and} \quad a_1 \wedge \cdots \wedge a_m \wedge b = 0.$$

Under our assumption that $a_1 \wedge \cdots \wedge a_m \neq 0$, the solution of (7.9) is unique. To see this, let $\{a_i'\}_{i=1}^m$ be the *reciprocal frame* for the frame $\{a_i\}_{i=1}^m$. Recall that this means

$$a_i \cdot a_j' = \begin{cases} 1 & \text{if } i = j, \\ 0 & \text{if } i \neq j. \end{cases}$$

If we then dot (7.9) with a_i', we obtain

$$\chi_i = a_i' \cdot b. \tag{7.10}$$

This gives us our solution.

Cramer's rule

A situation encountered in essentially all introductions to systems of linear equations is the one in which $m = n$ and $\det(A) \neq 0$; the solution is typically given by Cramer's rule. We now derive Cramer's rule which the reader can compare with (7.10).

We assume as before that

$$a_1 \wedge \cdots \wedge a_m \neq 0 \quad \text{and} \quad a_1 \wedge \cdots \wedge a_m \wedge b = 0.$$

On each side of Equation (7.9), we take the wedge product on the right by the term $A_i = (-1)^{i-1} a_1 \wedge \cdots \wedge \widehat{a_i} \wedge \cdots \wedge a_m$. We see that

$$a_j \wedge A_i = \begin{cases} a_1 \wedge \cdots \wedge a_m = \det(A) e_1 \wedge \cdots \wedge e_m & \text{if } j = i, \\ 0 & \text{if } j \neq i \end{cases}$$

and

$$b \wedge A_i = a_1 \wedge \cdots \wedge b \wedge \cdots \wedge a_m = \det(a_1, \ldots, b \ldots a_m) e_1 \wedge \cdots \wedge e_m$$

where b occurs in the ith position. The appearance of the determinants in the last two equations can be justified by referring back to Proposition 7.1 or Proposition 3.8. Therefore, taking the wedge product of Equation (7.9) and dividing by $\det(A) e_1 \wedge \cdots \wedge e_m$ yields

$$\chi_i = \frac{\det(a_1, \ldots, b, \ldots, a_m)}{\det(a_1, \ldots, a_i, \ldots, a_m)}$$

which is Cramer's rule.

7.1.2 Simplexes and geometric algebra

In what follows, we will denote points of $\mathbb{R}^n$ by capital letters A, B, C, A_i, etc. This means we can treat them as vectors and add them or multiply them by scalars.

The simplex S determined by $A_0, A_1, \ldots, A_k$, points in $\mathbb{R}^n$, is

$$S = \left\{ \tau_0 A_0 + \cdots + \tau_k A_k : \tau_i \in \mathbb{R}, 0 \leq \tau_i, \sum_{i=0}^k \tau_i = 1 \right\}.$$

S is the smallest convex set containing $A_0, \ldots, A_k$. We say that $A_0, \ldots, A_k$ are the *vertices* of S. Then 1-, 2-, and 3-simplexes are, respectively, a straight line, a triangle, and a tetrahedron.

Simplexes as simple k-vectors

Given $k+1$ points $A_0, A_1, \ldots, A_k$ in $\mathbb{R}^n$, then following [27] and [35], we introduce the notation

$$A_0 A_1 \cdots A_k \overset{\text{def.}}{=} \frac{1}{k!} (A_1 - A_0) \wedge \cdots \wedge (A_k - A_0). \tag{7.11}$$

The factor $1/k!$ is introduced because it is the ratio of the volume of the simplex represented by $A_0 A_1 \cdots A_k$ to that of the parallelepiped with edges $A_1 - A_0$, $A_2 - A_0$, $\ldots, A_k - A_0$.

Notice that on the the right-hand side of (7.11), we treat $A_i - A_0$ as a vector. The notation on the left-hand side of (7.11) is a generalization of the symbol PQ that denotes the vector from P to Q, namely $Q - P$. The result in the general case of (7.11) is a simple k-vector which is visualized as an oriented k-simplex with vertices $A_0, \ldots, A_k$. Notice that we can also write

$$A_0 A_1 \cdots A_k = \frac{1}{k!} (A_0 A_1) \wedge (A_0 A_2) \wedge \cdots \wedge (A_0 A_k) \tag{7.12}$$

where each $A_0 A_i$ is a vector in $\mathbb{R}^n$. (See Figure 7.1.) It would appear that A_0 plays a very special role in this definition, but Equation (7.13) will make it clear that this is not so.

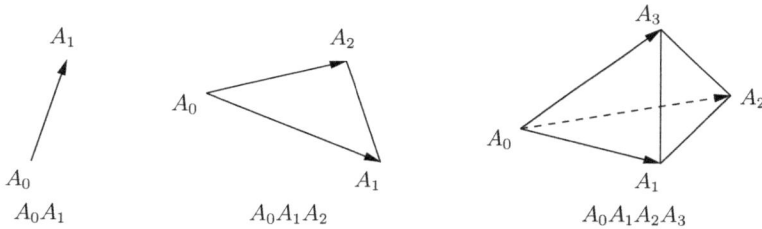

Figure 7.1: k-vectors of the form $A_0 \cdots A_k$.

The magnitude of $A_0 \cdots A_k$, that is, $|A_0 \cdots A_k|$, is the k-dimensional volume of the simplex. We say that $A_0, \ldots, A_k$ are *linearly independent points* provided that $A_0 \cdots A_k \neq 0$, that is, provided the vectors $A_1 - A_0, \ldots, A_k - A_0$ are linearly independent.

The following results are established in [27] and [35]:

Proposition 7.3. *For a simplex $A_0 \cdots A_k$ we have the following:*

1. If we interchange A_i and A_j, where $i < j$, then

$$A_0 \cdots A_j \cdots A_i \cdots A_k = -A_0 \cdots A_i \cdots A_j \cdots A_k. \tag{7.13}$$

2. If $k \geq 2$, then

$$\sum_{i=0}^{k} (-1)^i A_0 A_1 \cdots \widehat{A_i} \cdots A_k = 0, \tag{7.14}$$

where $\widehat{A_i}$ denotes omission of A_i.

Proof. 1. First consider the case where $i = 0$ and $j = 1$. Notice that

$$A_0A_1\cdots A_k = \frac{1}{k!}(A_0A_1)\wedge(A_0A_1+A_1A_2)\wedge\cdots\wedge(A_0A_1+A_1A_k)$$

$$= -\frac{1}{k!}(A_1A_0)\wedge(A_1A_2)\wedge\cdots\wedge(A_1A_k)$$

$$= -A_1A_0A_2\cdots A_k.$$

The cases where $j > 1$ follow similarly. For all other values of i, where $j > i \geq 1$, the proof is an obvious consequence of (7.12).

2. Notice that

$$A_1A_2\cdots A_k = \frac{1}{(k-1)!}(A_1A_2)\wedge(A_1A_3)\wedge\cdots\wedge(A_1A_k)$$

$$= \frac{1}{(k-1)!}(A_1A_0+A_0A_2)\wedge(A_1A_0+A_0A_3)\wedge\cdots\wedge(A_1A_0+A_0A_k).$$

When we multiply this out, because $(A_1A_0)\wedge(A_1A_0) = 0$, we must have

$$A_1A_2\cdots A_k = \frac{1}{(k-1)!}\Big\{(A_0A_2)\wedge(A_0A_3)\wedge\cdots\wedge(A_0A_k)$$

$$+ \sum_{i=2}^{k}(A_0A_2)\wedge\cdots\wedge(A_0A_{i-1})\wedge(A_1A_0)\wedge(A_0A_{i+1})\wedge\cdots\wedge(A_0A_k)\Big\}$$

$$= A_0A_2A_3\cdots A_k$$

$$+ \frac{1}{(k-1)!}\Big\{\sum_{i=2}^{k}(-1)^{i-1}(A_0A_1)\wedge(A_0A_2)\wedge\cdots\wedge\widehat{(A_0A_i)}\wedge\cdots\wedge(A_0A_k)\Big\}$$

$$= \sum_{i=1}^{k}(-1)^{i-1}A_0A_1\cdots\widehat{A_i}\cdots A_k.$$

$\square$

Corollary 7.2. *If there is a repeated point in $A_0\cdots A_k$, that is, if $A_i = A_j$ for some $i \neq j$, then $A_0\cdots A_k = 0$.*

The law of vector addition is usually understood in terms of a triangle as in Figure 7.2. Part 2 of Proposition 7.3 can be interpreted as a generalization of the law of vector addition. It says, in effect, that if the $(k-1)$-dimensional faces of a k-simplex are thought of as $(k-1)$-vectors with the proper orientations, then the sum of these "face-vectors" is 0.

Example 7.3. From Proposition 7.3, the "law of vector addition" for 2-simplexes is

$$A_1A_2A_3 - A_0A_2A_3 + A_0A_1A_3 - A_0A_1A_2 = 0.$$

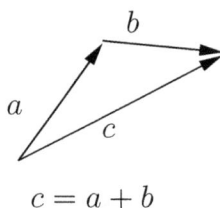

$$c = a + b$$

Figure 7.2: Law of vector addition.

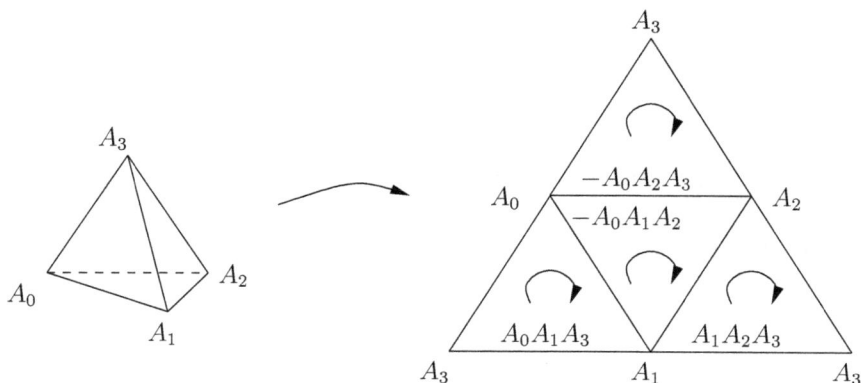

Figure 7.3: Vector addition law for 2-vectors.

We provide a picture of how this works in Figure 7.3 by "unfolding" a 3-simplex $A_0A_1A_2A_3$ into the plane and indicating with circular arrows on the "unfolded" version the orientation of each 2-dimensional face of the 3-simplex.

We say that an n-simplex $A_0A_1 \cdots A_n$ is *orthogonal* if, for some labelling of its vertices, the angle between the faces $A_0 \cdots \widehat{A}_j \cdots A_n$ and $A_0 \cdots \widehat{A}_l \cdots A_n$ is $\pi/2$ whenever $1 \leq j, l \leq n$ and $j \neq l$. Under these circumstances, we call $A_1 \cdots A_n$ the *oblique face* of the simplex.

The following is proved in [27] and here is left as an exercise:

Proposition 7.4 (Pythagorean theorem for simplexes). *If $A_0A_1 \cdots A_n$, $n \geq 2$, is an orthogonal n-simplex with oblique face $A_1 \cdots A_n$, then*

$$|A_1 \cdots A_n|^2 = \sum_{i=1}^{n} |A_0 \cdots \widehat{A}_i \cdots A_n|^2.$$

7.1.3 Barycentric coordinates

Definition 7.3. Suppose that $A_0, A_1, \ldots, A_k$ are points in $\mathbb{R}^n$ and V is the hyperplane determined by them. If $A \in V$, then we say that $\lambda_0, \ldots, \lambda_k$ are *barycentric coordinates* of A with respect to the ordered $(k+1)$-tuple $(A_0, \ldots, A_k)$ provided

$$A = \lambda_0 A_0 + \cdots + \lambda_k A_k$$

where $\lambda_0 + \cdots + \lambda_k = 1$.

For example, we know that $(\lambda_0, \ldots, \lambda_i, \ldots, \lambda_k) = (0, \ldots, 1, \ldots, 0)$, where $\lambda_i = 1$ and all other λ_js are 0, are the barycentric coordinates of the vertex A_i. Or if we look at the case where $k = 1$, then $A_0 A_1$ is a directed line segment, and its midpoint has barycentric coordinates $(\lambda_0, \lambda_1) = (\frac{1}{2}, \frac{1}{2})$.

Remark 7.1. If $A_0, \ldots, A_k$ are independent points, then the barycentric coordinates are unique.

Remark 7.2. The points A *inside* the simplex $A_0 \cdots A_k$ are the ones for which each $\lambda_i > 0$. A point will be *outside* the simplex if at least one of its barycentric coordinates is negative.

Remark 7.3. There is a physical interpretation of the barycentric coordinates in the case of a 2-simplex. If we think of $A_0 A_1 A_2$ as a rigid plate of negligible mass and place masses μ_0, μ_1, μ_2 at the vertices A_0, A_1, A_2 respectively, then the point with barycentric coordinates

$$\lambda_i = \frac{\mu_i}{\mu_0 + \mu_1 + \mu_2}$$

will be the *balance point* of the plate.

Proposition 7.5. *Suppose that* $A_0, A_1, \ldots, A_k$ *are points in* $\mathbb{R}^n$ *and* $A_0 = \lambda_0 B + \lambda_1 C$ *where* $\lambda_0 + \lambda_1 = 1$. *Then*

$$A_0 \cdots A_k = \lambda_0 (BA_1 \cdots A_k) + \lambda_1 (CA_1 \cdots A_k).$$

The proof is very simple, and of course a similar decomposition can be constructed using any A_i, not just A_0.

The next result gives a geometric interpretation of barycentric coordinates.

Proposition 7.6. *Let* $A_0, A_1, \ldots, A_k$ *be linearly independent points in* $\mathbb{R}^n$, *and let* V *be the* k-dimensional hyperplane passing through those points. If the barycentric coordinates of a point A in V are given by*

$$A = \lambda_0 A_0 + \cdots + \lambda_k A_k,$$

then for each i,

$$\lambda_i = (-1)^{i-1} \frac{AA_0 \cdots \widehat{A_i} \cdots A_k}{A_0 \cdots A_k} \tag{7.15}$$

where the quotient in (7.15) is to be understood as the geometric product

$$(AA_0 \cdots \widehat{A_i} \cdots A_k)(A_0 \cdots A_k)^{-1}.$$

Proof. By Proposition 7.5, we have

$$AA_0\cdots\widehat{A_i}\cdots A_k = \lambda_0\,(A_0A_0A_1\cdots\widehat{A_i}\cdots A_k)+\cdots$$
$$+\lambda_k\,(A_kA_0A_1\cdots\widehat{A_i}\cdots A_k).$$

It follows from Corollary 7.2 that all of these terms except possibly one will be zero and that we have

$$AA_0\cdots\widehat{A_i}\cdots A_k = \lambda_i\,(A_iA_0\cdots\widehat{A_i}\cdots A_k)$$
$$= (-1)^{i-1}\lambda_i\,(A_0\cdots A_i\cdots A_k).$$

The desired result is then immediate. □

Remark 7.4. Since (as is easily shown) we have

$$\lambda_i = (A_0\cdots A\cdots A_k)(A_0\cdots A_i\cdots A_k)^{-1} = (A_0\cdots A_i\cdots A_k)^{-1}(A_0\cdots A\cdots A_k),$$

where it is understood that A takes the place of A_i, then we may rewrite (7.15) as

$$\lambda_i = \frac{A_0\cdots A\cdots A_k}{A_0\cdots A_i\cdots A_k}.$$

Exercises 7.1.

1. If a is a simple k-vector in $\mathbb{R}^n$, then by the *subspace determined by a* we mean

$$\mathrm{sub}(a) \stackrel{\text{def.}}{=} \{x\in\mathbb{R}^n : x\wedge a = 0\}. \tag{7.16}$$

   Suppose a and b are p- and q-blades in $\mathbb{R}^n$ respectively ($1\leq p,q\leq n$). Show that if $\mathrm{sub}(a)\subseteq\mathrm{sub}(b)$, then $a\cdot_L b = ab$ and $b\cdot_R a = ba$.

2. If $A_0,A_1,\ldots,A_k$ are points of $\mathbb{R}^n$, show that

$$(A_0\cdots A_k)^{\dagger} = (-1)^{\frac{k(k-1)}{2}}A_0\cdots A_k.$$

3. Use Equation (7.10) to solve the system

$$3\chi_1+2\chi_2 = 1$$
$$\chi_1-4\chi_2 = 2$$

   for χ_1 and χ_2.

4. Prove Proposition 7.4.

5. Justify Remark 7.1.

6. Prove Proposition 7.5.

7. Justify Remark 7.4.

7.2 Geometric derivatives again

We are concerned here with multivector-valued functions defined on manifolds, $f \colon \mathcal{M} \to \mathbb{G}^n$, and with their directional derivatives, derivatives with respect to coordinates, and geometric derivatives.

Before beginning, we introduce a basic and useful concept:

7.2.1 Orthogonal projection onto a manifold

Given a multivector field f on a manifold $\mathcal{M}$, we are often concerned to consider the portion of f which is, at any given point, parallel to or tangent to the manifold.

Definition 7.4. If f is a multivector field on $\mathcal{M}$, then by $P_{\mathcal{M}}(f)$, we mean the multivector field on $\mathcal{M}$ whose value at $x_0 \in \mathcal{M}$ is given by

$$\big(P_{\mathcal{M}}(f)\big)(x_0) = \underline{P}_{x_0}\big(f(x_0)\big)$$

where $P_{x_0} \colon \mathbb{R}^n \to T_{x_0}\mathcal{M}$ is the orthogonal projection of $\mathbb{R}^n$ onto the tangent space $T_{x_0}\mathcal{M}$ and $\underline{P}_{x_0}$ is the outermorphism which operates on multivectors.

It follows from Proposition 5.21 that if $\mathcal{M}$ is a p-manifold and v is a p-blade that is tangent to $\mathcal{M}$ at x_0 or if $w(x_0)$ is the orientation of $\mathcal{M}$ at x_0, then

$$\big(P_{\mathcal{M}}(f)\big)(x_0) = (f(x_0) \cdot_L v)v^{-1} = \big(f(x_0) \cdot_L w(x_0)\big)w(x_0)^\dagger. \tag{7.17}$$

Example 7.4. Let $\mathcal{M}$ be the 1-manifold $\chi_1^2 + \chi_2^2 = 1$ in $\mathbb{R}^2$ and let $f(x_0) = e_1$ (the first of the standard basis vectors e_1, e_2) for all $x_0 \in \mathcal{M}$. See Figure 7.4. At every point

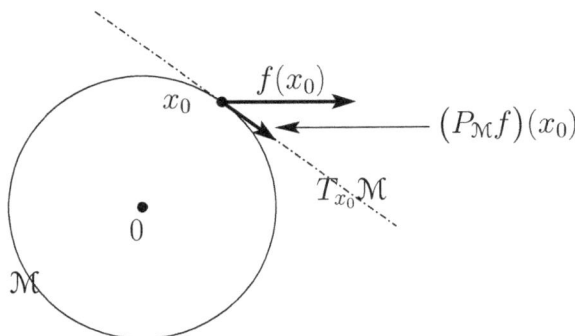

Figure 7.4: $P_{\mathcal{M}}$ operating on f

$x_0 = (\chi_{01}, \chi_{02})$ on the circle, let the orientation of $\mathcal{M}$ be $w(x_0) = -\chi_{02}e_1 + \chi_{01}e_2$, the counterclockwise unit tangent vector. Then using Equation (7.17), it can be computed that

$$\big(P_{\mathcal{M}}f\big)(x_0) = \chi_{02}^2 e_1 - \chi_{01}\chi_{02}e_2.$$

In particular, we have

$$\left(P_{\mathcal{M}}f\right)(1,0) = 0 \quad \text{and} \quad \left(P_{\mathcal{M}}f\right)(0,1) = e_1,$$

the validity of which should be geometrically obvious without calculation.

7.2.2   Derivatives, submanifolds, and coordinate changes

One of the most useful and basic tricks in calculus is rewriting a derivative which was calculated in terms of one set of coordinates in terms of a new set of coordinates. We would like to extend the standard formulas for doing this to the setting of multivector-valued functions on manifolds. At the same time we are often concerned with the situation where one manifold lies inside another, $\mathcal{M} \subseteq \mathcal{N}$, as occurs for example if we have a curve lying in a manifold or a surface bounding a region of one dimension higher. In this section, we shall combine both sets of concerns.

We now make a standing assumption that $\mathcal{M}$ and $\mathcal{N}$ are $\mathcal{C}^1$ p- and q-manifolds respectively in $\mathbb{R}^n$ such that

1. $\mathcal{M} \subseteq \mathcal{N}$,

2. x_0 is a point in $\mathcal{M}$,

3. $\mathcal{M}$ has $\mathcal{C}^1$ coordinates $(\chi_1, \ldots, \chi_p)$ and $\mathcal{N}$ has $\mathcal{C}^1$ coordinates $(\xi_1, \ldots, \xi_q)$ associated with respective charts x and y that both cover x_0.

4. For $i = 1, \ldots, p$ and $j = 1, \ldots, q$, we set

$$u_i = \frac{\partial x}{\partial \chi_i}(x_0) \quad \text{and} \quad v_j = \frac{\partial y}{\partial \xi_j}(x_0),$$

and we let $\{m_i\}_{i=1}^p$ be the reciprocal frame for $\{u_i\}_{i=1}^p$. Thus $\{u_i\}_{i=1}^p$ and $\{m_i\}_{i=1}^p$ are each bases for $T_{x_0}\mathcal{M}$. Notice that $m_i = d\chi_i(x_0)$ for $i = 1, \ldots, p$.

Since $T_{x_0}\mathcal{M} \subseteq T_{x_0}\mathcal{N}$, it is easily seen that we can extend $\{u_i\}_{i=1}^p$ to a basis $\{u_i\}_{i=1}^q$ for $T_{x_0}\mathcal{N}$ such that

$$u_i \cdot u_j = 0 \text{ and } |u_j| = 1 \text{ for } i = 1, \ldots, p \text{ and } j = p+1, \ldots, q.$$

If we set $m_i = u_i$ for $i = p+1, \ldots, q$, then it is easily checked that $\{m_i\}_{i=1}^q$ is the reciprocal frame for $\{u_i\}_{i=1}^q$. From this point on, it is important to keep in mind that $u_{p+1}, \ldots, u_q$ (hence $m_{p+1}, \ldots, m_q$) are orthogonal to $T_{x_0}\mathcal{M}$. We shall treat this particular construction as part of our standing assumption in this section.

Note on derivatives with respect to coordinates: We recall that if f is a function on $\mathcal{N}$, then by the notation $\partial f / \partial \xi_i$ we mean

$$\frac{\partial f}{\partial \xi_j}(x_0) = \partial_{e_j}(f \circ y)(r_0)$$

where y is the chart that induces the coordinates $(\xi_1, \ldots, \xi_q)$ on $\mathcal{N}$ and $x_0 = y(r_0)$. It is may be a bit less than clear what we should mean by $\partial f / \partial \chi_i$ since the chart x applies only to $\mathcal{M}$, a subset of $\mathcal{N}$; and further, it is not immediately clear how to compare $\partial f / \partial \xi_j$ and $\partial f / \partial \chi_i$ since we are compelled to look at different charts to compute them. However we know by Proposition 6.3 that

$$\frac{\partial f}{\partial \xi_j}(x_0) = \partial_{v_j} f(x_0) \quad \text{where } v_j = \frac{\partial y}{\partial \xi_j}(x_0),$$

and here everything is defined on $\mathcal{N}$. At points of $\mathcal{M}$, we therefore turn $\partial f / \partial \chi_i$ into a function defined on $\mathcal{N}$ by setting

$$\frac{\partial f}{\partial \chi_i}(x_0) = \partial_{u_i} f(x_0) \quad \text{where } u_j = \frac{\partial x}{\partial \chi_i}(x_0).$$

This agrees with the way we define $\partial f / \partial \chi_i$ on $\mathcal{M}$.

We gather together in the next proposition certain calculational results. Some are extensions of familiar formulas from introductory calculus.

Proposition 7.7. *The following hold:*

1. *If ϕ is a real-valued $\mathcal{C}^1$ function on $\mathcal{N}$, then $P_{\mathcal{M}}(\vec{\nabla}_{\mathcal{N}} \phi) = \vec{\nabla}_{\mathcal{M}} \phi$.*

2. *If v is a tangent vector to $\mathcal{N}$, then $v \cdot d\xi_i = \partial_v \xi_i$.*

3. $\dfrac{\partial \xi_j}{\partial \chi_i} = d\xi_j \cdot \dfrac{\partial x}{\partial \chi_i}.$

4. *If v is a tangent vector to $\mathcal{N}$ and ϕ is a $\mathcal{C}^1$ real-valued function on $\mathcal{N}$, then*

$$\left(\vec{\nabla}_{\mathcal{N}} \phi \right) \cdot v = \partial_v \phi.$$

5. $P_{\mathcal{M}}(d\xi_j) = \displaystyle\sum_{i=1}^{p} \frac{\partial \xi_j}{\partial \chi_i} d\chi_i.$

6. $d\chi_i = \displaystyle\sum_{j=1}^{q} \frac{\partial \chi_i}{\partial \xi_j} P_{\mathcal{M}}(d\xi_j).$

7. *If f is a $\mathcal{C}^1$ multivector field on $\mathcal{N}$, then*

$$\frac{\partial f}{\partial \chi_i} = \sum_{j=1}^{q} \frac{\partial \xi_j}{\partial \chi_i} \frac{\partial f}{\partial \xi_j}.$$

Example 7.5. Let $\mathcal{M}$ be the unit circle $\chi_1^2 + \chi_2^2 = 1$ in $\mathbb{R}^2$. We take $\mathcal{N}$ to be $\mathbb{R}^2$, so $\mathcal{M}$ is a submanifold of $\mathcal{N} = \mathbb{R}^2$.

Let x be the parametrization of $\mathbb{R}^2$ given by $x(\chi_1, \chi_2) = \chi_1 e_1 + \chi_2 e_2$. Then

$$\frac{\partial x}{\partial \chi_1} = e_1 = d\chi_1 \quad \text{and} \quad \frac{\partial x}{\partial \chi_2} = e_2 = d\chi_2.$$

We parametrize $\mathcal{M}$ by the map $y(\theta) = \cos(\theta) e_1 + \sin(\theta) e_2$. Then

$$\frac{\partial y}{\partial \theta} = -\sin(\theta) e_1 + \cos(\theta) e_2 = d\theta.$$

If we apply Part 3 of Proposition 7.7, then at any point $y(\theta) = \big(\cos(\theta), \sin(\theta)\big)$ of $\mathcal{M}$, we obtain

$$\frac{\partial \chi_1}{\partial \theta} = e_1 \cdot \frac{\partial y}{\partial \theta} = -\sin(\theta),$$

$$\frac{\partial \chi_2}{\partial \theta} = e_2 \cdot \frac{\partial y}{\partial \theta} = \cos(\theta).$$

Similarly, using Part 5 of Proposition 7.7, we obtain

$$P_{\mathcal{M}}(d\chi_1) = \frac{\partial \chi_1}{\partial \theta} d\theta = \sin^2(\theta) e_1 - \sin(\theta)\cos(\theta) e_2,$$

$$P_{\mathcal{M}}(d\chi_2) = \frac{\partial \chi_2}{\partial \theta} d\theta = -\sin(\theta)\cos(\theta) e_1 + \cos^2(\theta) e_2.$$

Proof of Proposition 7.7. Part 1: We know that

$$\vec{\nabla}_{\mathcal{N}} \phi(x_0) = \sum_{i=1}^{q} \partial_{u_i} \phi(x_0) m_i.$$

Since m_i is tangent to $T_{x_0}\mathcal{M}$ for $i = 1, \ldots, p$ and orthogonal for $i = p+1, \ldots, q$, we see that

$$P_{\mathcal{M}}\big(\vec{\nabla}_{\mathcal{N}}\phi\big)(x_0) = \sum_{i=1}^{p} \partial_{u_i} \phi(x_0) m_i.$$

But this last expression is just the formula for $\vec{\nabla}_{\mathcal{M}} \phi(x_0)$.

(Notice that $\vec{\nabla}_{\mathcal{M}} \phi = \overleftarrow{\nabla}_{\mathcal{M}} \phi$, so we have this result for both right and left geometric derivatives.)

Part 2: Keep in mind that all calculations are being carried out at the point x_0 and that $\{(\partial y / \partial \xi_j)(x_0)\}_{j=1}^{q}$ is a basis for $T_{x_0}\mathcal{N}$. We see that

$$v_j \cdot d\xi_i = \delta_{ij} \quad \text{since } v_j = \frac{\partial y}{\partial \xi_j}.$$

We also know from Propositions 6.3 and 6.4 that

$$\partial_{v_j} \xi_i = \frac{\partial \xi_i}{\partial \xi_j} = \delta_{ij}.$$

Thus $v_j \cdot d\xi_i = \partial_{v_j}\xi_i$. Since a tangent vector v can be written in the form

$$v = \sum_{j=1}^{q} \lambda_j \frac{\partial y}{\partial \xi_j}$$

and each of the expressions $v \cdot d\xi_i$ and $\partial_v \xi$ is linear in v, it follows that $v \cdot d\xi_i = \partial_v \xi_i$.

Part 3: We know that $\partial x / \partial \chi_j$ is a vector tangent to $\mathcal{M}$, hence to $\mathcal{N}$, and $\partial_{u_j}\xi_i = \partial \xi_i / \partial \chi_j$ by Proposition 6.3. The result then follows immediately from part 2.

Part 4: We know that

$$\vec{\nabla}_{\mathcal{N}}\phi = \sum_{i=1}^{q} \frac{\partial \phi}{\partial \xi_i} d\xi_i,$$

so

$$(\vec{\nabla}_{\mathcal{N}}\phi) \cdot \frac{\partial y}{\partial \xi_i} = \frac{\partial \phi}{\partial \xi_i}.$$

By Proposition 6.3, we also have

$$\partial_{v_i}\phi = \frac{\partial \phi}{\partial \xi_i} \quad \text{where } v_i = \frac{\partial y}{\partial \xi_i}.$$

Thus the identity holds if $v = v_i = \partial y / \partial \xi_i$. Since v can be written in the form

$$v = \sum_{i=1}^{q} \lambda_i \frac{\partial y}{\partial \xi_i},$$

the general identity holds by linearity.

Part 5: By Corollary 6.1, $d\xi_j = \vec{\nabla}_{\mathcal{N}}\xi_j$, so

$$d\xi_j(x_0) = \sum_{i=1}^{q} \partial_{u_i}\xi_j(x_0) m_i.$$

Since m_i is orthogonal to $T_{x_0}\mathcal{M}$ for $i = p+1, \ldots, q$ and appealing again to the definition of geometric derivative, we see that

$$\begin{aligned}
P_{\mathcal{M}}(d\xi_j)(x_0) &= \sum_{i=1}^{p} \partial_{u_i}\xi_j(x_0) m_i \\
&= \vec{\nabla}_{\mathcal{M}}\xi_j(x_0) \\
&= \sum_{i=1}^{p} \frac{\partial \xi_j}{\partial \chi_i}(x_0) d\chi_i(x_0)
\end{aligned}$$

which was the desired result.

Part 6: Using Corollary 6.1 and part 3, we see that

$$d\chi_i = \vec{\nabla}_{\mathcal{M}}\chi_i = P_{\mathcal{M}}(\vec{\nabla}_{\mathcal{N}}\chi_i).$$

Then

$$d\chi_i = P_{\mathcal{M}}\left(\sum_{j=1}^{q} \frac{\partial \chi_i}{\partial \xi_j} d\xi_j\right) = \sum_{j=1}^{q} \frac{\partial \chi_i}{\partial \xi_j} P_{\mathcal{M}}(d\xi_j).$$

Part 7: Let ψ be a $\mathcal{C}^1$ real-valued function on $\mathcal{N}$. Since f can be written as a sum of terms of the form ψe_l, it will suffice to prove the formula for ψ.

Part 1 tells us that $P_{\mathcal{M}}(\vec{\nabla}_{\mathcal{N}} \psi) = \vec{\nabla}_{\mathcal{M}} \psi$, that is,

$$P_{\mathcal{M}}\left(\sum_{j=1}^{q} \frac{\partial \psi}{\partial \xi_j} d\xi_j\right) = \sum_{i=1}^{p} \frac{\partial \psi}{\partial \chi_i} d\chi_i.$$

We dot this last equation with $\partial x / \partial \chi_i$ and obtain

$$\frac{\partial \psi}{\partial \chi_i} = \sum_{j=1}^{q} \frac{\partial \psi}{\partial \xi_j}\left(P_{\mathcal{M}}(d\xi_j) \cdot \frac{\partial x}{\partial \chi_i}\right). \tag{7.18}$$

The vector $\partial x / \partial \chi_i$ is tangent to $\mathcal{M}$, so if we dot it with $d\xi_j$, only the components of $d\xi_j$ that are tangent to $\mathcal{M}$ will make a contribution to the product. Using this fact and then appealing to part 3, we see that

$$P_{\mathcal{M}}(d\xi_j) \cdot \frac{\partial x}{\partial \chi_i} = d\xi_j \cdot \frac{\partial x}{\partial \chi_i} = \frac{\partial \xi_j}{\partial \chi_i}.$$

So Equation (7.18) becomes

$$\frac{\partial \psi}{\partial \chi_i} = \sum_{j=1}^{q} \frac{\partial \xi_j}{\partial \chi_i} \frac{\partial \psi}{\partial \xi_j}. \qquad \qquad \square$$

Remark 7.5. Notice that if $\mathcal{M} = \mathcal{N}$, then Parts 5, 6, and 7 of Proposition 7.7 amount to Proposition 6.5 which we earlier left unproved. Also it is implicit in what we have shown that there might be other formulas we might want to find. Could we, for example, find a nice formula for $\partial \chi_i / \partial \xi_j$? Such questions will be better handled after we introduce the notion of the normal identity.

There is an interesting implication to part 2 of Proposition 7.7. If v is a unit vector, then

$$\partial_v \xi_i = v \cdot d\xi_i = |d\xi_i| \cos(\theta)$$

where θ is the angle between v and $d\xi_i$. Since we can interpret $\partial_v \xi_i$ as the rate of change of ξ_i in the direction v, this is telling us that the coordinate ξ_i on the manifold is increasing most rapidly in the direction indicated by $d\xi_i$.

Example 7.6. Let our manifold $\mathcal{M}$ be $\mathbb{R}^2$. We have a natural coordinate system (χ_1, χ_2) on $\mathcal{M}$ inherited from $\mathbb{R}^2$. We define a chart $y \colon \mathbb{R}^2 \to \mathcal{M}$ by

$$y(\xi_1, \xi_2) = (\xi_1 + \xi_2)e_1 + \xi_2 e_2 = \chi_1 e_1 + \chi_2 e_2,$$

and this induces a second coordinate system (ξ_1, ξ_2) on $\mathcal{M}$. The two systems are related at any given point of $\mathcal{M}$ by

$$\xi_1 = \chi_1 - \chi_2 \quad \text{and} \quad \xi_2 = \chi_2.$$

If, at any point of $\mathcal{M}$, we calculate the tangent basis induced by the coordinates (ξ_1, ξ_2) and the associated reciprocal basis, we obtain

$$\frac{\partial y}{\partial \xi_1} = e_1 \qquad\qquad\qquad d\xi_1 = e_1 - e_2$$

$$\frac{\partial y}{\partial \xi_2} = e_1 + e_2 \qquad\qquad\qquad d\xi_2 = e_2.$$

If we now sketch on $\mathcal{M}$ lines on which ξ_1 and ξ_2 are held constant and superimpose the vectors $d\xi_1$ and $d\xi_2$ (see Figure 7.5), we see that $d\xi_i$ points in the direction in which ξ_i increases most rapidly.

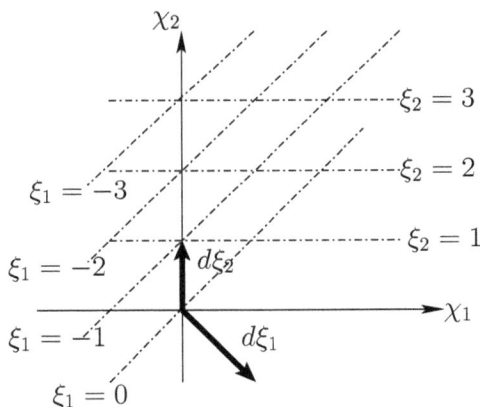

Figure 7.5: $d\xi_i$ indicates the direction of maximal ξ_i increase

7.2.3 A "shrinking" result

Recall that when the derivative of a real-valued function ϕ is first introduced in calculus, it is defined by something like this:

$$\phi'(\tau) = \lim_{\lambda \to 0} \frac{\phi(\tau + \lambda) - \phi(\tau)}{\lambda}.$$

This can be thought of as the limit of

$$\frac{\text{change of } \phi \text{ over an interval}}{\text{length of the interval}}$$

as the length of the interval shrinks to 0. We are going to get a similar characterization of $\vec{\nabla}_{\mathcal{M}} f(x)$.

First, however, we describe a more general result, a useful tool that shows what happens to the quotient of an integral over a cell by the volume of the cell while "shrinking" the cell to a point.

Let $\mathcal{M}$ be a $\mathcal{C}^1$ m-cell in $\mathbb{R}^n$ and suppose that x_0 is an interior point of $\mathcal{M}$. Given a parametrization $x \colon \mathcal{I}^m \to \mathcal{M}$ of $\mathcal{M}$, there is some $t_0 = (\tau_1, \ldots, \tau_m) \in (0,1)^m$ such that $x_0 = x(t_0)$. For sufficiently small positive δ, we see that the cube

$$\mathcal{I}^m_\delta \overset{\text{def.}}{=} [\tau_1 - \delta, \tau_1 + \delta] \times \cdots \times [\tau_m - \delta, \tau_m + \delta]$$

lies in $(0,1)^m$. We then set

$$\mathcal{M}^\delta = x(\mathcal{I}^m_\delta)$$

and call this *the δ-zone of x_0 with respect to the parametrization x.* (See Figure 7.6.) Clearly, if by x^δ we mean the restriction of x to $\mathcal{I}^m_\delta$, then we may treat $x^\delta \colon \mathcal{I}^m_\delta \to \mathcal{M}^\delta$

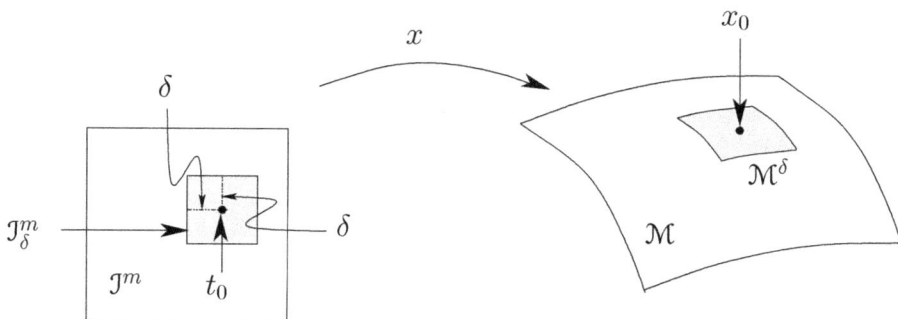

Figure 7.6: Construction of $\mathcal{M}^\delta$

as a parametrization of the m-cell $\mathcal{M}^\delta$; and if we let $\delta \to 0^+$, we can think of $\mathcal{M}^\delta$ as "shrinking" to x_0.

Proposition 7.8. *Let $\mathcal{M}$, x_0, x, and t_0 be as above. If $f \colon \mathcal{M} \to \mathbb{G}^n$ is a continuous multivector field, then*

$$\lim_{\delta \to 0^+} \frac{1}{vol(\mathcal{M}^\delta)} \int_{\mathcal{M}^\delta} f = f(x_0).$$

That is, the average value of a continuous function f on a cell goes to $f(x_0)$ if one lets the cell shrink to x_0.

Proof of Proposition 7.8. Let ϕ be a continuous, real-valued function on $\mathcal{M}$. Choose $\varepsilon > 0$. We can find $\delta > 0$ so small that $|\phi(x) - \phi(x_0)| < \varepsilon$ for all $x \in \mathcal{M}^\delta$. (The fact that we can take this last step can be regarded as following from what is known as the $\varepsilon - \delta$ definition of continuity of ϕ at x_0—see any introductory work on mathematical analysis, for example [24] or [35]. In line with our "cavalier" treatment of limits and continuity, we have not invoked the $\varepsilon - \delta$ definition of continuity till now.) Notice that

$$\int_{\mathcal{M}^\delta} \phi(x_0) = \phi(x_0) \int_{\mathcal{M}^\delta} 1 = \phi(x_0) \operatorname{vol}(\mathcal{M}^\delta).$$

Then

$$\left| \left(\frac{1}{\operatorname{vol}(\mathcal{M}^\delta)} \int_{\mathcal{M}^\delta} \phi \right) - \phi(x_0) \right| = \frac{1}{\operatorname{vol}(\mathcal{M}^\delta)} \left| \int_{\mathcal{M}^\delta} (\phi - \phi(x_0)) \right|$$

$$\leq \frac{1}{\operatorname{vol}(\mathcal{M}^\delta)} \int_{\mathcal{M}^\delta} |\phi - \phi(x_0)|$$

$$< \frac{1}{\operatorname{vol}(\mathcal{M}^\delta)} \varepsilon \operatorname{vol}(\mathcal{M}^\delta) = \varepsilon.$$

We see from this that we can make the difference between $\left(\int_{\mathcal{M}^\delta} \phi \right) / \operatorname{vol}(\mathcal{M}^\delta)$ and $\phi(x_0)$ arbitrarily small, that is, less than any given ε, provided we make δ sufficiently small. That is,

$$\lim_{\delta \to 0^+} \frac{1}{\operatorname{vol}(\mathcal{M}^\delta)} \int_{\mathcal{M}^\delta} \phi = \phi(x_0).$$

As every continuous multivector field has the form

$$f = \sum_{k=0}^{n} \sum_{i_1 < \cdots < i_k} \phi_{i_1 \ldots i_k} (e_{i_1} \cdots e_{i_k})$$

where each $\phi_{i_1 \ldots i_k}$ is continuous, we see that a similar result holds for any continuous f. $\qquad\square$

Now to our characterization of $\vec{\nabla}_{\mathcal{M}} f$.

Notice that by the Fundamental Theorem, we have

$$\int_{\mathcal{M}^\delta} (\vec{\nabla}_{\mathcal{M}} f) w = \int_{\partial \mathcal{M}^\delta} f \, \vec{\partial} w$$

where w may be taken to be the orientation of $\mathcal{M}$ and hence, by restriction, of $\mathcal{M}^\delta$. When we combine this observation with Proposition 7.8, we obtain a new description of $\vec{\nabla}_{\mathcal{M}}$:

Corollary 7.3. *Let $\mathcal{M}$, x_0, x, and t_0 be as Proposition 7.8 except that $\mathcal{M}$ is now required to be $\mathcal{C}^2$. Let w be the orientation of $\mathcal{M}$ and f be a $\mathcal{C}^1$ multivector field on $\mathcal{M}$. Then*

$$(\vec{\nabla}_{\mathcal{M}} f(x_0)) \, w(x_0) = \lim_{\delta \to 0^+} \frac{1}{vol(\mathcal{M}^\delta)} \int_{\partial \mathcal{M}^\delta} f \, \vec{\partial} w \qquad (7.19)$$

or, equivalently,

$$\vec{\nabla}_{\mathcal{M}} f(x_0) = \lim_{\delta \to 0^+} \left(\int_{\partial \mathcal{M}^\delta} f \, \vec{\partial} w \right) \frac{w(x_0)^\dagger}{vol(\mathcal{M}^\delta)}. \qquad (7.20)$$

We can interpret Equation (7.19) as saying

the rate of change of f at x_0 with respect to orientation w

$$= \lim_{\delta \to 0^+} \frac{1}{\mathrm{vol}(\mathcal{M}^\delta)} \int_{\partial \mathcal{M}^\delta} f \, \vec{\partial} w$$

where the integral of $f \, \vec{\partial} w$ over $\partial \mathcal{M}^\delta$ amounts to the change of f over the "infinitesimal" cell $\mathcal{M}^\delta$. It has been suggested by Hestenes in [17] and [22] that (7.20) be used as the definition of the geometric derivative (though Hestenes works with $\overleftarrow{\nabla}_{\mathcal{M}} f$ rather than $\vec{\nabla}_{\mathcal{M}} f$). In (7.20), the expression $w(x_0)^\dagger / \mathrm{vol} \mathcal{M}^\delta$ can be thought of as playing the role of division by an oriented $\mathcal{M}^\delta$.

From this point on, when using this idea of "shrinking" a cell to an interior point x_0, we feel free (if it is convenient) to skip mention of the parametrization, to drop the symbol $\mathcal{M}^\delta$, and to indicate what is going on by writing

$$\lim_{\mathcal{M} \to x_0} \frac{1}{\mathrm{vol}(\mathcal{M})} \int_{\partial \mathcal{M}} f.$$

In particular, we have

the rate of change of f at x_0 with respect to orientation w

$$= \lim_{\mathcal{M} \to x_0} \frac{1}{\mathrm{vol}(\mathcal{M})} \int_{\partial \mathcal{M}} f \, \vec{\partial} w$$

and

$$\vec{\nabla}_{\mathcal{M}} f(x_0) = \left(\lim_{\mathcal{M} \to x_0} \frac{1}{\mathrm{vol}(\mathcal{M})} \int_{\partial \mathcal{M}} f \, \vec{\partial} w \right) w(x_0)^\dagger. \qquad (7.21)$$

7.2.4   Curl and divergence

In vector analysis in three dimensions, there are several operations on real-valued functions and vector fields that are of great utility. Suppose ϕ is a real-valued function and $f = \phi_1 e_1 + \phi_2 e_2 + \phi_3 e_3$, a vector field, each defined on some subset of $\mathbb{R}^3$ and differentiable as often as needed. Then the *gradient*, *divergence*, *curl*, and *Laplacian* are given by

$$\mathrm{grad}\,\phi = \nabla\phi \stackrel{\text{def.}}{=} \frac{\partial \phi}{\partial \chi_1} e_1 + \frac{\partial \phi}{\partial \chi_2} e_2 + \frac{\partial \phi}{\partial \chi_3} e_3,$$

$$\mathrm{div}\,f = \nabla \cdot f \stackrel{\text{def.}}{=} \frac{\partial \phi_1}{\partial \chi_1} + \frac{\partial \phi_2}{\partial \chi_2} + \frac{\partial \phi_3}{\partial \chi_3},$$

$$\mathrm{curl}\,f = \nabla \times f \stackrel{\text{def.}}{=} \left(\frac{\partial \phi_3}{\partial \chi_2} - \frac{\partial \phi_2}{\partial \chi_3} \right) e_1 + \left(\frac{\partial \phi_1}{\partial \chi_3} - \frac{\partial \phi_3}{\partial \chi_1} \right) e_2 \qquad (7.22)$$

$$+ \left(\frac{\partial \phi_2}{\partial \chi_1} - \frac{\partial \phi_1}{\partial \chi_2} \right) e_3,$$

$$\Delta\phi = \nabla^2 \phi \stackrel{\text{def.}}{=} \frac{\partial^2 \phi}{\partial \chi_1^2} + \frac{\partial^2 \phi}{\partial \chi_2^2} + \frac{\partial^2 \phi}{\partial \chi_3^2}.$$

We have already begun to encounter these operations in Examples 6.18 and 6.19. It is possible to generalize them to the setting of tensor fields, but we shall not be concerned with that. We shall instead concentrate on curl and divergence because of their intimate connection with the geometric derivative.

We try to give a little intuition for divergence and curl as defined in (7.22) by the following examples:

Example 7.7. A typical physical example of divergence arises from considering the expansion of a cloud of gas in a vacuum. We may picture the molecules of gas as all crowded together in the region where the cloud had its origin. (This might have been an explosion, a volcanic vent, or a leak.) The molecules are all pushing to get away from one another, but at the point of the cloud's origin, they cannot move swiftly because of all the other molecules clustered about them, hemming them in. As one moves farther away, one finds that the molecules are moving more swiftly because they are less crowded.

We give an (overly simplified) picture of this sort of thing in Figure 7.7. We treat

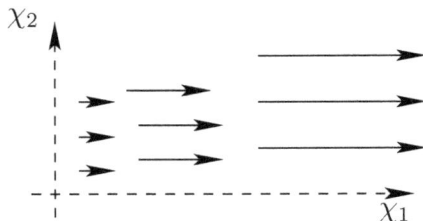

Figure 7.7: Example of divergence

the "expansion of the cloud" as though it is occurring in the first quadrant of $\mathbb{R}^2$ and only in the horizontal direction. The fact that the arrows grow longer as one moves to the right shows that the molecules are accelerating as one moves in that direction.

For an algebraic expression that captures this sort of behavior, we might try $f(\chi_1, \chi_2) = \chi_1 e_1$. We see by computation that $\operatorname{div} f = 1$ and $\operatorname{curl} f = 0$. We have positive divergence because the cloud is expanding.

Example 7.8. Picture paddles affixed to an axis in such a way that they can revolve freely about the axis. One imagines holding the axis and dipping the paddle end in a flowing stream. Either the paddles will rotate in the stream or they will not. We think of curl as being a measure of this tendency to rotate.

To construct a simple example, let us draw a vector field in the upper half of $\mathbb{R}^2$. (See Figure 7.8.) We think of it as representing the flow of a river with the χ_1-axis as the riverbank. The flow is is from left to right, and as one approaches the riverbank, the water moves ever more slowly and ultimately stops when $\chi_2 = 0$. It seems reasonable that if one dipped the paddle wheel in this river, it would rotate because the water would strike the paddles farther from the riverbank more forcefully than those close to the bank.

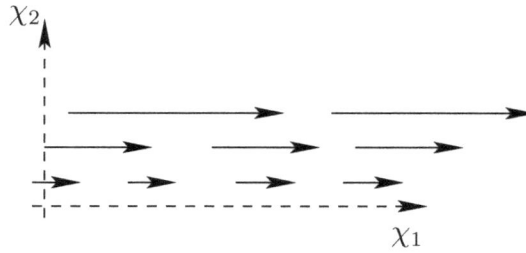

Figure 7.8: Example of curl

We can capture this sort of behavior in an expression for the vector field such as $f(\chi_1, \chi_2, \chi_3) = \chi_2 \, e_1$. This has the property that the vectors are always directed to the right and they grow in magnitude as one moves away from the riverbank. We calculate that $\text{curl} f = -e_3$ and $\text{div} f = 0$. Notice that though f acts as a vector field in the $\chi_1 \chi_2$-plane, we get an answer in the third dimension. Speaking imprecisely, this is because we act as though ∇ is a 1-vector in $\mathbb{R}^3$, $\nabla = (\partial/\partial \chi_1)e_1 + (\partial/\partial \chi_2)e_2 + (\partial/\partial \chi_3)e_3$, and curl is a vector product as defined in introductory calculus: $\text{curl} f = \nabla \times f$.

We now give a very general definition of curl and divergence for the geometric calculus setting:

Definition 7.5. Let $\mathcal{M}$ be a $\mathcal{C}^1$ manifold in $\mathbb{R}^n$ and f a $\mathcal{C}^1$ multivector field on $\mathcal{M}$. We then define the *right* (or *right-hand*) *divergence* and *curl* of f on $\mathcal{M}$ by

$$\overrightarrow{\text{div}}_{\mathcal{M}} f \;\overset{\text{def.}}{=}\; \sum_{k=1}^{n} \left\langle \overrightarrow{\nabla}_{\mathcal{M}} \langle f \rangle_k \right\rangle_{k-1}$$

$$\overrightarrow{\text{curl}}_{\mathcal{M}} f \;\overset{\text{def.}}{=}\; \sum_{k=0}^{n} \left\langle \overrightarrow{\nabla}_{\mathcal{M}} \langle f \rangle_k \right\rangle_{k+1}.$$

We define *left* (or *left-hand*) *divergence* and *curl* of f, that is $\overleftarrow{\text{div}}_{\mathcal{M}} f$ and $\overleftarrow{\text{curl}}_{\mathcal{M}} f$, by replacing $\overrightarrow{\nabla}_{\mathcal{M}}$ with $\overleftarrow{\nabla}_{\mathcal{M}}$.

Two things to notice about this definition: First, it looks nothing like the definitions of Equation (7.22). Second, this definition is *coordinate-free*, that is, in contrast to (7.22), it has no dependence on $\chi_1, \ldots, \chi_n$. This is an appealing property since if one is solving a physical problem, though coordinates are useful for measurements and computations, there are no coordinates painted on space-time.

Example 7.9. Let $\mathcal{M}$ be $\mathbb{R}^n$ and take ϕ to be a $\mathcal{C}^1$, real-valued function on $\mathbb{R}^n$. (As noted earlier, when we indicate geometric derivatives on this particular manifold, instead of using symbols such as $\overrightarrow{\nabla}_{\mathcal{M}}$ or $\overleftrightarrow{\nabla}_{\mathbb{R}^n}$, we shall simply write $\overrightarrow{\nabla}$ or $\overleftrightarrow{\nabla}$ or $\overrightarrow{\text{curl}}$ etc. as needed). We take $\{e_i\}_{i=1}^n$ as our basis for $\mathbb{R}^n$, and it is its own reciprocal basis, so

$$\overrightarrow{\nabla}\phi = \overleftarrow{\nabla}\phi = \sum_{i=1}^n \frac{\partial \phi}{\partial \chi_i} e_i.$$

This is an n-dimensional version of the gradient in (7.22). Since ϕ is grade 0 and $\overrightarrow{\nabla}\phi$ and $\overleftarrow{\nabla}\phi$ are grade 1, by Definition 7.5,

$$\overrightarrow{\text{curl}}\,\phi = \overleftarrow{\text{curl}}\,\phi = \text{grad}\,\phi.$$

Since there are no grade -1 terms,

$$\overrightarrow{\text{div}}\,\phi = \overleftarrow{\text{div}}\,\phi = 0.$$

Example 7.10. Let $f = \sum_{i=1}^n \phi_i e_i$ be a $\mathcal{C}^1$ vector field on some open subset of $\mathbb{R}^n$, each of the ϕ_i being understood to be a real-valued function. Using the facts that $e_i e_j = e_i \cdot e_j + e_i \wedge e_j$ and $e_i \cdot e_j = \delta_{ij}$, we calculate

$$\overrightarrow{\nabla}f = \sum_{j=1}^n \frac{\partial f}{\partial \chi_i} e_j = \sum_{i,j=1}^n \frac{\partial \phi_i}{\partial \chi_j} e_i e_j$$
$$= \sum_{i=1}^n \frac{\partial \phi_i}{\partial \chi_i} + \sum_{i<j} \left(\frac{\partial \phi_i}{\partial \chi_j} - \frac{\partial \phi_j}{\partial \chi_i} \right) e_i \wedge e_j.$$

Then by Definition 7.5,

$$\overrightarrow{\text{div}}f = \left\langle \overrightarrow{\nabla}\langle f \rangle_1 \right\rangle_0 = \sum_{i=1}^n \frac{\partial \phi_i}{\partial \chi_i},$$

which is the usual definition of divergence in vector analysis; and

$$\overrightarrow{\text{curl}}f = \left\langle \overrightarrow{\nabla}\langle f \rangle_1 \right\rangle_2 = \sum_{i<j} \left(\frac{\partial \phi_i}{\partial \chi_j} - \frac{\partial \phi_j}{\partial \chi_i} \right) e_i \wedge e_j,$$

which looks like the formula of (7.22) except that it is a 2-vector instead of a 1-vector.

Let us take the particular case of $n = 3$ in this example. Set $e = e_1 e_2 e_3$, the standard orientation of $\mathbb{R}^3$. $\overrightarrow{\text{curl}}f$ is a 2-vector, so the geometric product $(\overrightarrow{\text{curl}}f)\,e$ will be a 1-vector. By Proposition 5.9, it follows from $(\overrightarrow{\text{curl}}f)\,e$ being a 1-vector that $\overrightarrow{\text{curl}}f$ must be a simple 2-vector, and from the discussion of Section 5.4, that 1-vector must be orthogonal to $\overrightarrow{\text{curl}}f$. We note finally that by computation, we have $(\overrightarrow{\text{curl}}f)\,e = \text{curl}f$, where the 1-vector field on the right is that of (7.22). So at any given point, $\text{curl}f$ is a 1-vector that is orthogonal to the 2-vector $\overrightarrow{\text{curl}}f$.

Example 7.11. Now we get to the Laplacian. Let ϕ be a $\mathcal{C}^2$ real-valued function on an open subset of $\mathbb{R}^n$. By $\overrightarrow{\nabla}^2$ we mean two applications of $\overrightarrow{\nabla}$, one followed by the other.

$$\overrightarrow{\nabla}^2\phi = \overrightarrow{\nabla}\left(\sum_{i=1}^n \frac{\partial \phi}{\partial \chi_i} e_i \right)$$

$$= \sum_{i,j=1}^{n} \frac{\partial^2 \phi}{\partial \chi_j \partial \chi_i} e_i e_j$$

$$= \sum_{i=1}^{n} \frac{\partial^2 \phi}{\partial \chi_i^2} + \sum_{i,j=1}^{n} \frac{\partial^2 \phi}{\partial \chi_j \partial \chi_i} e_i \wedge e_j.$$

Because of the equality of mixed partial derivatives and the fact that $e_i \wedge e_j = -e_j \wedge e_i$, we have

$$\sum_{i,j=1}^{n} \frac{\partial^2 \phi}{\partial \chi_j \partial \chi_i} e_i \wedge e_j = 0.$$

Thus $\vec{\nabla}^2 \phi = \Delta \phi$, the Laplacian of ϕ.

We now establish some general properties of divergence and curl. We suppose $\mathcal{M}$ is a $\mathcal{C}^1$ manifold in $\mathbb{R}^n$ and f is a $\mathcal{C}^1$ multivector field on $\mathcal{M}$ taking values in $\mathbb{G}^n$.

This first result shows us how to switch between right-hand and left-hand versions of divergence and curl.

Proposition 7.9. $\overrightarrow{\mathrm{div}}_{\mathcal{M}}(f^\dagger) = (\overleftarrow{\mathrm{div}}_{\mathcal{M}} f)^\dagger$ *and* $\overrightarrow{\mathrm{curl}}_{\mathcal{M}}(f^\dagger) = (\overleftarrow{\mathrm{curl}}_{\mathcal{M}} f)^\dagger$.

Proof. Beginning with divergence in Definition 7.5 and applying Propositions 5.6 and 6.8, we see that

$$\langle \vec{\nabla}_{\mathcal{M}} \langle f^\dagger \rangle_k \rangle_{k-1} = \langle \vec{\nabla}_{\mathcal{M}} (\langle f \rangle_k^\dagger) \rangle_{k-1}$$
$$= \langle (\overleftarrow{\nabla}_{\mathcal{M}} \langle f \rangle_k)^\dagger \rangle_{k-1}$$
$$= \langle \overleftarrow{\nabla}_{\mathcal{M}} \langle f \rangle_k \rangle_{k-1}^\dagger.$$

This establishes the formula for divergence. We obtain the one for curl similarly. $\square$

The next result gives formulas for computing divergence and curl and shows the exact relation of these operations to the geometric derivative.

Proposition 7.10. *Suppose $\mathcal{M}$ has a frame of tangent vectors $\{u_i\}_{i=1}^{p}$ (at least locally defined) and reciprocal frame $\{m_i\}_{i=1}^{p}$. Then*

$$\overrightarrow{\mathrm{div}}_{\mathcal{M}} f = \sum_{i=1}^{p} (\partial_{u_i} f) \cdot_R m_i \quad \textit{and} \quad \overrightarrow{\mathrm{curl}}_{\mathcal{M}} f = \sum_{i=1}^{p} (\partial_{u_i} f) \wedge m_i,$$

$$\overleftarrow{\mathrm{div}}_{\mathcal{M}} f = \sum_{i=1}^{p} m_i \cdot_L (\partial_{u_i} f) \quad \textit{and} \quad \overleftarrow{\mathrm{curl}}_{\mathcal{M}} f = \sum_{i=1}^{p} m_i \wedge (\partial_{u_i} f),$$

$$\vec{\nabla}_{\mathcal{M}} f = \overrightarrow{\mathrm{div}}_{\mathcal{M}} f + \overrightarrow{\mathrm{curl}}_{\mathcal{M}} f \quad \textit{and} \quad \overleftarrow{\nabla}_{\mathcal{M}} f = \overleftarrow{\mathrm{div}}_{\mathcal{M}} f + \overleftarrow{\mathrm{curl}}_{\mathcal{M}} f.$$

Proof. Since the map $f \mapsto \partial_{u_i} f$ does not change grade, we have $\langle \partial_{u_i} f \rangle_k = \partial_{u_i} \langle f \rangle_k$. Then

$$\vec{\nabla}_{\mathcal{M}} \langle f \rangle_k = \sum_{i=1}^{p} \langle \partial_{u_i} f \rangle_k m_i$$

$$= \sum_{i=1}^{p} \langle \partial_{u_i} f \rangle_k \cdot_R m_i + \sum_{i=1}^{p} \langle \partial_{u_i} f \rangle_k \wedge m_i$$

where we have used Proposition 5.14 for the last step. Since the dot product lowers grade and the wedge product raises it, we must have

$$\left\langle \vec{\nabla}_{\mathcal{M}} \langle f \rangle_k \right\rangle_{k-1} = \sum_{i=1}^{p} \langle \partial_{u_i} f \rangle_k \cdot_R m_i,$$

$$\left\langle \vec{\nabla}_{\mathcal{M}} \langle f \rangle_k \right\rangle_{k+1} = \sum_{i=1}^{p} \langle \partial_{u_i} f \rangle_k \wedge m_i.$$

Then

$$\vec{\mathrm{div}}_{\mathcal{M}} f = \sum_{k=1}^{n} \left\langle \vec{\nabla}_{\mathcal{M}} \langle f \rangle_k \right\rangle_{k-1}$$

$$= \sum_{i=1}^{p} \left(\sum_{k=1}^{n} \langle \partial_{u_i} f \rangle_k \right) \cdot_R m_i$$

$$= \sum_{i=1}^{p} (\partial_{u_i} f) \cdot_R m_i.$$

The derivation for curl is similar. The fact that $\vec{\nabla}_{\mathcal{M}} f$ is the sum of the divergence and the curl should be obvious at this point.

The left-hand versions of this result follow from Propositions 7.9 and 5.13. $\quad\square$

Remark 7.6. A notation of approximately the form $\nabla \cdot f$ and $\nabla \wedge f$ (or, a little more closely, $\partial \cdot f$ and $\partial \wedge f$) is often used in the geometric algebra literature for divergence and curl (see [22] and [8]). One can see that it is suggested by the result above, though it is generally used for the left-hand versions of the operators.

Remark 7.7. If we have coordinates $\chi_1, \dots, \chi_p$ on $\mathcal{M}$, then we can, for example, write the right-hand divergence and curl as

$$\vec{\mathrm{div}}_{\mathcal{M}} f = \sum_{i=1}^{p} \frac{\partial f}{\partial \chi_i} \cdot_R d\chi_i \quad \text{and} \quad \vec{\mathrm{curl}}_{\mathcal{M}} f = \sum_{i=1}^{p} \frac{\partial f}{\partial \chi_i} \wedge d\chi_i. \tag{7.23}$$

Similarly for the left-hand operations.

One can prove a version of the product rule for curl:

Proposition 7.11. *Suppose that f and g are $\mathcal{C}^1$ multivector fields on a $\mathcal{C}^1$ p-manifold $\mathcal{M}$. If g is a grade k field, then*

$$\vec{\mathrm{curl}}_{\mathcal{M}}(f \wedge g) = (-1)^k (\vec{\mathrm{curl}}_{\mathcal{M}} f) \wedge g + f \wedge (\vec{\mathrm{curl}}_{\mathcal{M}} g).$$

If f is a grade k field, then

$$\overleftarrow{\mathrm{curl}}_{\mathcal{M}}(f \wedge g) = (\overleftarrow{\mathrm{curl}}_{\mathcal{M}} f) \wedge g + (-1)^k f \wedge (\overleftarrow{\mathrm{curl}}_{\mathcal{M}} g).$$

Proof. We establish only the first equation.

Let $(\chi_1, \ldots, \chi_p)$ be a local coordinate patch on $\mathcal{M}$.

$$\overrightarrow{\mathrm{curl}}_{\mathcal{M}}(f \wedge g) = \sum_{i=1}^{p} \frac{\partial (f \wedge g)}{\partial \chi_i} \wedge d\chi_i$$

$$= \sum_{i=1}^{p} \left(\frac{\partial f}{\partial \chi_i} \wedge g \wedge d\chi_i \right) + \sum_{i=1}^{p} \left(f \wedge \frac{\partial g}{\partial \chi_i} \wedge d\chi_i \right)$$

$$= (-1)^k \sum_{i=1}^{p} \left(\frac{\partial f}{\partial \chi_i} \wedge d\chi_i \right) \wedge g + f \wedge \sum_{i=1}^{p} \left(\frac{\partial g}{\partial \chi_i} \wedge d\chi_i \right)$$

$$= (-1)^k (\overrightarrow{\mathrm{curl}}_{\mathcal{M}} f) \wedge g + f \wedge (\overrightarrow{\mathrm{curl}}_{\mathcal{M}} g). \qquad \square$$

7.2.5   Some geometric derivatives

We will now spend a little time calculating geometric derivatives of functions that look as though they might be interesting or useful.

Identity functions on a manifold

We know that the identity function on a manifold $\mathcal{M}$ where $\mathcal{M} \subseteq \mathbb{R}^n$ is not necessarily unique when one considers its $\mathcal{C}^k$ extension to the $\mathbb{R}^n$. However geometric derivatives of the identity function are uniquely determined since they can be calculated using only points of $\mathcal{M}$.

Proposition 7.12. *Suppose that $\mathcal{M}$ is a $\mathcal{C}^1$ p-manifold and $I_{\mathcal{M}}$ is a $\mathcal{C}^1$ function which is the identity on $\mathcal{M}$. Then*

$$\overrightarrow{\nabla}_{\mathcal{M}} I_{\mathcal{M}} = \overleftarrow{\nabla}_{\mathcal{M}} I_{\mathcal{M}} = p.$$

Proof. Suppose we are carrying out our computations at a point $x_0 \in \mathcal{M}$. We choose an orthonormal basis $\{u_i\}_{i=1}^{p}$ for $T_{x_0} \mathcal{M}$ and we know it is its own reciprocal basis. From Proposition 6.1, we have $\partial_{u_i} I_{\mathcal{M}} = u_i$. It follows that

$$\overrightarrow{\nabla}_{\mathcal{M}} I_{\mathcal{M}} = \overleftarrow{\nabla}_{\mathcal{M}} I_{\mathcal{M}} = \sum_{i=1}^{p} |u_i|^2 = p. \qquad \square$$

Linear transformations

If V is a vector subspace of some $\mathbb{R}^n$ and $f \colon V \to V$ is a linear transformation, there is a nice connection between geometric derivatives and the *trace* of f.

If A is a square matrix with entries α_{ij}, by the *trace* of A, written $\mathrm{tr}(A)$, we mean the sum of its diagonal elements, $\sum_i \alpha_{ii}$. If A is the matrix of a linear transformation f, we can take $\mathrm{tr}(f)$ to be $\mathrm{tr}(A)$. Of course, f will be represented by an infinity of matrices because each particular A is tied to a choice of bases for the vector spaces involved. However it can be shown that if $f \colon V \to V$ and A is the matrix of f with respect to a basis $\{u_i\}$ while B is its matrix with respect to a second basis $\{v_i\}$, then

$\text{tr}(A) = \text{tr}(B)$. (See Exercise 1.) The trace of f is, in this sense, independent of its matrix representation.

Proposition 7.13. *If f is a linear transformation $f : V \to V$ where V is a p-dimensional vector subspace of $\mathbb{R}^n$, then*

$$\frac{1}{2}\overrightarrow{\nabla}_V f + \frac{1}{2}\overleftarrow{\nabla}_V f \;=\; \overrightarrow{\text{div}}_V f \;=\; \overleftarrow{\text{div}}_V f \;=\; tr(f).$$

Proof. Let $\{u_i\}_{i=1}^p$ be an orthonormal basis for V and suppose the matrix for f with respect to this basis has entries λ_{ij}. This amounts to

$$f(u_i) \;=\; \sum_{j=1}^p \lambda_{ji} u_j.$$

For any point $x \in V$, by Proposition 2.4, we have

$$\partial_{u_i} f(x) \;=\; [f'(x)]u_i \;=\; f(u_i).$$

Keeping in mind that

$$u_j u_i \;=\; u_j \cdot u_i + u_j \wedge u_i \;=\; \delta_{ji} + u_j \wedge u_i,$$

we calculate

$$\begin{aligned}
\frac{1}{2}\overrightarrow{\nabla}_V f + \frac{1}{2}\overleftarrow{\nabla}_V f &= \frac{1}{2}\sum_{i=1}^p (\partial_{u_i} f)\, u_i + \frac{1}{2}\sum_{i=1}^p u_i\, (\partial_{u_i} f) \\
&= \frac{1}{2}\sum_{i,j=1}^p \lambda_{ij} u_j u_i + \frac{1}{2}\sum_{i,j=1}^p \lambda_{ij} u_i u_j \\
&= \sum_{i=1}^p \lambda_{ii} + \frac{1}{2}\sum_{i \neq j} \lambda_{ji}\big((u_j \wedge u_i) + (u_i \wedge u_j)\big).
\end{aligned}$$

The second term of the last line is clearly zero, hence

$$\frac{1}{2}\overrightarrow{\nabla}_V f + \frac{1}{2}\overleftarrow{\nabla}_V f \;=\; \sum_{i=1}^p \lambda_{ii} \;=\; tr(f).$$

We leave it to the reader to show that

$$\overrightarrow{\text{div}}_V f \;=\; \overleftarrow{\text{div}}_V f \;=\; tr(f). \qquad \square$$

The function $\mathbf{x}/|\mathbf{x}|^{\mathbf{n}}$

We now introduce a special function $G(x)$ defined for all points x of $\mathbb{R}^n$ other than the origin:

$$G(x) \stackrel{\text{def.}}{=} \frac{x}{|x|^n}. \tag{7.24}$$

This is a useful function and will be a starring player in the Clifford-Cauchy integral formula. (See Proposition 7.25.)

We outline the relevant calculations and leave the details to the reader:

Assume we use the coordinates $(\chi_1, \ldots, \chi_n)$ for $x \in \mathbb{R}^n$. If we set $r = |x|$, then we can show that

$$\frac{\partial}{\partial \chi_i}(r^n) = nr^{n-2}\chi_i.$$

Using the fact that

$$G(x) = \frac{1}{r^n} \sum_{i=1}^{n} \chi_i e_i,$$

we calculate that

$$\frac{\partial G}{\partial \chi_i} = \sum_{j=1}^{n} \frac{\partial}{\partial \chi_i}\left(\frac{\chi_j}{r^n}\right) e_j$$

$$= \frac{1}{r^n} e_i - \frac{n\chi_i}{r^{n+2}} \sum_{j=1}^{n} \chi_j e_j$$

$$= \frac{1}{r^n}\left(e_i - \frac{n\chi_i}{r^2} x\right).$$

Then

$$\vec{\nabla} G = \sum_{i=1}^{n} \frac{\partial G}{\partial \chi_i} e_i$$

$$= \frac{n}{r^n} - \frac{n}{r^{n+2}} \sum_{i,j=1}^{n} \chi_i \chi_j e_j e_i$$

$$= \frac{n}{r^n} - \frac{n}{r^{n+2}} r^2 = 0.$$

A similar result holds for $\overleftarrow{\nabla} G$. Thus we have

Proposition 7.14. *If* $G(x) = x/|x|^n$ *on* $\mathbb{R}^n - \{0\}$, *then* $\vec{\nabla} G = \overleftarrow{\nabla} G = 0$.

This example hints at an important concept:

Definition 7.6. A multivector field f defined on an open subset of $\mathbb{R}^n$ is called *right holomorphic* provided $\vec{\nabla} f = 0$ and *left holomorphic* if $\overleftarrow{\nabla} f = 0$. (The terms *monogenic* and *analytic* are also used.)

The orientation and spur of a manifold

In a different direction, an obvious question is to ask whether we can calculate the rate of change of orientation of a manifold. We have in mind something roughly of the form $(\vec{\nabla}_{\mathcal{M}} w) w$ where w is the orientation of the manifold.

Example 7.12. Let $\mathcal{A}$ be an oriented $\mathcal{C}^1$ arc in $\mathbb{R}^n$ with a parametrization $x : [0, \alpha] \to \mathcal{A}$ that agrees with the orientation $w = w_{\mathcal{A}}$ of $\mathcal{A}$. We may, without loss of generality, suppose that the parameter σ of $x = x(\sigma)$ is arclength. Then $|dx/d\sigma| = 1$ and

$$w = w_{\mathcal{A}} = \frac{\frac{dx}{d\sigma}}{\left|\frac{dx}{d\sigma}\right|} = \frac{dx}{d\sigma}.$$

To compute $\vec{\nabla}_{\mathcal{A}} w$ we need a basis for the tangent space at an arbitrary point x_0; this is just $\{w(x_0)\} = \{(dx/d\sigma)(\sigma_0)\}$ where $x_0 = x(\sigma_0)$. The reciprocal basis is $\{w(x_0)\}$. Using Proposition 6.3, we see that

$$\vec{\nabla}_{\mathcal{A}} w(x_0) = \left[(\partial_{w(x_0)} w)(x_0)\right] w(x_0)$$
$$= \frac{d^2 x}{d\sigma^2}(\sigma_0) \, w(x_0).$$

So

$$\vec{\nabla}_{\mathcal{A}} w = \frac{d^2 x}{d\sigma^2} \, w.$$

If we think of $x = x(\sigma)$ as describing the motion of a particle moving with a constant speed of 1 along the arc $\mathcal{A}$, then the rate of change of w with respect to w is

$$(\vec{\nabla}_{\mathcal{A}} w) \, w = \frac{d^2 x}{d\sigma^2} = \text{acceleration.}$$

It is easily checked that at every point of $\mathcal{A}$, this is a vector that is orthogonal to $\mathcal{A}$.

Example 7.13. Suppose $\rho_0 > 0$ and $m \geq 1$. We fix a point x_0 on the sphere $\rho_0 S^m$ which is centered on the origin of $\mathbb{R}^{m+1}$ and has radius ρ_0. We wish to compute the rate of change of the orientation of the sphere at x_0.

We first specify the orientation of $\rho_0 S^m$. (See Figure 7.9.)

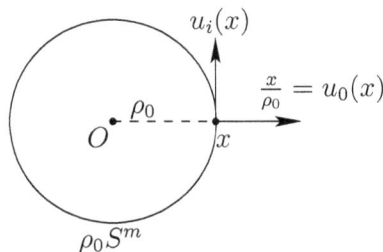

Figure 7.9: Setting up a frame at $x \in \rho_0 S^m$

We know that $\rho_0 S^m \subseteq \mathbb{R}^{m+1}$ and every $\mathbb{R}^k$ has a standard orientation. For every $x \in \rho_0 S^m$, we know that x/ρ_0 is a unit vector that points outward from $\rho_0 S^m$ and is orthogonal to the sphere at x. Let us set $u_0(x) = x/\rho_0$. We next find an orthonormal basis $\{u_i(x)\}_{i=1}^m$ for $T_x\rho_0 S^m$, the tangent space to the sphere at x. It is automatically true that $\{u_i(x)\}_{i=0}^m$ is an orthonormal basis for $\mathbb{R}^{m+1}$. Further, we may make our choice in such a way that $u = u_0 u_1 \cdots u_m$ is the standard orientation for $\mathbb{R}^{m+1}$, namely, $u = e_1 e_2 \cdots e_{m+1}$.

We know that $\rho_0 S^m$ is the boundary of the ball $\rho_0 B^{m+1}$ and u is the natural orientation for $\rho_0 B^{m+1}$. We take the orientation w of $\rho_0 S^m$ to be the orientation induced on the boundary of the solid ball $\rho_0 B^{m+1}$. From our discussion of induced orientation and its relation to the unit outward normal vector on the boundary of a manifold (see Equation (6.18)), this means that

$$w(x) = \frac{x}{\rho_0} u = u_1(x) \cdots u_m(x). \qquad (7.25)$$

We see from (7.25) that w extends to the linear map $x \mapsto (x/\rho_0)u$ on $\mathbb{R}^{m+1}$. Therefore we can use an argument like that in the proof of Proposition 2.4 to compute

$$\partial_{u_i(x_0)} w(x_0) = \frac{u_i(x_0)}{\rho_0} u.$$

Next, if we introduce $\mathcal{M}$ as another name for $\rho_0 S^m$, and since $\{u_i(x_0)\}_{i=1}^m$ is its own reciprocal frame, it follows that

$$\left(\vec{\nabla}_{\mathcal{M}} w\right)(x_0) = \sum_{i=1}^m \frac{u_i(x_0)}{\rho_0} u u_i(x_0) = (-1)^m \frac{m}{\rho_0} u.$$

Let us now introduce the symbol

$$N(x) = \frac{x}{\rho_0}$$

to emphasize the fact that x/ρ_0 is the unit outward normal vector from $\rho_0 S^m$ at the point x. We then see that the geometric derivative of w at x_0 is

$$\left(\vec{\nabla}_{\mathcal{M}} w\right)(x_0) = (-1)^m \frac{m}{\rho_0} N(x_0) w(x_0).$$

Finally, we compute the rate of change of w with respect to $w^\dagger$ on the sphere $\mathcal{M} = \rho_0 S^m$:

$$\left(\vec{\nabla}_{\mathcal{M}} w\right) w^\dagger = (-1)^m \frac{m}{\rho_0} N.$$

(We compute it with respect to $w^\dagger$ rather than w because the result is simpler.)

Remark 7.8. The reader may be bothered by the fact that the rate of change of the orientation of a sphere, considered as a vector, points sometimes into and sometimes out of the sphere, depending on the dimension. This is an artifact of our choice to work with $w^\dagger$. We can calculate the rate of change of w with respect to $(-1)^k w$ for any k that catches our fancy. For example, if we draw some pictures of a tangent line rolling around a circle or a tangent plane rolling around on a 2-sphere, we may well be led to feel that the vector that represents rate of change on a sphere should always point inward. We can obtain that effect by remembering that N is the unit vector that points outward and then calculating the rate of change of w with respect to $(-1)^{m-1} w^\dagger$:

$$\left(\vec{\nabla}_{\rho_0 S^m} w\right)(-1)^{m-1} w^\dagger = -\frac{m}{\rho_0} N.$$

In our last two examples, the rate of change of orientation of a manifold in the sense of something like $\left(\vec{\nabla}_{\mathcal{M}} w\right) w$ is a 1-vector orthogonal to $\mathcal{M}$. It turns out this is always true. Furthermore, the vector has the magnitude of the *mean curvature* of the manifold, a fundamental geometric property.

Let us go into some details though the actual proof will be tucked into Appendix A.

In what follows, we assume $\mathcal{M}$ is a $\mathcal{C}^2$ p-manifold in $\mathbb{R}^n$ and that $\mathcal{M}$ has orientation w. We further suppose the existence of a $\mathcal{C}^2$ chart $x : U \to \mathcal{M}$ (where U is an open subset of $\mathbb{R}^p$) which agrees with w and induces coordinates $(\tau_1, \ldots, \tau_p)$ on at least a portion of $\mathcal{M}$.

From the chart x, we construct a frame of independent vectors $\{\partial x / \partial \tau_i\}_{i=1}^p$ which serves as a basis for each tangent space $T_{x_0} \mathcal{M}$ for which x_0 is in the range of x. The corresponding reciprocal frame is $\{d\tau_i\}_{i=1}^p$. We can compute the rate of change of a tangent vector $\partial x / \partial \tau_i$ by taking second derivatives, $\partial^2 x / \partial \tau_j \partial \tau_i$. It turns out that what is of interest to us is the component of such second derivatives which is orthogonal to $\mathcal{M}$:

$$\left.\frac{\partial^2 x}{\partial \tau_i \partial \tau_j}\right|_{\perp \mathcal{M}} \stackrel{\text{def.}}{=} \left(\frac{\partial^2 x}{\partial \tau_i \partial \tau_j} \wedge w\right) w^\dagger. \tag{7.26}$$

(See Proposition 5.21 for this formula.) When we say that it is orthogonal to $\mathcal{M}$, we mean, of course, that for any $x_0 \in \mathcal{M}$ at which we can perform the relevant calculations, the vector $\left((\partial^2 x / \partial \tau_j \partial \tau_i)|_{\perp \mathcal{M}}\right)(x_0)$ is orthogonal to $T_{x_0} \mathcal{M}$.

We are then able to prove the following:

Proposition 7.15. *Let $\mathcal{M}$ be a $\mathcal{C}^2$ p-manifold in $\mathbb{R}^n$ with orientation w. Using the notation and coordinates above, we have*

1. $\dfrac{\partial w}{\partial \tau_i} = \displaystyle\sum_{j=1}^{p} \left(\left.\frac{\partial^2 x}{\partial \tau_i \partial \tau_j}\right|_{\perp \mathcal{M}}\right)(d\tau_j) w$ *and*

2. $\vec{\nabla}_{\mathcal{M}} w = (-1)^{p-1} \left(\displaystyle\sum_{i,j=1}^{p} \left(\left.\frac{\partial^2 x}{\partial \tau_i \partial \tau_j}\right|_{\perp \mathcal{M}}\right)(d\tau_i \cdot d\tau_j)\right) w.$

We give the proof in Section A.7 of Appendix A.

The proposition suggests the following definition:

Definition 7.7. Let $\mathcal{M}$ be a $\mathcal{C}^2$ p-manifold in $\mathbb{R}^n$ with orientation w, and suppose that $(\tau_1, \ldots, \tau_p)$ are local coordinates induced on $\mathcal{M}$ by a $\mathcal{C}^2$ parametrization $x : U \to \mathcal{M}$ that agrees with w. Then on the coordinate patch $(\tau_1, \ldots, \tau_p)$,

$$N_{\mathcal{M}} \stackrel{\text{def.}}{=} \sum_{i,j=1}^{p} \left.\frac{\partial^2 x}{\partial \tau_i \partial \tau_j}\right|_{\perp \mathcal{M}} (d\tau_i \cdot d\tau_j). \tag{7.27}$$

An equivalent global definition is

$$N_{\mathcal{M}} \stackrel{\text{def.}}{=} (-1)^{p-1}\big(\vec{\nabla}_{\mathcal{M}} w\big)w^{\dagger} = -w^{\dagger}\big(\vec{\nabla}_{\mathcal{M}} w\big). \tag{7.28}$$

$N_{\mathcal{M}}$ is called the *spur* of $\mathcal{M}$. (See [22] for an extensive discussion of the spur and the related concept of the shape operator.)

From (7.27), we see that $N_{\mathcal{M}}$ is a 1-vector that is orthogonal to $\mathcal{M}$ at its base point and that by Equation (7.28) it is independent of our parametrization x of $\mathcal{M}$. The second result of Proposition 7.15 is just $\vec{\nabla}_{\mathcal{M}} w = (-1)^{p-1}N_{\mathcal{M}} w$, so that gives us the first equality in (7.28). Since $N_{\mathcal{M}}$ is orthogonal to the p-blade w, we have $N_{\mathcal{M}} w = (-1)^p w N_{\mathcal{M}}$, and from that, we get the second equality in (7.28).

Notice that if we replace w by $-w$ in (7.28), we still get the same value of $N_{\mathcal{M}}$, so $N_{\mathcal{M}}$ is independent of our choice of orientation. Indeed, $N_{\mathcal{M}}$ exists even if $\mathcal{M}$ is a non-orientable manifold since one can always define an orientation *locally*; the fact that (7.28) is invariant under change of sign of w shows that even in this case, $N_{\mathcal{M}}$ exists and is unique.

Equation (7.28) relates $N_{\mathcal{M}}$ to the right-hand geometric derivative. Applying properties of reversion and the fact that we can replace w by $w^{\dagger}$ in (7.28) gives us the relation to the left-hand geometric derivative:

$$N_{\mathcal{M}} = (-1)^{p-1} w^{\dagger}\big(\overleftarrow{\nabla}_{\mathcal{M}} w\big) = -\big(\overleftarrow{\nabla}_{\mathcal{M}} w\big)w^{\dagger}. \tag{7.29}$$

Equation (7.28) tells us that we may interpret $N_{\mathcal{M}}$ as a rate of change of w with respect to some orientation of $\mathcal{M}$.

Example 7.14. If $\mathcal{M}$ is a vector subspace of $\mathbb{R}^n$ with orientation w, since w is a constant, we have $\vec{\nabla}_{\mathcal{M}} w = 0$. Thus $N_{\mathcal{M}} = 0$.

Example 7.15. Let $\mathcal{A}$ be an oriented $\mathcal{C}^1$ arc in $\mathbb{R}^n$ with a parametrization $x : [0, \alpha] \to \mathcal{A}$ that agrees with the orientation $w = w_{\mathcal{A}}$ of $\mathcal{A}$. Let σ be arclength measured along $\mathcal{A}$ from the initial point. We may interpret this setup as a particle moving along the arc with a speed of 1; in this case, σ is both arclength and time. From Example 7.12, we can see that

$$N_{\mathcal{A}} = -w\big(\vec{\nabla}_{\mathcal{A}} w\big) = \frac{d^2 x}{d\sigma^2} = \text{acceleration.}$$

Notice that since speed is a constant here, acceleration has nothing to do with change of speed but rather with how rapidly the particle is changing direction.

Example 7.16. Let ρ_0 be a positive real number. By $\rho_0 S^{n-1}$ we mean the unit sphere in $\mathbb{R}^n$ with the origin as center and radius ρ_0. We set $\mathcal{M} = \rho_0 S^{n-1}$ and following the setup of Example 7.13, we take the orientation w of $\mathcal{M}$ at an arbitrary point x to be

$$w(x) = \frac{x}{\rho_0} e_1 \cdots e_n.$$

From Example 7.13, we know that

$$(\vec{\nabla}_{\mathcal{M}} w)(x) = (-1)^{n-1} \frac{(n-1)x}{\rho_0^2} w(x) = w(x) \frac{(n-1)x}{\rho_0^2}.$$

It can thus be seen that

$$N_{\mathcal{M}}(x) = -w(x)^\dagger (\vec{\nabla}_{\mathcal{M}} w)(x) = -\frac{(n-1)x}{\rho_0^2}.$$

Exercises 7.2.

1. Suppose that $f: V \to V$ is a linear transformation of a p-dimensional vector subspace of $\mathbb{R}^n$ into itself. Let $\{u_i\}_{i=1}^p$ and $\{v_i\}_{i=1}^p$ be two bases for V. Suppose that $A = (\alpha_{ij})_{p \times p}$ is the matrix of f with respect to $\{u_i\}_{i=1}^p$ and $B = (\beta_{ij})_{p \times p}$ is the matrix with respect to $\{v_i\}_{i=1}^p$.

   (a) Show there is a $p \times p$ matrix T such that $B = TAT^{-1}$.
   (b) Show that $\operatorname{tr}(A) = \operatorname{tr}(B)$.

2. Let $f: \mathbb{R}^2 \to \mathbb{G}^2$ be the function $f(\chi_1, \chi_2) = \chi_1 + \chi_2 e_1 e_2$. Show that f is left holomorphic but not right holomorphic.

3. Suppose that U is an open subset of $\mathbb{R}^n$ and f is a $\mathcal{C}^1$ multivector-valued function on U, $f: U \to \mathbb{G}^n$. Let U have the orientation $w = e_1 \cdots e_n$. Show the following:

   (a) f is right holomorphic on U if and only if $\int_{\partial \mathcal{M}} f \, \vec{\partial} w = 0$ for every $\mathcal{C}^2$ n-cell $\mathcal{M} \subseteq U$.
   (b) f is left holomorphic on U if and only if $\int_{\partial \mathcal{M}} \overleftarrow{\partial} w \, f = 0$ for every $\mathcal{C}^2$ n-cell $\mathcal{M} \subseteq U$.

   (Hint: Corollary 7.3.)

4. Let a be a k-blade in $\mathbb{R}^n$ and suppose that $x \in \mathbb{R}^n$. Let $f: \mathbb{R}^n \to \mathbb{R}^n$ be the linear transformation $f(x) = axa^{-1}$. Find the trace of f. (Hint: Look at the answer to Exercise 6 in Section 5.4.)

5. Let $\mathcal{N}$ be the manifold $\mathbb{R}^3$ with parametrization $y(\chi_1, \chi_2, \chi_3) = \sum_{i=1}^3 \chi_i e_i$. We then have

$$\frac{\partial y}{\partial \chi_i} = e_i = d\chi_i$$

at every point of $\mathbb{R}^3$. Let $\mathcal{M}$ be the unit 2-sphere centered at the origin, that is, the submanifold defined by $\chi_1^2 + \chi_2^2 + \chi_3^2 = 1$. We place coordinates (θ, ϕ) on $\mathcal{M}$ by the map

$$x(\theta, \phi) = \sin(\theta) \cos(\phi) e_1 + \sin(\theta) \sin(\phi) e_2 + \cos(\theta) e_3.$$

(This is not a real coordinatization unless one restricts the values of (θ, ϕ) to certain sets to insure x is one-to-one and x^{-1} is $\mathcal{C}^1$.) Use Proposition 7.7 to compute $\partial \chi_i / \partial \theta$, $\partial \chi_i / \partial \phi$, and $P_{\mathcal{M}}(d\chi_i)$ on $\mathcal{M}$ as functions of θ and ϕ for $i = 1, 2, 3$.

6. Let $\mathcal{M}$ be the 1-manifold $\chi_1^2 + \chi_2^2 = 1$ in $\mathbb{R}^2$ and let f be a vector field on $\mathcal{M}$ whose value at $x_0 \in \mathcal{M}$ is $f(x_0) = \phi_1(x_0)e_1 + \phi_2(x_0)e_2$ where ϕ_1, ϕ_2 are real-valued functions. Show that at every point $x_0 = (\chi_{01}, \chi_{02})$ of $\mathcal{M}$ we have

$$
\begin{aligned}
P_{\mathcal{M}}(f)(x_0) = {} & \left(\phi_1(x_0)\chi_{02}^2 - \phi_2(x_0)\chi_{01}\chi_{02} \right) e_1 \\
& + \left(-\phi_1(x_0)\chi_{01}\chi_{02} + \phi_2(x_0)\chi_{01}^2 \right) e_2.
\end{aligned}
$$

7. Let $\mathcal{M}$ be the unit 2-sphere centered at the origin in $\mathbb{R}^3$. It is thus defined by the equation $\chi_1^2 + \chi_2^2 + \chi_3^2 = 1$ where (χ_1, χ_2, χ_3) are the standard coordinates of $\mathbb{R}^3$. Consider the constant vector field e_1 on $\mathbb{R}^3$. Given a point $x = (\chi_1, \chi_2, \chi_3)$ on $\mathcal{M}$, find $P_{\mathcal{M}}(e_1)(x)$ in terms of χ_1, χ_2, χ_3.

8. Derive the curl formula of Proposition 7.9.

9. Let $\mathcal{M}$ be the 2-manifold in $\mathbb{R}^3$ consisting of the points (χ_1, χ_2, χ_3) that satisfy $\chi_3 = e^{\chi_1}$. Find the spur $N_{\mathcal{M}}$.

10. Given a $\mathcal{C}^1$ p-manifold $\mathcal{M}$ and a $\mathcal{C}^1$ identity function $I_{\mathcal{M}}$ on $\mathcal{M}$, what are $\overrightarrow{\operatorname{div}}_{\mathcal{M}}(I_{\mathcal{M}})$ and $\overrightarrow{\operatorname{curl}}_{\mathcal{M}}(I_{\mathcal{M}})$?

11. Suppose we have a $\mathcal{C}^2$ p-manifold $\mathcal{M}$ and a chart x that induces coordinates $\chi_1, \dots, \chi_p$ on $\mathcal{M}$. Then $\{\partial x/\partial \chi_i\}_{i=1}^{p}$ is a frame of tangent vectors on $\mathcal{M}$. Show that for the tangent vector field $\partial x/\partial \chi_i$ we have

$$
\overrightarrow{\operatorname{div}}_{\mathcal{M}}\left(\frac{\partial x}{\partial \chi_i} \right) = \frac{\partial}{\partial \chi_i}\left[\ln\left(\left| \frac{\partial x}{\partial \chi_1} \wedge \cdots \wedge \frac{\partial x}{\partial \chi_p} \right| \right) \right].
$$

12. If f is a $\mathcal{C}^1$ multivector field on the $\mathcal{C}^1$ manifold $\mathcal{M}$ and c is a constant multivector, show that

$$
\begin{aligned}
\overrightarrow{\operatorname{curl}}_{\mathcal{M}}(cf) &= c\left(\overrightarrow{\operatorname{curl}}_{\mathcal{M}}f \right), & \overrightarrow{\operatorname{curl}}_{\mathcal{M}}(fc) &= \left(\overrightarrow{\operatorname{curl}}_{\mathcal{M}}f \right) c, \\
\overrightarrow{\operatorname{div}}_{\mathcal{M}}(cf) &= c\left(\overrightarrow{\operatorname{div}}_{\mathcal{M}}f \right), & \overrightarrow{\operatorname{div}}_{\mathcal{M}}(fc) &= \left(\overrightarrow{\operatorname{div}}_{\mathcal{M}}f \right) c.
\end{aligned}
$$

7.3 A taste of analysis and geometry

Now that we have built some machinery, we shall delve a little further into the depths of geometric calculus. We shall consider an alternate form of the Fundamental Theorem of geometric calculus, one in which the spur of a manifold, $N_{\mathcal{M}}$, surprisingly appears; we shall say a little more about divergence and curl; and we shall exhibit a connection between the spur of a manifold and the classical concept of mean curvature.

7.3.1 An alternate form of the Fundamental Theorem

Let $\mathcal{M}$ be a $\mathcal{C}^2$ p-cell in $\mathbb{R}^n$ with orientation w. Suppose f and g are $\mathcal{C}^1$ multivector fields on $\mathcal{M}$. Then the standard, right-handed version of the Fundamental Theorem of geometric calculus (Theorem 6.1) is this:

$$\int_{\partial \mathcal{M}} f(\overrightarrow{\partial} w) g = \int_{\mathcal{M}} (\overrightarrow{\nabla}_{\mathcal{M}} f) w g + (-1)^{p-1} \int_{\mathcal{M}} f w (\overleftarrow{\nabla}_{\mathcal{M}} g). \qquad (7.30)$$

Recall from Equation (6.18) that the outward normal unit vector n to $\partial \mathcal{M}$ is

$$n = (\overrightarrow{\partial} w) w^\dagger = w^\dagger (\overleftarrow{\partial} w). \qquad (7.31)$$

(It should be kept in mind that although n is normal (orthogonal) to $\partial \mathcal{M}$, it is tangent to $\mathcal{M}$.) By $N_{\mathcal{M}}$, the spur of $\mathcal{M}$, we mean (see Equation (7.28))

$$N_{\mathcal{M}} = -w^\dagger (\overrightarrow{\nabla}_{\mathcal{M}} w) = (-1)^{p-1} (\overrightarrow{\nabla}_{\mathcal{M}} w) w^\dagger. \qquad (7.32)$$

$N_{\mathcal{M}}$ is a $\mathcal{C}^1$ 1-vector field such that $N_{\mathcal{M}}(x_0)$ is orthogonal to $T_{x_0}^1 \mathcal{M}$ when $x_0 \in \mathcal{M}$ and can be interpreted as a rate of change of w over $\mathcal{M}$.

Proposition 7.16. *Given the hypotheses above, Equation (7.30), the right-handed version of the Fundamental Theorem for a $\mathcal{C}^2$ cell, is equivalent to*

$$\int_{\partial \mathcal{M}} f n g = \int_{\mathcal{M}} \left((\overrightarrow{\nabla}_{\mathcal{M}} f) g + f (\overleftarrow{\nabla}_{\mathcal{M}} g) + f N_{\mathcal{M}} g \right). \qquad (7.33)$$

Proof. Substitute $w^\dagger g$ for g in (7.30). This gives us

$$\int_{\partial \mathcal{M}} f n g = \int_{\mathcal{M}} (\overrightarrow{\nabla}_{\mathcal{M}} f) g + (-1)^{p-1} \int_{\mathcal{M}} f w [\overleftarrow{\nabla}_{\mathcal{M}} (w^\dagger g)]. \qquad (7.34)$$

We calculate that

$$\begin{aligned}
w[\overleftarrow{\nabla}_{\mathcal{M}} (w^\dagger g)] &= w \sum_{i=1}^{p} d\tau_i \frac{\partial (w^\dagger g)}{\partial \tau_i} \\
&= w \sum_{i=1}^{p} d\tau_i \left(\frac{\partial w}{\partial \tau_i} \right)^\dagger g + w \sum_{i=1}^{p} d\tau_i w^\dagger \left(\frac{\partial g}{\partial \tau_i} \right) \\
&= \left(\sum_{i=1}^{p} \left(\frac{\partial w}{\partial \tau_i} \right) d\tau_i w^\dagger \right)^\dagger g + (-1)^{p-1} w w^\dagger \sum_{i=1}^{p} d\tau_i \left(\frac{\partial g}{\partial \tau_i} \right) \\
&= \left((\overrightarrow{\nabla}_{\mathcal{M}} w) w^\dagger \right)^\dagger g + (-1)^{p-1} \overleftarrow{\nabla}_{\mathcal{M}} g \\
&= (-1)^{p-1} N_{\mathcal{M}} g + (-1)^{p-1} \overleftarrow{\nabla}_{\mathcal{M}} g.
\end{aligned}$$

Substituting this result into (7.34) gives us (7.33). $\qquad \square$

Given this result, we deduce a pointwise version and a relation between $N_{\mathcal{M}}$ and the outward unit normal n to the boundary of $\mathcal{M}$:

Corollary 7.4. *Assuming the hypotheses of Proposition 7.16, the following holds for a $\mathcal{C}^2$ cell $\mathcal{M}$:*

1. *If x_0 is an interior point of $\mathcal{M}$, then*

$$\left[\left(\vec{\nabla}_{\mathcal{M}}f\right)(x_0)\right]g(x_0) + f(x_0)\left[\left(\overleftarrow{\nabla}_{\mathcal{M}}g\right)(x_0)\right] + f(x_0)N_{\mathcal{M}}(x_0)g(x_0)$$

$$= \lim_{\mathcal{M}\to x_0} \frac{1}{vol(\mathcal{M})} \int_{\partial\mathcal{M}} fng. \tag{7.35}$$

2. *If x_0 is an interior point of $\mathcal{M}$, then*

$$N_{\mathcal{M}}(x_0) = \lim_{\mathcal{M}\to x_0} \frac{1}{vol(\mathcal{M})} \int_{\partial\mathcal{M}} n.$$

Proof. 1. We simply apply to (7.33) the result (Proposition 7.8) that for h continuous on $\mathcal{M}$ we have

$$h(x_0) = \lim_{\mathcal{M}\to x_0} \frac{1}{vol(\mathcal{M})} \int_{\mathcal{M}} h.$$

2. Set $f = g = 1$ in (7.35). □

Remark 7.9. We showed in Proposition 7.16 that Equation (7.33) is equivalent to the right-hand version of the Fundamental Theorem, and it must also be equivalent to the left-hand version. Yet Equation (7.33) has no "handedness."

7.3.2 Some intuition about divergence and curl

We can use the results of the last section to get a little more of a geometric picture of divergence and curl. We first derive formulas for these quantities:

Proposition 7.17. *Suppose that f is a $\mathcal{C}^1$ multivector field on a $\mathcal{C}^2$ p-cell $\mathcal{M}$. Let n be the unit outward normal vector to $\partial\mathcal{M}$, $N_{\mathcal{M}}$ the spur of $\mathcal{M}$, and x_0 an interior point of $\mathcal{M}$. Then*

$$\left(\overrightarrow{\text{div}}_{\mathcal{M}}f\right)(x_0) + f(x_0)\cdot_R N_{\mathcal{M}}(x_0) = \lim_{\mathcal{M}\to x_0} \frac{1}{vol(\mathcal{M})} \int_{\partial\mathcal{M}} f\cdot_R n,$$

$$\left(\overrightarrow{\text{curl}}_{\mathcal{M}}f\right)(x_0) + f(x_0)\wedge N_{\mathcal{M}}(x_0) = \lim_{\mathcal{M}\to x_0} \frac{1}{vol(\mathcal{M})} \int_{\partial\mathcal{M}} f\wedge n. \tag{7.36}$$

Proof. Recall that if x is a local parametrization of $\mathcal{M}$ with associated coordinates $(\tau_1,\ldots,\tau_p)$, then

$$\overrightarrow{\text{div}}_{\mathcal{M}}f = \sum_{i=1}^{p} \frac{\partial f}{\partial \tau_i}\cdot d\tau_i,$$

$$\overrightarrow{\text{curl}}_{\mathcal{M}}f = \sum_{i=1}^{p} \frac{\partial f}{\partial \tau_i}\wedge d\tau_i,$$

$$\vec{\nabla}_{\mathcal{M}}f = \overrightarrow{\text{div}}_{\mathcal{M}}f + \overrightarrow{\text{curl}}_{\mathcal{M}}f.$$

We also remember that $ab = a \cdot_R b + a \wedge b$ whenever b is a 1-vector. Now set $g = 1$ in Equation (7.35) and notice that we have

$$
\begin{aligned}
&\left(\overrightarrow{\mathrm{div}}_{\mathcal{M}} f \right)(x_0) + f(x_0) \cdot_R N_{\mathcal{M}}(x_0) \\
&+ \left(\overrightarrow{\mathrm{curl}}_{\mathcal{M}} f \right)(x_0) + f(x_0) \wedge N_{\mathcal{M}}(x_0) \\
&= \lim_{\mathcal{M} \to x_0} \frac{1}{\mathrm{vol}(\mathcal{M})} \int_{\partial \mathcal{M}} (f \cdot_R n) + \lim_{\mathcal{M} \to x_0} \frac{1}{\mathrm{vol}(\mathcal{M})} \int_{\partial \mathcal{M}} (f \wedge n).
\end{aligned} \tag{7.37}
$$

We know that the grade map $h \mapsto \langle h \rangle_k$ is linear and that the operations of concern to us behave "nicely" with respect to the grade operation. For example,

$$
\left\langle \frac{\partial h}{\partial \tau_i} \right\rangle_k = \frac{\partial \langle h \rangle_k}{\partial \tau_i} \quad \text{and} \quad \left\langle \int_{\mathcal{M}} h \right\rangle_k = \int_{\mathcal{M}} \langle h \rangle_k.
$$

Then because we can decompose f thus, $f = \sum_{k=0}^{n} \langle f \rangle_k$, it suffices to establish our result in the case where $f = \langle f \rangle_k$. We construct a table showing various expressions and their grades:

f	$\overrightarrow{\mathrm{div}}_{\mathcal{M}} f$	$\overrightarrow{\mathrm{curl}}_{\mathcal{M}} f$	$f \cdot_R N_{\mathcal{M}}$	$f \wedge N_{\mathcal{M}}$	$f \cdot_R n$	$f \wedge n$
k	$k-1$	$k+1$	$k-1$	$k+1$	$k-1$	$k+1$

Equating terms of the same grade in (7.37) yields the equations of (7.36). $\square$

We need one more step to get a pictorial glimpse of the meaning of divergence and curl:

Corollary 7.5. *Suppose that f is a $\mathcal{C}^1$ multivector field on a $\mathcal{C}^2$ p-cell $\mathcal{M}$ in $\mathbb{R}^p$, that n is the unit outward normal vector to the boundary of $\mathcal{M}$, and that x_0 is an interior point of $\mathcal{M}$. Then*

$$
\begin{aligned}
\mathrm{div}_{\mathcal{M}} f(x_0) &= \lim_{\mathcal{M} \to x_0} \frac{1}{vol(\mathcal{M})} \int_{\partial \mathcal{M}} f \cdot_R n, \\
\mathrm{curl}_{\mathcal{M}} f(x_0) &= \lim_{\mathcal{M} \to x_0} \frac{1}{vol(\mathcal{M})} \int_{\partial \mathcal{M}} f \wedge n.
\end{aligned} \tag{7.38}
$$

Proof. Since $\mathcal{M}$ and $\mathbb{R}^p$ have the same dimension, $N_{\mathcal{M}}$ must be orthogonal to $\mathbb{R}^p$. This is only possible if $N_{\mathcal{M}} = 0$. Equation (7.36) immediately gives the desired result. $\square$

The limits of (7.38) are very suggestive to the imagination. Think of f as being decomposed into components that are orthogonal and parallel to $\partial \mathcal{M}$; denote them $f_{\perp}$ and $f_{\parallel}$ respectively. (This is certainly true if f is a 1-vector field but a little trickier if it is a general multivector field. See Figure 7.10.) Let us think of f being associated with fluid flow.

At every point x_0 on the boundary of $\mathcal{M}$, $f(x_0) \cdot_R n(x_0) = f_{\perp}(x_0) \cdot_R n(x_0)$, and this dot product can be thought of as a measure of the tendency of the fluid to spray out of $\mathcal{M}$ or into it. (See Figure 7.11.) Then we see that $\mathrm{div}_{\mathcal{M}} f(x_0)$ is a measure of the change,

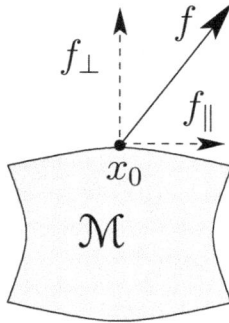

Figure 7.10: Decomposition of f Figure 7.11: f and n

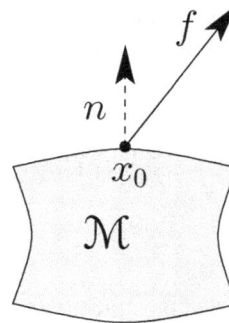

on an "infinitesimal" cell containing x_0, of that tendency of the fluid to "diverge out" or "diverge in."

In a similar way, we see that $f(x_0) \wedge n(x_0) = f_{\parallel}(x_0) \wedge n(x_0)$ is a measure of the tendency of the fluid flow to "hug" the boundary of $\mathcal{M}$ at x_0. Thus $\mathrm{curl}_{\mathcal{M}} f(x_0)$ is a measure of the change, on an "infinitesimal" cell containing x_0, of that tendency of the fluid to "hug" the boundary of the cell.

7.3.3 Mean curvature and the spur

The Weingarten map and mean curvature

Suppose that $\mathcal{M}$ is a $\mathcal{C}^2$ p-manifold in $\mathbb{R}^{p+1}$ and $\mathcal{M}$ has orientation w. (When a manifold $\mathcal{M}$ lies in a space or manifold of one dimension higher, it is usually described as a *hypersurface*, though we have not followed this dimensional convention when talking of hyperplanes.)

We assume the existence of a $\mathcal{C}^2$ unit vector field n on $\mathcal{M}$ such that n is orthogonal to $\mathcal{M}$ at every point where it is defined. More specifically, given any $x_0 \in \mathcal{M}$, then $|n(x_0)| = 1$ and $n(x_0)$ is orthogonal to the tangent space $T_{x_0}\mathcal{M}$. (Later we shall give an explicit construction of n.)

We now describe the *Weingarten map* and use it to define the mean curvature of $\mathcal{M}$. Let x_0 be a point of $\mathcal{M}$ and suppose that $v \in T_{x_0}\mathcal{M}$. The Weingarten map is defined on $T_{x_0}\mathcal{M}$ and is given by

$$W(v) \stackrel{\text{def.}}{=} \partial_v n(x_0). \qquad (7.39)$$

We know that $1 = n \cdot n$, and if we apply the directional derivative operator ∂_v to the last equation (under the assumption that all calculations are being carried out at the point x_0), then we obtain

$$0 = 2n \cdot (\partial_v n) = 2n \cdot W(v).$$

Since n is orthogonal to $\mathcal{M}$ and the manifold is of one dimension less than the ambient $\mathbb{R}^{p+1}$, it follows that $W(v)$ must be tangent to $\mathcal{M}$ at x_0. Thus we have a Weingarten map $W: T_{x_0}\mathcal{M} \to T_{x_0}\mathcal{M}$ for every $x_0 \in \mathcal{M}$.

Since the directional derivative operator ∂_v is linear in v, it follows that W is a linear transformation of each tangent space into itself.

We next show that W is *self-adjoint*.

Let $x: U \to \mathbb{R}^{p+1}$ be a $\mathcal{C}^2$ p-chart on $\mathcal{M}$ that covers x_0. We assume x induces coordinates $(\chi_1, \ldots, \chi_p)$ on $\mathcal{M}$ and we set $u_i = \partial x / \partial \chi_i$. This gives us a tangent frame on an open neighborhood of x_0 in $\mathcal{M}$. If we recall Remark 6.1, we see that

$$\partial_{u_j} u_i = \frac{\partial^2 x}{\partial \chi_j \partial \chi_i}.$$

Each u_i is tangent to $\mathcal{M}$, so $n \cdot u_i = 0$. If we apply the directional derivative operator ∂_{u_j} to this last equation, we obtain

$$0 = \left(\partial_{u_j} n\right) \cdot u_i + n \cdot \left(\partial_{u_j} u_i\right).$$

From this we deduce

$$
\begin{aligned}
\left(\partial_{u_j} n\right) \cdot u_i &= -n \cdot \left(\partial_{u_j} u_i\right) \\
&= -n \cdot \frac{\partial^2 x}{\partial \chi_j \partial \chi_i} \\
&= -n \cdot \frac{\partial^2 x}{\partial \chi_i \partial \chi_j} \\
&= \left(\partial_{u_i} n\right) \cdot u_j.
\end{aligned}
$$

It follows that for arbitrary u_i and u_j we have

$$W(u_i) \cdot u_j = \left(\partial_{u_i} n\right) \cdot u_j = \left(\partial_{u_j} n\right) \cdot u_i = W(u_j) \cdot u_i.$$

We conclude by linearity that $W(u) \cdot v = u \cdot W(v)$ for all $u, v \in T_{x_0}\mathcal{M}$. That is, W is self-adjoint.

A result from linear algebra says that W must therefore have an orthogonal set of eigenvectors $v_1, \ldots, v_p$ with associated real eigenvalues κ_i; so $W(v_i) = \kappa_i v_i$. Of course, $\{v_i\}_{i=1}^p$ is a basis for $T_{x_0}\mathcal{M}$. It is clear that the associated reciprocal basis is $\{v_i / |v_i|^2\}_{i=1}^p$.

We now define the *mean curvature of* $\mathcal{M}$ to be

$$H \stackrel{\text{def.}}{=} \frac{1}{p}\left(\kappa_1 + \cdots + \kappa_p\right). \tag{7.40}$$

(See, for example, [29] or [44].) Some authorities prefer the negative of our definition.

Example 7.17. Consider the p-sphere $\mathcal{S}$ in $\mathbb{R}^{p+1}$ having the origin as its center and radius $\rho_0 > 0$. Its equation is $\chi_1^2 + \cdots + \chi_p^2 = \rho_0^2$.

We define a unit normal vector field n to $\mathcal{S}$ by setting $n(x) = (1/\rho_0)x$ for all $x \in \mathcal{S}$. Let v be a tangent vector to $\mathcal{M}$ at x. The map $x \mapsto (1/\rho_0)x$, considered as a map on

$\mathbb{R}^{p+1}$, is linear, so by Proposition 2.4, $W(v) = \partial_v n(x) = (1/\rho_0)v$. This tells us that every $v \in T_x \mathcal{S}$ is an eigenvector with eigenvalue $1/\rho_0$. Choosing any orthonormal basis for $T_x \mathcal{S}$ and using it to compute the mean curvature at x, we obtain

$$H = \frac{1}{p}\left(\frac{1}{\rho_0} + \cdots + \frac{1}{\rho_0}\right) = \frac{1}{\rho_0}.$$

Mean curvature and the spur

We carry forward from the last section the assumptions and notations introduced there.

We now specify a particular unit normal vector field n to $\mathcal{M}$. Let $e_1, \ldots, e_{p+1}$ be the standard basis for $\mathbb{R}^{p+1}$, the ambient space for $\mathcal{M}$. Set $e = e_1 \cdots e_{p+1}$; this is the usual orientation for $\mathbb{R}^{p+1}$. Set $n = e^\dagger w$. We know from the discussion in Section 5.4 that n is a 1-vector field that is orthogonal to $\mathcal{M}$. It must also be $\mathcal{C}^2$ since e is constant while w is $\mathcal{C}^2$.

It follows that for a tangent vector v to $\mathcal{M}$ at x_0, we have $\partial_v n = e^\dagger(\partial_v w)$, hence

$$\partial_v w = e(\partial_v n) = eW(v). \tag{7.41}$$

Let $\{v_i\}_{i=1}^p$ be the set of orthogonal eigenvectors of W described before along with associated eigenvalues κ_i and the corresponding reciprocal frame $\{v_i/|v_i|^2\}_{i=1}^p$. From Equation (7.41), we must have

$$\partial_{v_i} w = \kappa_i e v_i.$$

This permits computation of the geometric derivative of w:

$$\vec{\nabla}_{\mathcal{M}} w = \sum_{i=1}^p \kappa_i e v_i \frac{v_i}{|v_i|^2} = \sum_{i=1}^p \kappa_i e.$$

From Equation (7.28), we have

$$N_{\mathcal{M}} = -w^\dagger(\vec{\nabla}_{\mathcal{M}} w) = -\sum_{i=1}^p \kappa_i w^\dagger e.$$

Since $n = e^\dagger w$, by Proposition 5.6, we have $n = n^\dagger = w^\dagger e$. Thus

$$N_{\mathcal{M}} = -\sum_{i=1}^p \kappa_i n = -p\left(\frac{1}{p}\sum_{i=1}^p \kappa_i\right)n = -pHn.$$

If we had chosen the opposite sign for the definition of H, then we would have obtained a plus instead of a minus on right-hand side of the last equation.

We conclude the discussion with a result that is independent of our choice of sign for either H or n. Keep in mind that n is a unit vector.

Proposition 7.18. *If $\mathcal{M}$ is a $\mathcal{C}^2$ p-cell in $\mathbb{R}^{p+1}$ with mean curvature H and spur $N_{\mathcal{M}}$, then $|N_{\mathcal{M}}| = p|H|$.*

Remark 7.10. A surface $\mathcal{M}$ in $\mathbb{R}^3$ is defined to be a *minimal surface* provided its mean curvature is zero. A physical example would be a soap film with boundary on a wire frame. Thus such a surface is minimal if and only if $N_{\mathcal{M}} = 0$.

Exercises 7.3.

1. Let $\mathcal{M}$ be a $\mathcal{C}^1$ p-cell ($p \geq 2$) with spur $N_{\mathcal{M}}$ and outward unit normal tangent vector n from $\partial \mathcal{M}$. Let $(\chi_1, \ldots, \chi_p)$ be $\mathcal{C}^1$ coordinates on $\mathcal{M}$. Show that at every interior point x_0 of $\mathcal{M}$ we have

$$\lim_{\mathcal{M} \to x_0} \frac{1}{\operatorname{vol}(\mathcal{M})} \int_{\partial \mathcal{M}} \chi_i \chi_j n = \left(\chi_j \, d\chi_i + \chi_i \, d\chi_j + \chi_i \chi_j N_{\mathcal{M}} \right)(x_0).$$

2. Let $\mathcal{M}$ be a $\mathcal{C}^1$ p-cell ($p \geq 2$) with outward unit normal tangent vector n from $\partial \mathcal{M}$. Show that if f is a $\mathcal{C}^1$ multivector-valued function on $\mathcal{M}$ and x_0 is an interior point of $\mathcal{M}$, then

$$\left(\vec{\nabla}_{\mathcal{M}} f \right)(x_0) = \lim_{\mathcal{M} \to x_0} \frac{1}{\operatorname{vol}(\mathcal{M})} \int_{\partial \mathcal{M}} (f - f(x_0)) \, n \quad \text{and}$$

$$\left(\overleftarrow{\nabla}_{\mathcal{M}} f \right)(x_0) = \lim_{\mathcal{M} \to x_0} \frac{1}{\operatorname{vol}(\mathcal{M})} \int_{\partial \mathcal{M}} n \, (f - f(x_0)).$$

3. Let $\mathcal{M}$ be a $\mathcal{C}^2$ p-manifold in $\mathbb{R}^{p+1}$ with orientation w. Set $e = e_1 e_2 \cdots e_{p+1}$, the orientation for $\mathbb{R}^{p+1}$, and $n = e^\dagger w$. Then n is a 1-vector field that is orthogonal to $\mathcal{M}$. Show that for the spur of $\mathcal{M}$ we have

$$N_{\mathcal{M}} = -n \left(\vec{\nabla}_{\mathcal{M}} n \right).$$

4. The spur of a manifold is a well-defined vector field, so it is natural to ask what its geometric derivative is. Here is one answer: Let $\mathcal{M}$ be a $\mathcal{C}^2$ p-cell. Assume that x_0 is an interior point of $\mathcal{M}$, that n is the outward unit normal vector field to $\partial \mathcal{M}$ that is tangent to $\mathcal{M}$, and that H is the mean curvature function on $\mathcal{M}$. Show that

$$\left(\vec{\nabla}_{\mathcal{M}} N_{\mathcal{M}} \right)(x_0) = \left(\lim_{\mathcal{M} \to x_0} \frac{1}{\operatorname{vol}(\mathcal{M})} \int_{\partial \mathcal{M}} N_{\mathcal{M}} n \right) - p^2 \, |H(x_0)|^2.$$

5. Assume $\mathcal{M}$ is a $\mathcal{C}^2$ p-manifold in $\mathbb{R}^{p+1}$ consisting of the points $(\chi_1, \ldots, \chi_{p+1})$ satisfying $\phi(\chi_1, \ldots, \chi_{p+1}) = 0$ where ϕ is a real-valued function. (Example: The ellipse $3\chi_1^2 + 4\chi_2^2 = 1$ in $\mathbb{R}^2$ is generated by the function $\phi(\chi_1, \chi_2) = 3\chi_1^2 + 4\chi_2^2 - 1$.)

   Recall that by $\vec{\nabla}\phi$ we mean the geometric derivative of ϕ with respect to the manifold $\mathbb{R}^{p+1}$ rather than $\vec{\nabla}_{\mathcal{M}}\phi$ and that $\vec{\nabla}\phi$ is the same as what we call the gradient of ϕ. Assume that $|\vec{\nabla}\phi|$ is a positive constant on $\mathcal{M}$.

(a) Give at least two examples of manifolds and associated ϕ such that $|\vec{\nabla}\phi|$ is a positive constant on the manifold.

(b) Prove that

$$N_{\mathcal{M}} = -\frac{\vec{\nabla}\phi}{|\vec{\nabla}\phi|^2}\,\vec{\nabla}_{\mathcal{M}}(\vec{\nabla}\phi).$$

(c) Apply the formula above to $\mathcal{M}$ given by the equation $\chi_1^2 + \cdots + \chi_{p+1}^2 = \rho_0^2$ (where ρ_0 is a positive constant) and calculate $N_{\mathcal{M}}$ at the point $x_0 = (0,\ldots,0,\rho_0)$.

6. We consider how to compute the spur for another large class of manifolds:

Let $\mathcal{M}$ be the p-manifold in $\mathbb{R}^{p+1}$ defined by the equation $\chi_{p+1} = \psi(\chi_1,\ldots,\chi_p)$ where ψ is a $\mathcal{C}^2$ real-valued function and $\chi_1,\ldots,\chi_{p+1}$ are the standard coordinates of $\mathbb{R}^{p+1}$. (Example in $\mathbb{R}^3$: $\chi_3 = \chi_1^2 + \chi_2^2$. This is a paraboloid centered on the χ_3-axis.)

Let us induce coordinates $(\chi_1,\ldots,\chi_p)$ on $\mathcal{M}$ via the chart

$$x(\chi_1,\ldots,\chi_p) = \sum_{i=1}^{p} \chi_i\,e_i + \psi(\chi_1,\ldots,\chi_p)\,e_{p+1}$$

and then introduce the notations

$$u_i \overset{\text{def.}}{=} \frac{\partial x}{\partial \chi_i} \quad \text{and} \quad \psi_i \overset{\text{def.}}{=} \frac{\partial \psi}{\partial \chi_i} \quad (\text{where } i = 1,\ldots,p)$$

for tangent vectors to $\mathcal{M}$ and first partials of ψ. Let $\{d\chi_i\}_{i=1}^{p}$ be the reciprocal frame for $\{u_i\}_{i=1}^{p}$.

Show that the following hold:

(a) $u_i \cdot u_j = \delta_{ij} + \psi_i\psi_j$.

(b) The vector

$$n = \left(\sum_{i=1}^{p} \psi_i e_j\right) - e_{p+1}$$

is orthogonal to $\mathcal{M}$. (That is, when evaluated at a point x_0 of $\mathcal{M}$, it is orthogonal to $T_{x_0}\mathcal{M}$.)

(c) $d\chi_i = u_i - \sum_{j=1}^{p} \frac{\psi_i\psi_j}{|n|^2} u_j = e_i - \frac{\psi_i}{|n|^2} n.$

(d) $d\chi_i \cdot d\chi_j = \delta_{ij} - \frac{1}{|n|^2} \psi_i\psi_j.$

(e) $\dfrac{\partial^2 x}{\partial\chi_i \partial\chi_j} = \dfrac{\partial^2 \psi}{\partial\chi_i \partial\chi_j} e_{p+1}.$

(f) If we want the component of the vector in the last item that is orthogonal to $\mathcal{M}$, it is given by

$$\frac{\partial^2 x}{\partial \chi_i \partial \chi_j}\bigg|_{\perp \mathcal{M}} = -\frac{1}{|n|^2} \frac{\partial^2 \psi}{\partial \chi_i \partial \chi_j} n.$$

(g) The spur of the manifold is given by the formulas

$$
\begin{aligned}
N_{\mathcal{M}} &= \left(\frac{1}{|n|^4} \sum_{i,j=1}^{p} \psi_i \psi_j \frac{\partial^2 \psi}{\partial \chi_i \partial \chi_j} - \frac{1}{|n|^2} \sum_{i=1}^{p} \frac{\partial^2 \psi}{\partial \chi_i^2} \right) n \\
&= -\frac{1}{|n|^2} \left[\sum_{i,j=1}^{p} \frac{\partial^2 \psi}{\partial \chi_i \partial \chi_j} \left(\delta_{ij} - \frac{1}{|n|^2} \psi_i \psi_j \right) \right] n \\
&= -\frac{1}{|n|^2} \left[\sum_{i,j=1}^{p} \frac{\partial^2 \psi}{\partial \chi_i \partial \chi_j} (d\chi_i \cdot d\chi_j) \right] n.
\end{aligned}
$$

7.4 Complex analysis

To a rough first approximation, one can think of complex analysis as redoing calculus but with real numbers and real variables replaced by complex numbers and complex variables. However new, strange, and interesting things happen in complex analysis that do not appear in calculus.

We want to give a feel for how complex analysis could arise within the setting of geometric calculus. This shall be no more than a taste, a small one. We shall go as far as the Cauchy integral theorem but most of the field will be left out (for example, anything about power series or Laurent series).

Complex analysis is only the beginning of a vaster field of study referred to as Clifford analysis (see, for example, [2] and [38]).

7.4.1 Complex numbers and quaternions

Suppose we are working in $\mathbb{G}^n$ with $n \geq 2$. Let u_1, u_2 be orthonormal vectors in $\mathbb{R}^n$ and set $\mathbf{i} = u_1 u_2$. Notice that $\mathbf{i}^2 = -1$. This is exactly the way we expect the unit imaginary of the complex numbers to behave. (See the discussion given in [10] of the occurrence of square roots of -1 in geometric algebras.)

More generally, set $z_i = \alpha_i + \beta_i \mathbf{i}$ for $i = 1, 2$, where α_i, β_i are real numbers, and we then calculate, using the multivector addition and geometric product of $\mathbb{G}^n$, that

$$
\begin{aligned}
z_1 + z_2 &= (\alpha_1 + \beta_1 \mathbf{i}) + (\alpha_2 + \beta_2 \mathbf{i}) = (\alpha_1 + \alpha_2) + (\beta_1 + \beta_2)\mathbf{i}, \\
z_1 z_2 &= (\alpha_1 \alpha_2 - \beta_1 \beta_2) + (\alpha_1 \beta_2 + \alpha_2 \beta_1)\mathbf{i}
\end{aligned}
$$

which is exactly the usual sum and product of complex numbers. So we see that we have, in effect, a copy of the complex numbers embedded in $\mathbb{G}^n$. (To be sure, if the dimension of $\mathbb{R}^n$ is greater than 2, than there are *many* candidates for the unit imaginary

i.) There is even ready machinery for the conjugate of a complex number; it is the reversion of our complex number:

$$\overline{\alpha + \beta \mathbf{i}} = \alpha - \beta \mathbf{i} = (\alpha + \beta \mathbf{i})^{\dagger}.$$

This means that for any complex number of the type we have just constructed, its magnitude is $|z| = \sqrt{zz^{\dagger}}$, and this is exactly the same thing as its magnitude considered as an element of $\mathbb{G}^2$. (See Definition 5.4.)

The field of complex numbers, $\mathbb{C}$, is usually thought of as an extension of the real numbers. But there is a still larger "number field," the *quaternions*, and this too can be found in $\mathbb{G}^n$. To see how this works, consider $\mathbb{G}^n$ where $n \geq 3$. Let u_1, u_2, u_3 be orthonormal vectors in $\mathbb{R}^n$ and set

$$\mathbf{i} = u_3 u_2,$$
$$\mathbf{j} = u_1 u_3,$$
$$\mathbf{k} = u_2 u_1.$$

By the quaternions, $\mathbb{H}$, we shall mean the set of multivectors of the form $\alpha + \beta \mathbf{i} + \gamma \mathbf{j} + \delta \mathbf{k}$ where $\alpha, \beta, \gamma, \delta$ are arbitrary real numbers. Equipped with the operations of multivector addition and the geometric product, $\mathbb{H}$ is a field except that multiplication is not commutative. The generating elements $\mathbf{i}, \mathbf{j}, \mathbf{k}$ satisfy

$$\mathbf{i}^2 = \mathbf{j}^2 = \mathbf{k}^2 = \mathbf{ijk} = -1.$$

Just as there is a discipline of complex analysis, there is a field of quaternionic analysis. Both of these are contained in the field of Clifford analysis. We shall not investigate the properties of quaternions, but for those who are interested, there is an accessible paper, [6], *The Quaternion Calculus*.

7.4.2   Translations between $\mathbb{C}$ and $\mathbb{G}^2$

For the sake of simplicity and particularity, in what follows we shall confine our attention to $\mathbb{R}^2$ and $\mathbb{G}^2$. We set $\mathbf{i} = e_1 e_2$ where $\{e_1, e_2\}$ is the standard basis of $\mathbb{R}^2$. We take the complex numbers to be

$$\mathbb{C} \overset{\text{def.}}{=} \{\alpha + \beta \mathbf{i} : \alpha, \beta \in \mathbb{R}\}.$$

This means we have $\mathbb{C} \subseteq \mathbb{G}^2$. We note also that when we write $\vec{\nabla} f$, we mean $\vec{\nabla}_{\mathbb{R}^2} f$.

When we are doing complex analysis over $\mathbb{C}$ and geometric calculus over $\mathbb{R}^2$, we have different pictures in mind and tend to talk about different (but related) objects and ideas.

For instance, a typical point z in $\mathbb{C}$ is $z = \chi_1 + \chi_2 \mathbf{i}$ while in $\mathbb{R}^2$ it is $x = \chi_1 e_1 + \chi_2 e_2$.

In $\mathbb{C}$, we are interested in complex-valued functions $F(z)$ while in the $\mathbb{G}^2$ setting, we are interested in multivector-valued functions $f(x)$ where $x \in \mathbb{R}^2$.

When we want to differentiate $F(z)$, we write down $F'(z)$ and say that F is *analytic* or *holomorphic*. Now $F'(z)$ is neither a directional nor a geometric derivative; we are doing something different. (We explain exactly what in Equation (7.43).) Also note the occurrence of the word *holomorphic*. We attached a meaning to this term earlier in Definition 7.6, and later we shall see how that definition is connected to the differentiability of $F(z)$.

We also know that a fundamental construct in complex analysis is the integral $\int_{\mathcal{M}} F(z)\,dz$ where $\mathcal{M}$ is a curve or closed curve in the plane. However this last integral is not exactly the same thing as $\int_{\mathcal{M}} f$ in the sense in which we have explained it in our exposition of geometric calculus.

Because of such differences, our first concern will be to construct a means of translating between concepts in the complex setting and in geometric calculus. Here is one particular way to do it, a way which is adapted to the use of the right-hand geometric derivative operator $\overrightarrow{\nabla}$ for multivector fields on the manifold $\mathbb{R}^2$:

Notice that if we have $z = \chi_1 + \chi_2 \mathbf{i}$, then $x = e_1 z$ is a point of $\mathbb{R}^2$. We have

$$z = \chi_1 + \chi_2 \mathbf{i} \quad \Longleftrightarrow \quad x = e_1 z = \chi_1 e_1 + \chi_2 e_2.$$

As for $F : \mathbb{C} \to \mathbb{C}$, we must be able to write $F(z) = \omega_1(x) + \omega_2(x)\mathbf{i}$ where ω_1 and ω_2 are real-valued functions. We note that $F(z) = F(e_1 x)$ is now a function on $\mathbb{R}^2$ but is still complex-valued. To turn it into a vector-valued function $f(x)$, we replace $F(z)$ by $F(z)e_1$. Then

$$F(z) = \omega_1(x) + \omega_2(x)\mathbf{i} \quad \Longleftrightarrow \quad f(x) = F(z)e_1 = \omega_1(x)e_1 - \omega_2(x)e_2.$$

So our first and most basic translation rules are these:

$$\begin{aligned} x = e_1 z \quad &\text{and} \quad z = e_1 x \\ f(x) = F(z)e_1 \quad &\text{and} \quad F(z) = f(x)e_1. \end{aligned} \tag{7.42}$$

7.4.3 Derivatives

Let us see what differentiability of $F(z)$ corresponds to in the geometric calculus setting.
Let $F(z)$ and $f(x)$ be as in (7.42). By the *derivative of F at z* we mean

$$F'(z) \stackrel{\text{def.}}{=} \lim_{z_0 \to z} \frac{F(z_0) - F(z)}{z_0 - z} \tag{7.43}$$

provided the limit exists.

It is understood, when computing this limit, that z_0 may approach z from any direction whatsoever provided only that z_0 is always an element of $\mathbb{C}$. Let us write $F(z) = \omega_1(x) + \omega_2(x)\mathbf{i}$. If $z_0 = z + \lambda = (\chi_1 + \lambda) + \chi_2 \mathbf{i}$ where λ is a real number, then

$$\frac{F(z_0) - F(z)}{z_0 - z} \xrightarrow[\lambda \to 0]{} \frac{\partial \omega_1}{\partial \chi_1}(x) + \frac{\partial \omega_2}{\partial \chi_1}(x)\mathbf{i}.$$

If, on the other hand, we have $z_0 = z + \lambda \mathbf{i} = \chi_1 + (\chi_2 + \lambda)\mathbf{i}$, where λ is again a real number, then

$$\frac{F(z_0) - F(z)}{z_0 - z} \xrightarrow[\lambda \to 0]{} -\frac{\partial \omega_1}{\partial \chi_2}(x)\mathbf{i} + \frac{\partial \omega_2}{\partial \chi_2}(x).$$

This tells us that if F has a derivative at z, then ω_1 and ω_2 must satisfy the *Cauchy-Riemann equations*:

$$\frac{\partial \omega_1}{\partial \chi_1}(x) = \frac{\partial \omega_2}{\partial \chi_2}(x) \quad \text{and} \quad \frac{\partial \omega_2}{\partial \chi_1}(x) = -\frac{\partial \omega_1}{\partial \chi_2}(x). \tag{7.44}$$

Recall that $f(x) = \omega_1(x)e_1 - \omega_2(x)e_2$. Let us assume f is $\mathcal{C}^1$ and compute $\vec{\nabla}f(x)$.

$$\begin{aligned}
\vec{\nabla}f(x) &= \partial_{e_1}f(x)e_1 + \partial_{e_2}f(x)e_2 \\
&= \frac{\partial \omega_1}{\partial \chi_1}(x) - \frac{\partial \omega_2}{\partial \chi_2}(x) + \left(\frac{\partial \omega_1}{\partial \chi_2}(x) + \frac{\partial \omega_2}{\partial \chi_1}(x)\right)\mathbf{i}.
\end{aligned}$$

This result leads us to the following:

Proposition 7.19. *Suppose that x, z, f, and F satisfy (7.42) and that $F(z) = \omega_1(x) + \omega_2(x)\mathbf{i}$. If ω_1 and ω_2 are $\mathcal{C}^1$ real-valued functions, then the following are equivalent:*

1. $\vec{\nabla}f(x) = 0$.

2. The Cauchy-Riemann equations hold for ω_1, ω_2 at x.

3. $F'(z)$ *exists.*

Proof. The equivalence of the first and second items is immediate, and we also know that the existence of $F'(z)$ implies the Cauchy-Riemann equations. The only thing we must prove is that if the Cauchy-Riemann equations hold at x, then $F'(z)$ exists.

Given the Cauchy-Riemann equations at x, we choose a complex number $z_0 \neq z$ but still in the domain of F. We then define

$$\begin{aligned}
z_1 &= z_0 - z, \\
x_0 &= e_1 z_0, \\
x_1 &= e_1 z_1 = x_0 - x.
\end{aligned}$$

Since $f(x) = \omega_1(x)e_1 - \omega_2(x)e_2$ and ω_1 and ω_2 are $\mathcal{C}^1$, we see that f must be differentiable at x and that we can write

$$f(x_0) - f(x) = f'(x)x_1 + g(x_1) \tag{7.45}$$

where $g(x_1)/|x_1| \to 0$ as $x_1 \to 0$, that is, as $x_0 \to x$. Let us now set

$$F^*(z) \stackrel{\text{def.}}{=} \frac{\partial \omega_1}{\partial \chi_1}(x) + \frac{\partial \omega_2}{\partial \chi_1}(x)\mathbf{i}.$$

Since $x = e_1 z$, this is a well-defined complex number, and we shall show that it is $F'(z)$. We multiply (7.45) on the right by e_1 and obtain

$$F(z_0) - F(z) = \left(f'(x)x_1\right)e_1 + g(x_1)e_1.$$

We are to understand that the notation $\left(f'(x)x_1\right)e_1$ means that we are to apply the linear transformation $f'(x)$ to the vector x_1 and then multiply the result in $\mathbb{G}^2$ by e_1. Let us set $x_1 = \xi_1 e_1 + \xi_2 e_2$ so that $z_1 = \xi_1 + \xi_2 \mathbf{i}$. It follows by straightforward computation and by calling upon the Cauchy-Riemann equations where necessary, that

$$
\left(f'(x)x_1\right)e_1 = \xi_1 \frac{\partial \omega_1}{\partial \chi_1}(x) + \xi_2 \frac{\partial \omega_1}{\partial \chi_2}(x) + \left(\xi_1 \frac{\partial \omega_2}{\partial \chi_1}(x) + \xi_2 \frac{\partial \omega_2}{\partial \chi_2}(x)\right)\mathbf{i}
$$
$$
= F^*(z)z_1.
$$

Thus

$$
F(z_0) - F(z) = F^*(z)z_1 + g(x_1)e_1.
$$

We divide by z_1:

$$
\frac{F(z_0) - F(z)}{z_0 - z} = F^*(z) + g(x_1)e_1 z_1^{-1}.
$$

Since $z_1 = e_1 x_1$, we can show that $|e_1 z_1^{-1}| = 1/|x_1|$ and thus $|g(x_1)e_1 z_1^{-1}| \to 0$ as $x_1 \to 0$. We then see that

$$
\frac{F(z_0) - F(z)}{z_0 - z} \to F^*(z) \quad \text{as } z_0 \to z.
$$

Therefore $F'(z)$ exists. $\qquad\qquad\qquad\qquad\qquad\qquad\qquad\qquad\qquad\qquad\quad\square$

The reader may recall that we introduced *right holomorphic* and *left holomorphic* in Definition 7.6. Proposition 7.19 gives us a means of generating examples of right holomorphic functions $f(x)$ on $\mathbb{R}^2$. This follows from the fact that many differentiable complex functions $F(z)$ can be obtained from elementary functions such as χ^k, e^χ, $\sin(\chi)$, etc. by replacing χ with z and showing they have the "same" derivatives as their real-variable originals.

Example 7.18. We give some simple instances of complex variable functions $F(z)$ with the corresponding $f(x)$ according to the translation rule (7.42). We leave it to the reader to check (a straightforward calculation) that in each case $\overrightarrow{\nabla}f = 0$.

1. If $F(z) = z$, then the corresponding f is $f(x) = \chi_1 e_1 - \chi_2 e_2$. It is easily checked that $F'(z) = 1$.

2. For $F(z) = z^2$, we have $f(x) = \left(\chi_1^2 - \chi_2^2\right)e_1 - 2\chi_1\chi_2 e_2$. Of course $F'(z) = 2z$.

3. By $e^z = e^{\chi_1 + \chi_2 \mathbf{i}}$ we understand $e^{\chi_1}\left(\cos(\chi_2) + \mathbf{i}\sin(\chi_2)\right)$. So for $F(z) = e^z$, the corresponding $f(x)$ is $e^{\chi_1}\left(\cos(\chi_2)e_1 - \sin(\chi_2)e_2\right)$. The derivative of $F(z)$ (in the complex variable sense) is $F'(z) = e^z$.

Remark 7.11. The reader is asked to show in Exercise 1 of Section 7.4 that a $\mathcal{C}^1$ vector-valued function f on $\mathbb{R}^2$ is right holomorphic if and only if it is left holomorphic. It follows that if $F(z)$ and $f(x)$ are connected according to the translation rule (7.42), then to say that F is differentiable, f is right holomorphic, or f is left holomorphic are all equivalent condititons. We thus feel free to use the phrase $F(z)$ *is holomorphic* to mean $F(z)$ is differentiable in the complex variable sense.

It now becomes reasonable, since we have a means of translating between the realms of geometric calculus and complex analysis, to ask if we can express $F'(z)$ in terms of f and x. Here is one way to do it:

Proposition 7.20. *Let z, x, F, and f be as in (7.42). If f is $\mathcal{C}^1$ and $F'(z)$ exists, then*

$$F'(z) = \frac{1}{2}\left(\vec{\nabla}(e_1 f e_1)(x)\right)^{\dagger}.$$

Proof. We write $F(z)$ in the form $F(z) = \omega_1(x) + \omega_2(x)\mathbf{i}$, so that $f(x) = \omega_1(x)e_1 - \omega_2(x)e_2$. Since $F'(z)$ exists, the Cauchy-Riemann equations hold for ω_1, ω_2. We know that one form for $F'(z)$ is

$$F'(z) = \frac{\partial \omega_1}{\partial \chi_1}(x) + \frac{\partial \omega_2}{\partial \chi_1}(x)\mathbf{i}.$$

Invoking the Cauchy-Riemann equations where necessary, it is straightforward to compute that

$$\vec{\nabla}(e_1 f e_1)(x) = 2\frac{\partial \omega_1}{\partial \chi_1}(x) - 2\frac{\partial \omega_2}{\partial \chi_1}(x)\mathbf{i}.$$

The desired result now follows. $\square$

We conclude this particular discussion with a result the reader is asked to establish in Exercises 3 and 4 of this section:

If z is a complex variable and $z = \chi_1 + \chi_2\mathbf{i}$, then we may consider any function of z to be a function of $x = (\chi_1, \chi_2) \in \mathbb{R}^2$,

$$F(z) = F\big(z(x))\big),$$

and we can compute its right and left geometric derivatives over $\mathbb{R}^2$,

$$\vec{\nabla}F(z) = \frac{\partial\big(F(z)\big)}{\partial \chi_1}e_1 + \frac{\partial\big(F(z)\big)}{\partial \chi_2}e_2,$$

$$\overleftarrow{\nabla}F(z) = e_1\frac{\partial\big(F(z)\big)}{\partial \chi_1} + e_2\frac{\partial\big(F(z)\big)}{\partial \chi_2}.$$

Proposition 7.21. *For functions of a complex variable z, the following hold:*

1. *For any polynomial $\phi(z) = \alpha_0 + \alpha_1 z + \cdots + \alpha_k z^k$ of degree $k \geq 1$ with real coefficients, we have $\overleftarrow{\nabla}\phi(z) = 0$.*

2. *Suppose $F(z) = \omega_1(x) + \omega_2(x)\mathbf{i}$ where ω_1, ω_2 are real-valued and $\mathcal{C}^1$. Then the Cauchy-Riemann equations hold for ω_1, ω_2 if and only if $\overleftarrow{\nabla}F(z) = 0$.*

7.4.4 Line integrals

Let $\mathcal{A}$ be a $\mathcal{C}^1$ arc in $\mathbb{R}^2$ with orientation w. We take a $\mathcal{C}^1$ parametrization $x : \mathcal{I} \to \mathcal{A}$ of $\mathcal{A}$ that agrees with the orientation. We can write this in the form

$$x = x(\tau) = \chi_1(\tau) e_1 + \chi_2(\tau) e_2$$

where χ_1 and χ_2 are real-valued $\mathcal{C}^1$ functions. Then $x'(\tau) = \chi_1'(\tau) e_1 + \chi_2'(\tau) e_2$ is a tangent vector to $\mathcal{A}$ which is a positive scalar multiple of the orientation. See Figure 7.12.

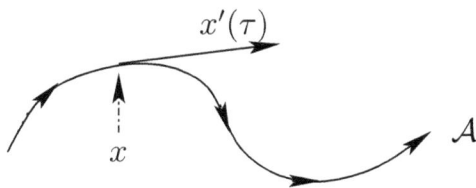

Figure 7.12: The "complex curve" $\mathcal{A}$

We turn the points of $\mathcal{A}$ into the set of complex numbers $e_1\mathcal{A}$. That is, the elements of $e_1\mathcal{A}$ are the complex numbers $z = e_1 x$ where $x \in \mathcal{A}$. We may parameterize this "complex curve" by writing

$$z = z(\tau) = e_1 x(\tau) = \chi_1(\tau) + \chi_2(\tau)\mathbf{i}.$$

At this point, we permit ourselves to be a bit fuzzy in our notation: We shall feel free to write $\mathcal{A}$ for both $e_1\mathcal{A}$ and $\mathcal{A}$. It should be clear from context whether we wish to regard points of $\mathcal{A}$ as complex numbers or points of $\mathbb{R}^2$.

Now let us introduce a complex-valued function on this complex curve, $F : \mathcal{A} \to \mathbb{C}$. The standard definition from complex analysis of the line integral of F over $\mathcal{A}$ is

$$\int_{\mathcal{A}} F(z)\, dz \stackrel{\text{def.}}{=} \int_0^1 F\big(z(\tau)\big)\, z'(\tau)\, d\tau \tag{7.46}$$

where $z'(\tau) = \chi_1'(\tau) + \chi_2'(\tau)\mathbf{i}$ and where we assume for convenience that $0 \leq \tau \leq 1$. In (7.46), the product of $F(z(\tau))$ and $z'(\tau)$ can be understood as either the product of two complex numbers or their geometric product; it makes no difference. When multiplied out, this product takes the form

$$F\big(z(\tau)\big)\, z'(\tau) = \psi_1(\tau) + \psi_2(\tau)\mathbf{i}$$

where $\psi_1(\tau)$ and $\psi_2(\tau)$ are real-valued functions, and we understand the integral of this with respect to τ to be

$$\int_0^1 F\big(z(\tau)\big)\, z'(\tau)\, d\tau = \int_0^1 \psi_1(\tau)\, d\tau + \mathbf{i}\left(\int_0^1 \psi_2(\tau)\, d\tau \right).$$

Each integral $\int_0^1 \psi_j(\tau)\,d\tau$ is to be evaluated in the manner of an integral in introductory calculus, so (7.46) does indeed reduce the integral of $F(z)$ along $\mathcal{A}$ to something well-defined. Notice that this definition requires us to know the orientation of $\mathcal{A}$.

Remark 7.12. We note that a similar result holds if we integrate not along arcs but around simple closed curves. A simple closed curve $\mathcal{S}$ can be treated as a sum of two arcs $\mathcal{A}_1$ and $\mathcal{A}_2$ having only their endpoints in common.

Example 7.19. Let $F(z) = 1/z$ and consider the circle S^1 in $\mathbb{R}^2$ with equation $\chi_1^2 + \chi_2^2 = 1$ and counterclockwise orientation. This is a simple closed curve rather than a mere curve, but (7.46) can be used here as well. Treating S^1 as a subset of $\mathcal{C}$, a parametrization is given by

$$z(\tau) = \cos(\tau) + \sin(\tau)\mathbf{i} \quad \text{where } 0 \leq \tau \leq 2\pi.$$

We see that

$$z'(\tau) = -\sin(\tau) + \cos(\tau)\mathbf{i} = \big(\cos(\tau) + \sin(\tau)\mathbf{i}\big)\mathbf{i} = z(\tau)\mathbf{i}.$$

Then

$$\int_{S^1} \frac{1}{z}\,dz = \int_0^{2\pi} \frac{1}{z(\tau)} z(\tau)\mathbf{i}\,d\tau = 2\pi\mathbf{i}.$$

We have the following connection between line integrals in the complex and geometric calculus domains:

Proposition 7.22. *Let $\mathcal{A}$ be a $\mathcal{C}^1$ curve in $\mathbb{R}^2$ with orientation w. Suppose $F : \mathcal{A} \to \mathbb{C}$ is a complex-valued function of a complex variable. Let z, x, F, and f be related as in (7.42). Then*

$$\int_{\mathcal{A}} F(z)\,dz = \int_{\mathcal{A}} f\,w,$$

assuming the integrals exist.

Proof. Let $x = x(\tau)$, where $0 \leq \tau \leq 1$, be a $\mathcal{C}^1$ parametrization of $\mathcal{A}$ that agrees with the orientation w. Set $z(\tau) = e_1 x(\tau)$ as described above. We then simply note that

$$\int_{\mathcal{A}} f\,w = \int_0^1 f(x(\tau))\,x'(\tau)\,d\tau$$
$$= \int_0^1 f(x(\tau))\,e_1 e_1 x'(\tau)\,d\tau$$
$$= \int_0^1 F(z(\tau))z'(\tau)\,d\tau. \qquad \square$$

Remembering that to say $F(z)$ is holomorphic means it is differentiable in the complex variables sense, and calling on Proposition 7.19, we can now write down a basic result of complex analysis, Cauchy's theorem:

Proposition 7.23. *If $F(z)$ is a complex-valued function of a complex variable, if F is holomorphic on the open set U, and if the closed set S is bounded by a simple closed curve and lies entirely in U, then*

$$\int_{\partial S} F(z)\, dz = 0.$$

The statement really requires more careful conditions on the simple, closed curve, such as its being C^1 or piecewise C^1. Notice that the proof we give is only for a special case. However the spirit of the thing is correct.

Proof of Proposition 7.23 when S is a C^2 2-cell. Let $f(x)$ be the function associated with $F(z)$ by translation rule (7.42) and let w be the orientation of S. Recall that we need to know the orientation of ∂S to evaluate $\int_{\partial S} F(z)\, dz$. We assign ∂S the orientation $\overrightarrow{\partial} w$. (It is easily seen that if we were to give it the opposite orientation, we would still get the same final answer.)

By Proposition 7.22 and the Fundamental Theorem,

$$\int_{\partial S} F(z)\, dz = \int_{\partial S} f\, \overrightarrow{\partial} w = \int_{S} \left(\overrightarrow{\nabla} f \right) w.$$

But by Proposition 7.19, $\overrightarrow{\nabla} f = 0$, so we are done. $\qquad\qquad\square$

Appealing to the versions of the Fundamental Theorem in Equation (6.23) and Example 6.20, we can immediately and trivially write down a generalization of Cauchy's theorem to p-cells $\mathcal{M}$ lying in $\mathbb{R}^n$ and having orientation w:

$$\begin{aligned}
\int_{\partial \mathcal{M}} f\, \overrightarrow{\partial} w &= 0 \quad \text{if } f \text{ is right holomorphic,} \\
\int_{\partial \mathcal{M}} \overleftarrow{\partial} w f &= 0 \quad \text{if } f \text{ is left holomorphic.}
\end{aligned} \tag{7.47}$$

7.4.5 Handwaving and annuli

Our goal from this point on will be to establish and generalize the Cauchy Integral Formula, a basic result of complex analysis. We need, however, some preliminary ideas.

We want to talk about integrating over certain n-dimensional regions called *annuli* and the boundaries of these annuli. We shall not do it rigorously for that would require prolonged topological considerations. However we hope the discussion will be reasonably clear and convincing.

Let D_1 and D_2 be two disks in $\mathbb{R}^2$ such that D_2 lies in the interior of D_1. If we remove the interior of D_2 from D_1, the resulting 2-dimensional region A is an example of a 2-dimensional *annulus*. (See Figure 7.13.)

Suppose we assign the orientations w, w_1, and w_2 to A, D_1, and D_2 respectively, and in each case these are $e_1 e_2$, the standard orientation of $\mathbb{R}^2$. Now $\overrightarrow{\partial} w_1$ and $\overrightarrow{\partial} w_2$ will be the counterclockwise orientations on ∂D_1 and ∂D_2 respectively. We see that $\partial A = \partial D_1 \cup \partial D_2$. Recall that the induced orientation of ∂A is $\overrightarrow{\partial} w = n w$ where n is the outward unit normal vector to ∂A. These induced orientations have the property that

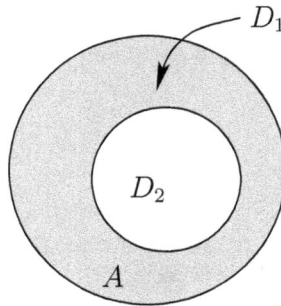

Figure 7.13: Example of 2-dimensional annulus

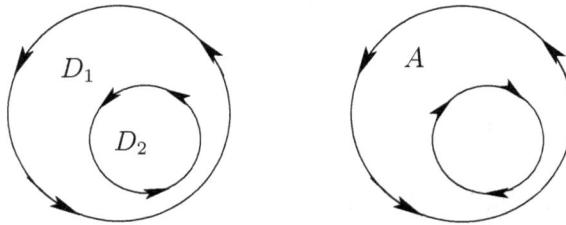

Figure 7.14: Induced orientations first of D_1, D_2 then of ∂A

$$\vec{\partial} w = \begin{cases} \vec{\partial} w_1 & \text{on } \partial D_1 \\ -\vec{\partial} w_2 & \text{on } \partial D_2. \end{cases} \tag{7.48}$$

See Figure 7.14.

Suppose that $f : A \to \mathbb{G}^2$. Assuming we can perform the integrations, we see that

$$\int_{\partial A} f \, \vec{\partial} w = \int_{\partial D_1} f \, \vec{\partial} w_1 - \int_{\partial D_2} f \, \vec{\partial} w_2.$$

Then if the Fundamental Theorem applies, we have

$$\int_A (\vec{\nabla} f) w = \int_{\partial D_1} f \, \vec{\partial} w_1 - \int_{\partial D_2} f \, \vec{\partial} w_2. \tag{7.49}$$

This scenario generalizes in two ways:

First, instead of considering disks within disks, we may state similar results for a greater variety of figures.

For example, we may suppose that D_1 is triangle and D_2 a 2-cell in its interior. See Figure 7.15. Equation (7.49) still holds with A the region between ∂D_1 and ∂D_2. We still consider A in these circumstances to be an annulus.

Second, we can generalize this construction to arbitrarily high dimensions and to "cell-like" sets with smaller "cell-like" sets extracted from their interiors.

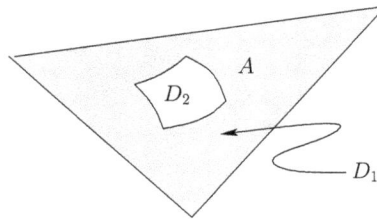

Figure 7.15: Another annulus

For example, if D_1 is an n-cell in $\mathbb{R}^n$ and D_2 is an n-dimensional ball in the interior of D_1, then we take A as before, namely D_1 minus the interior of D_2, and Equation (7.49) still holds. See Figure 7.16 for a picture of this when $n = 3$. It should be understood

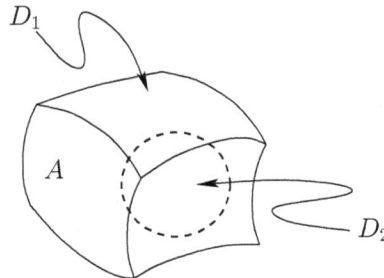

Figure 7.16: A 3-dimensional annulus: cell minus a ball

that A, D_1, and D_2 all have the same orientation, usually $w, w_1, w_2 = e_1 \cdots e_n$. We still refer to A as an annulus, an n-dimensional annulus.

We note that the regions A we have talked about in all these cases are *topological* annuli. That is, A is an n-dimensional (topological) annulus ($n \geq 2$) provided there is a one-to-one onto map $g : S^{n-1} \times \mathcal{I} \to A$ such that g and g^{-1} are both continuous. It is understood that S^{n-1} is the $(n-1)$-dimensional unit sphere centered at the origin and $\mathcal{I}$ is the unit interval in $\mathbb{R}$. The map g is called a *homeomorphism*.

For the sake of integrability and so that we can appeal to the Fundamental Theorem, we think of the sets A and D_2 as being composed of some of the cells which make up a "nice" partition of D_1. To get an idea of what we mean by this last statement, let us rework Figure 7.15 by dividing it into cells $\mathcal{M}_1, \ldots, \mathcal{M}_6$; see Figure 7.17. The annulus A is the shaded portion and D_2 has been relabeled $\mathcal{M}_6$.

Now here are two restrictions we put on our annuli:

Number one is that we consider only m-dimensional annuli lying in $\mathbb{R}^m$. That is, we want our annuli to have the same dimension as the ambient Euclidean space.

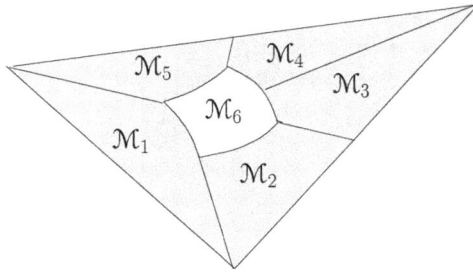

Figure 7.17: A "nice" partition of an annulus into cells

Number two is that we consider only cases where D_1 and D_2 are either closed solid m-dimensional balls or m-cells. When we say that D_i is a closed solid ball, we mean it is of the form

$$D_i = \{x \in \mathbb{R}^m : |x - x_0| \leq \rho\}$$

for some $x_0 \in \mathbb{R}^m$ which is its center and some $\rho > 0$ which is its radius.

As always, D_2 lies in the interior of D_1, and the annulus A is $D_1 - D_2$. (More precisely, it is the closure of this set which is slightly "larger" by virtue of including boundary points.) Note that $\partial A = \partial D_1 \cup \partial D_2$. We take the orientations of A, D_1, and D_2 to be w, w_1, and w_2 respectively and generally take each of them to be the restriction of $e = e_1 \cdots e_m$ to the appropriate set. The important thing to note from this set up is that Equations 7.48 and 7.49 hold. In particular, if f is right holomorphic on A, we have the very useful result

$$\int_{\partial D_1} f \, \overrightarrow{\partial} w_1 = \int_{\partial D_2} f \, \overrightarrow{\partial} w_2. \tag{7.50}$$

Suppose when considering our construction of an annulus A from D_1 and D_2, we let the unit outward normal vector from the boundary of each of these three manifolds be n_A, n_{D_1}, and n_{D_2} respectively. Appealing to Equation (7.48), we see that on ∂D_1 we have $n_A = n_{D_1}$, but on ∂D_2 we have $n_A = -n_{D_2}$. This tells us that if we have functions f and g on A, then

$$\int_{\partial A} f n_A g = \int_{\partial D_1} f n_{D_1} g - \int_{\partial D_2} f n_{D_2} g, \tag{7.51}$$

a result we shall shortly find to be useful.

7.4.6   Spheres and the function G

We now produce a surprising result involving the function $G(x) = x/|x|^n$ which, by Proposition 7.14, is both right and left holomorphic on the set $R^n - \{0\}$.

Proposition 7.24. *Let $\mathcal{M}$ be a $\mathcal{C}^2$ n-cell in $\mathbb{R}^n$ (where $n \geq 2$) and $\mathcal{M}$ has orientation $w = e_1 \cdots e_n$. If the origin is an interior point of $\mathcal{M}$, then*

$$\int_{\partial \mathcal{M}} G \, \overrightarrow{\partial} w = \int_{\partial \mathcal{M}} \overleftarrow{\partial} w \, G = vol(S^{n-1}) \, w$$

where S^{n-1} is the unit $(n-1)$-dimensional sphere centered at the origin. If $0 \notin \mathcal{M}$, then

$$\int_{\partial \mathcal{M}} G \, \overrightarrow{\partial} w = \int_{\partial \mathcal{M}} \overleftarrow{\partial} w \, G = 0.$$

Proof. We consider only right-hand induced orientation in this proof.

Notice that S^{n-1} is the boundary of the unit n-ball, B^n, that is the set of points x such that $|x| \le 1$. By $\text{vol}(S^{n-1})$ we mean of course $\int_{S^{n-1}} 1$. Let us give B^n the orientation $w = e_1 \cdots e_n$. The outward unit normal vector to $\partial B^n = S^{n-1}$ at $x \in S^{n-1}$ is $m(x) = x$. We take the orientation of S^{n-1} at x to be $\overrightarrow{\partial} w(x) = m(x) \, w = xw$. (See Equation (6.18).) Since $G(x) = x$ for $x \in S^{n-1}$, $|x| = 1$, and w is a constant, we then compute

$$\int_{S^{n-1}} G \, \overrightarrow{\partial} w = \int_{S^{n-1}} w = \text{vol}(S^{n-1}) w.$$

Next, given a scalar $\rho > 0$, we take ρS^{n-1} to be the set of ρx such that $x \in S^{n-1}$. That is, ρS^{n-1} is the $(n-1)$-dimensional sphere centered at the origin and having radius ρ. The outward unit normal vector to $x \in \rho S^{n-1}$ is $m(x) = x/|x|$, and we assign ρS^{n-1} the orientation $\overrightarrow{\partial} w(x) = m(x) \, w$ where w is as before.

Take ρ to satisfy $0 < \rho < 1$ and let A be the n-dimensional annulus between ρS^{n-1} and S^{n-1}. See Figure 7.18. Then Equation (7.49) holds with ∂D_1 and ∂D_2 equal to

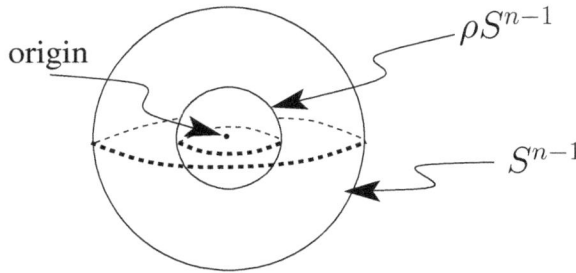

Figure 7.18: Annulus defined by S^{n-1} and ρS^{n-1}

S^{n-1} and ρS^{n-1} respectively. However $\overrightarrow{\nabla} G = 0$ on A since $0 \notin A$. Therefore we have

$$\int_{S^{n-1}} G \, \overrightarrow{\partial} w - \int_{\rho S^{n-1}} G \, \overrightarrow{\partial} w = 0.$$

This means that

$$\int_{\rho S^{n-1}} G \, \overrightarrow{\partial} w = \text{vol}(S^{n-1}) w.$$

Now let $\mathcal{M}$ be a $\mathcal{C}^2$ n-cell which has 0 as an interior point. There must exist $\rho > 0$ sufficiently small that ρS^{n-1} lies entirely in the interior of $\mathcal{M}$. (See Figure 7.16 again with ∂D_2 playing the role of ρS^{n-1} and D_1 corresponding to $\mathcal{M}$.) Let A be the annulus between $\partial \mathcal{M}$ and ρS^{n-1}. Since $0 \notin A$ and $\overrightarrow{\nabla}_{\mathcal{M}} G = 0$ on A, we can then show that

$$\int_{\partial \mathcal{M}} G \, \overrightarrow{\partial} w = \int_{\rho S^{n-1}} G \, \overrightarrow{\partial} w = \text{vol}(S^{n-1}) w.$$

Of course if $0 \notin \mathcal{M}$, then $\vec{\nabla}_{\mathcal{M}} G = 0$ on $\mathcal{M}$, so

$$\int_{\partial \mathcal{M}} G \, \vec{\partial} w = \int_{\mathcal{M}} (\vec{\nabla}_{\mathcal{M}} G) \, w = 0. \qquad\qquad \square$$

7.4.7   The Clifford-Cauchy integral formula

The Clifford-Cauchy integral formula can be thought of as a generalization of the Cauchy integral formula from complex analysis to the setting of geometric calculus. It says that if we have a holomorphic function on a cell, then the values of the function inside the cell are completely determined by its values on the boundary of the cell.

This is such a striking and non-intuitive result that even though it involves an epsilon-delta style proof, we shall lay out all the details before the reader in this section rather than hiding them in an appendix as we have often done before.

It is convenient to work with left-hand induced orientation on the boundary of the cell rather than the right-hand version, so we recall to the reader's mind Equation (6.19) which gives the useful relationship

$$\overleftarrow{\partial} w = wn$$

where n is the outward unit normal vector to the boundary of the cell.

We also need the following fact which depends on advanced calculus: If f is an integrable multivector function on a manifold $\mathcal{M}$, then

$$\left| \int_{\mathcal{M}} f \right| \leq \int_{\mathcal{M}} |f|. \tag{7.52}$$

If we were to establish it, we would need to more carefully specify the type of f with which we worked, though the fact that we consider only continuous f should be more than sufficient; and we should also need to more carefully pin down which integration theory we worked with, Riemann, Lebesgue, or some other. However such a development would take us too far afield, so we shall content ourselves with an assurance that (7.52) will hold for all f and manifolds that we consider.

Proposition 7.25 (Clifford-Cauchy integral formula for a cell). *Assume $\mathcal{M}$ is a $\mathcal{C}^2$ m-cell (where $m \geq 2$) in $\mathbb{R}^m$ with unit outward normal n on $\partial \mathcal{M}$. Assume x_0 is a point in the interior of $\mathcal{M}$ and that $f, g : \mathcal{M} \to \mathbb{G}^m$ are $\mathcal{C}^1$ multivector fields on $\mathcal{M}$. If f is right holomorphic and g is left holomorphic on $\mathcal{M}$, then*

$$f(x_0) = \frac{1}{vol(S^{m-1})} \int_{\partial \mathcal{M}} f(x) \, n(x) \, G(x - x_0) \, dx \quad and$$

$$g(x_0) = \frac{1}{vol(S^{m-1})} \int_{\partial \mathcal{M}} G(x - x_0) \, n(x) \, g(x) \, dx.$$

Remark 7.13. The dx in the integrals is not a vector or reciprocal vector. It merely indicates the variable with respect to which the integration is carried out to help clarify the fact that $G(x - x_0)$ is a translation of the function $G(x)$.

Proof of Proposition 7.25. For the sake of simplicity, we consider only the case where $x_0 = 0$. The general case is then established by applying a translation and the change of variables formula for integrals.

Let us establish the formula for f.

Choose $\varepsilon > 0$. We can next choose $\rho > 0$ so small that two conditions are satisfied:

1. ρS^{m-1} lies in the interior of $\mathcal{M}$.

2. Since f is continuous, we can further suppose ρ so small that if $x \in \rho S^{m-1}$, then $|f(x) - f(0)| < \varepsilon$.

Let A be the m-dimensional annulus between $\partial \mathcal{M}$ and ρS^{m-1}. We take the orientation on A, and hence on $\mathcal{M}$, to be $w = e_1 \cdots e_m$. By Proposition 7.16,

$$\int_{\partial A} fnG = \int_A \left((\overrightarrow{\nabla}_A f)G + f(\overleftarrow{\nabla}_A G) + f N_A G \right)$$

where n is the unit outward normal vector to ∂A and N_A is the spur of A. We know that $\overrightarrow{\nabla}_A f = 0$ and $\overleftarrow{\nabla}_A G = 0$, and since A is m-dimensional, N_A cannot be orthogonal to A without being orthogonal to $\mathbb{R}^m$, thus we also have $N_A = 0$. So $\int_{\partial A} fnG = 0$.

We know that $\partial A = \partial \mathcal{M} \cup \rho S^{m-1}$. We may think of ρS^{m-1} as the boundary of the solid m-dimensional ball ρB^m centered at 0 and having radius ρ. Now if, in the next equation, we understand that the occurrence of n on the left refers to the unit outward normal vector to $\mathcal{M}$ while on the right it refers to the unit outward normal vector to the solid ball ρB^m, then by virtue of Equation (7.51), we have

$$\int_{\partial \mathcal{M}} fnG = \int_{\rho S^{m-1}} fnG. \tag{7.53}$$

Since $n = w^\dagger \overleftarrow{\partial} w$, and $w^\dagger$ is the constant $e_m \cdots e_1$, we compute

$$\int_{\rho S^{m-1}} nG = w^\dagger \int_{\rho S^{m-1}} (\overleftarrow{\partial} w) G$$
$$= w^\dagger \operatorname{vol}(S^{m-1}) w = \operatorname{vol}(S^{m-1})$$

where we have appealed to Proposition 7.24. From this we can deduce

$$f(0)\operatorname{vol}(S^{m-1}) = \int_{\rho S^{m-1}} f(0)nG = \int_{\partial \mathcal{M}} f(0)nG \tag{7.54}$$

where the last step follows from Equation (7.53).

We put Equations (7.53) and (7.54) together to obtain

$$\left| \int_{\partial \mathcal{M}} fnG - f(0)\operatorname{vol}(S^{m-1}) \right| = \left| \int_{\rho S^{m-1}} \left(fnG - f(0)nG \right) \right|.$$

Recall that n on ρS^{m-1} is the unit outward directed normal vector to the boundary of ρB^m. Notice that if we evaluate n and G at $x \in \rho S^{m-1}$, we have $n(x) = x/|x|$ and

$G(x) = x/|x|^m$, so that $nG = 1/|x|^{m-1}$, a positive real number. From this we see that $|fnG - f(0)nG| = |f - f(0)|nG$, and applying (7.52), we deduce

$$\left| \int_{\rho S^{m-1}} fnG - f(0)nG \right| \leq \int_{\rho S^{m-1}} |f - f(0)| nG$$

$$\leq \varepsilon \int_{\rho S^{m-1}} nG = \varepsilon \operatorname{vol}(S^{m-1}).$$

Thus

$$\left| \int_{\partial \mathcal{M}} fnG - f(0) \operatorname{vol}(S^{m-1}) \right| \leq \varepsilon \operatorname{vol}(S^{m-1}).$$

Letting $\varepsilon \to 0$ gives us the first result of the proposition. The other one is obtained similarly. □

7.4.8 The Cauchy integral formula

We want to show that the Cauchy integral formula of complex analysis is a special case of the Clifford-Cauchy integral formula.

In what follows, it should be kept in mind that when we say a point $x = \chi_1 e_1 + \chi_2 e_2$ belongs to a set U in $\mathbb{R}^2$, then we shall feel free to say (inexactly and perhaps confusingly) that the associated complex number $z = e_1 x = \chi_1 + \chi_2 \mathbf{i}$ also belongs to U. What we really mean, of course, is that

$$z = e_1 x \in e_1 U = \{e_1 y : y \in U\}.$$

Let $\mathcal{M}$ be a $\mathcal{C}^2$ 2-cell in $\mathbb{R}^2$ with orientation $w = e_1 e_2 = \mathbf{i}$. Suppose that $F(z)$ is a complex-valued function of a complex variable on $\mathcal{M}$ and that it is $\mathcal{C}^1$ in the sense that if $F(z) = \omega_1(x) + \omega_2(x)\mathbf{i}$, then ω_1 and ω_2 are $\mathcal{C}^1$ functions. It is understood that z and x are related by $z = e_1 x$.

Corollary 7.6 (Cauchy's Integral Formula for a cell). *If z_0 is a complex number in the interior of the $\mathcal{C}^2$ 2-cell $\mathcal{M}$ and $F(z)$ is holomorphic on $\mathcal{M}$, then we have*

$$F(z_0) = \frac{1}{2\pi \mathbf{i}} \int_{\partial \mathcal{M}} \frac{F(z)}{z - z_0} dz.$$

Proof. In what follows, let $x, z, f(x), F(z)$ be associated by $z = e_1 x$ and $F(z) = f(x)e_1$. Also let $x_0 = e_1 z_0$. We intend, of course, that x_0 is an interior point of $\mathcal{M}$.

Consider the complex-valued function $F(z)/(z - z_0)$. By the translation rules (7.42), the corresponding function $h : \mathbb{R}^2 \to \mathbb{R}^2$ is

$$h(x) = \left(\frac{F(z)}{z - z_0} \right) e_1.$$

It then follows from Proposition 7.22 that

$$\int_{\partial \mathcal{M}} \frac{F(z)}{z - z_0} dz = \int_{\partial \mathcal{M}} \left(\frac{F(z)}{z - z_0} \right) e_1 \vec{\partial} w(x) dx.$$

Now

$$\frac{1}{z-z_0} = \frac{(z-z_0)^\dagger}{|z-z_0|^2} = \frac{(x-x_0)e_1}{|x-x_0|^2} = G(x-x_0)e_1,$$

and $\vec{\partial}w(x) = n(x)w = n(x)\mathbf{i}$. Therefore

$$F(z)\frac{1}{z-z_0}e_1\,\vec{\partial}w(x) = f(x)e_1G(x-x_0)e_1e_1n(x)\mathbf{i}$$
$$= f(x)e_1\,G(x-x_0)n(x)\mathbf{i}.$$

It is easily shown by direct calculation that if $a,b \in \mathbb{R}^2$, then $e_1ab = bae_1$. Thus

$$F(z)\frac{1}{z-z_0}e_1\,\vec{\partial}w(x) = f(x)n(x)G(x-x_0)e_1\mathbf{i}.$$

We know that $f(x)$ is right holomorphic on $\mathcal{M}$, so we are now able to invoke the Clifford-Cauchy integral formula:

$$\int_{\partial\mathcal{M}}\frac{F(z)}{z-z_0}\,dz = \left(\int_{\partial\mathcal{M}}f(x)n(x)G(x-x_0)\,dx\right)e_1\mathbf{i}$$
$$= 2\pi f(x_0)e_1\mathbf{i} = 2\pi F(z_0)\mathbf{i}$$

which is the desired result. $\qquad\square$

7.4.9 The fundamental theorem of algebra

This result, usually encountered in high school algebra, tends to be thought of as a purely algebraic statement rather than something involving complex analysis and the tools of calculus, but its proof depends on analysis. Its statement is that any polynomial of degree one or higher with real coefficients must have at least one zero in $\mathbb{C}$.

If we are given a real-valued function $\phi(\chi)$ of a real variable χ, we notice that it sometimes makes sense to replace χ by a multivector x. This is clearly the case if ϕ is a polynomial with real coefficients, $\phi(\chi) = \alpha_0 + \alpha_1\chi + \cdots + \alpha_k\chi^k$. We can then consider the function

$$x \mapsto \phi(x) = \alpha_0 + \alpha_1 x + \cdots + \alpha_k x^k,$$

where $x \in \mathbb{G}^n$. (Of course, by x^r, we mean the geometric product $xx\cdots x$ with r factors.)

From this point on, we shall confine our attention to the settings of $\mathbb{R}^2$ and $\mathbb{G}^2$. We shall be principally, but not exclusively, concerned with polynomials of the form $\phi(z) = \alpha_0 + \alpha_1 z + \cdots + \alpha_k z^k$ where z is a complex variable and each α_i is a real number.

We also make use of the fact that if we have a polynomial $\phi(z)$ with real coefficients and a complex variable z, then—considering $\phi(z) = \phi(\chi_1 + \chi_2\mathbf{i})$ as a function of $x = (\chi_1, \chi_2)$—we know that $\phi(z)$ is left holomorphic. That is $\overleftarrow{\nabla}\phi(z) = 0$. We have not proved this, but it follows from Exercise 3 of this section.

We need one other result before getting to the fundamental theorem of algebra:

Lemma 7.1. *If ϕ is a polynomial with real coefficients and z is a complex variable, then the reciprocal function $g(x) = g(\chi_1, \chi_2) = 1/\phi(z)$ is left holomorphic, at least at the points where $\phi(z) \neq 0$.*

Proof. Set $f(x) = \phi(z)$. We know f is left holomorphic from Proposition 7.21. Since $fg = 1$, we apply $\overset{\leftarrow}{\nabla}$ to this equation and obtain

$$e_1 \left(\frac{\partial f}{\partial \chi_1} g + f \frac{\partial g}{\partial \chi_1} \right) + e_2 \left(\frac{\partial f}{\partial \chi_2} g + f \frac{\partial g}{\partial \chi_2} \right) = 0.$$

Because $\overset{\leftarrow}{\nabla} f = 0$, we can rewrite this last equation as

$$e_1 f \frac{\partial g}{\partial \chi_1} + e_2 f \frac{\partial g}{\partial \chi_2} = 0. \tag{7.55}$$

Notice that $e_1 z = z^\dagger e_1$ and $e_2 z = z^\dagger e_2$ where $z^\dagger$ is the reversion of z. One shows by induction that $e_i z^k = (z^\dagger)^k e_i$ for $i = 1, 2$ and $k = 0, 1, 2, \ldots$. It follows from this that one can rewrite Equation (7.55) as $f^\dagger \overset{\leftarrow}{\nabla} g = 0$. The value of f at any point is a complex number, and $f^\dagger$ is what, in complex analysis, is called the conjugate value of f. At a given point, $f = 0$ if and only if $f^\dagger = 0$. Since we can always divide out nonzero complex numbers, we see that $\overset{\leftarrow}{\nabla} g = 0$ at all points at which $f \neq 0$. $\qquad \square$

The proof given below is an adaptation of [39] to the setting of geometric calculus.

Proposition 7.26 (The Fundamental Theorem of Algebra). *Every polynomial of degree at least 1 and with real coefficients has a complex zero.*

Proof. Suppose there is a polynomial ϕ with real coefficients and degree $m \geq 1$ such that ϕ does *not* have a complex zero. Without loss of generality, we may suppose that

$$\phi(z) = z^m + \beta_1 z^{m-1} + \cdots + \beta_m = z^m \left(1 + \frac{\beta_1}{z} + \cdots + \frac{\beta_m}{z^m} \right) \tag{7.56}$$

where $z = \chi_1 + \chi_2 e_1 e_2 \in \mathbb{G}^2$. We know from Lemma 7.1 that $f(x) = f(\chi_1, \chi_2) = 1/\phi(z)$ is left holomorphic and never zero on $\mathbb{R}^2$. Let ρS^1 be the circle in $\mathbb{R}^2$ of radius ρ and center 0. It then follows from the Clifford-Cauchy integral formula that

$$\frac{1}{2\pi} \int_{\rho S^1} \frac{1}{\rho} f = f(0) = \frac{1}{\phi(0)} \neq 0.$$

We see that

$$\left| 1 + \frac{\beta_1}{z} + \cdots + \frac{\beta_m}{z^m} \right| \to 1 \quad \text{as } |z| \to \infty.$$

So by Equation (7.56),

$$|\phi(z)| = |z|^m \left| 1 + \frac{\beta_1}{z} + \cdots + \frac{\beta_m}{z^m} \right| \geq \frac{|z|^m}{2}$$

for $|z|$ sufficiently large. Therefore we have

$$\frac{1}{|\phi(0)|} \leq \frac{1}{2\pi} \int_{\rho S^1} \frac{1}{\rho} |f|$$

$$\leq \frac{2}{\rho^m} \to 0 \quad \text{as } \rho \to \infty,$$

a contradiction. $\square$

Exercises 7.4.

1. Show that if $f(x) = \omega_1(x)e_1 + \omega_2(x)e_2$ is a $\mathcal{C}^1$ function on $\mathbb{R}^2$, then f is right holomorphic if and only if it is left holomorphic.

2. Using the magnitude or norm of $\mathbb{G}^2$ (see Definition 5.4), show that for two complex numbers z_1, z_2, we have $|z_1 z_2| = |z_1| |z_2|$.

3. Treating z as a function of $x = (\chi_1, \chi_2)$ where $z = \chi_1 + \chi_2 \mathbf{i}$, show that z^k is left holomorphic for $k = 1, 2, \dots$ but it is not necessarily right holomorphic.

4. Suppose $F(z) = \omega_1(x) + \omega_2(x)\mathbf{i}$ where ω_1, ω_2 are real-valued $\mathcal{C}^1$ functions on $\mathbb{R}^2$. Show the following are equivalent:

   (a) ω_1 and ω_2 satisfy the Cauchy-Riemann equations.

   (b) $\overleftarrow{\nabla} F(z) = 0$ (where $F(z)$ is treated as a function of $x \in \mathbb{R}^2$).

   (c) $F'(z) = \frac{1}{2}\left(\overrightarrow{\nabla} F(z) \right) e_1$.

5. Let $F(z) = z/|z|^2$ where z is a complex variable. Show that $F(z)$ is not differentiable in the complex variable sense, that is, $F'(z)$ does not exist.

6. Assume that x, z, $f(x)$, and $F(z)$ are connected by the translation rules 7.42. Let $\mathcal{M}$ be a $\mathcal{C}^2$ 2-cell in $\mathbb{R}^2$ with orientation $\mathbf{i} = e_1 e_2$, and assume we deal with $\mathcal{C}^1$ functions.

   (a) Show that

   $$\int_{\partial \mathcal{M}} F(z)\, dz = \int_{\mathcal{M}} \left(\overrightarrow{\nabla} f \right) \mathbf{i}.$$

   (b) Show that if $F(z) = \bar{z} = \chi_1 - \chi_2 \mathbf{i}$ (where $x = \chi_1 e_1 + \chi_2 e_2$), then

   $$\int_{\partial \mathcal{M}} F(z)\, dz = 2\,\text{area}(\mathcal{M})\,\mathbf{i}.$$

7.5   Differential forms

We want to see how we can think of the machinery of differential forms as fitting into geometric calculus. Logically, our construction is different from that of the standard theory of differential forms, but it permits us to perform the same calculations and obtain "isomorphic" results. There is also a certain overlap and a certain divergence in the notations used in differential form theory and those we have used for geometric calculus. We shall point these out as we go along.

The reader may wish to compare our discussion with that of Hestenes's article [21] or Section 6-4 of [22].

7.5.1   What shall we mean by a differential form?

Suppose f is a $\mathcal{C}^1$ multivector field on a $\mathcal{C}^1$ p-manifold $\mathcal{M}$ in $\mathbb{R}^n$. Recall that $P_{\mathcal{M}}(f)$ is the "orthogonal projection" of f onto $\mathcal{M}$. (See Definition 7.4.) That is, given $x_0 \in \mathcal{M}$ and an orientation w of $\mathcal{M}$ (at least an orientation on some neighborhood of x_0), then

$$\Big(P_{\mathcal{M}}(f)\Big)(x_0) = \Big(f(x_0) \cdot_L w(x_0)\Big)w(x_0)^{\dagger},$$

the orthogonal projection of $f(x_0)$ onto the tangent space $T_{x_0}\mathcal{M}$. Let us say that f is *tangent to* $\mathcal{M}$ provided $P_{\mathcal{M}}(f) = f$.

Definition 7.8. Given a k-vector field f on a $\mathcal{C}^1$ p-manifold $\mathcal{M}$, where $0 \le k \le p$, we say that f is a *differential k-form* (or more briefly a k-form) on $\mathcal{M}$ provided f is tangent to $\mathcal{M}$.

If we have coordinates $(\chi_1, \ldots, \chi_p)$ on $\mathcal{M}$, we know that at any given point of $\mathcal{M}$, the vectors $d\chi_1, \ldots, d\chi_p$ form a basis for the tangent space to the given point. It follows that f is a differential k-form precisely when it can be written in the form

$$f = \sum_{i_1 < \cdots < i_k} \phi_{i_1 \ldots i_k} \, d\chi_{i_1} \wedge \cdots \wedge d\chi_{i_k} \tag{7.57}$$

where each $\phi_{i_1 \ldots i_k}$ is a real-valued function.

Thus 1-forms on $\mathcal{M}$ have the form $\sum_{i=1}^{p} \phi_i \, d\chi_i$, the 0-forms are simply the real-valued functions ϕ on $\mathcal{M}$, and if we consider an orientation w of $\mathcal{M}$, since we can write it as $w = \psi \, d\chi_1 \wedge \cdots \wedge d\chi_p$ for some real-valude ψ, we see that w is p-form on $\mathcal{M}$.

The following are obvious consequences of our construction of differential form theory:

If f is a k-form on $\mathcal{M}$ and λ is a scalar, then λf is also a k-form.

If f and g are k-forms on $\mathcal{M}$, then the same is true of $f + g$.

If f is a k-form and g is an m-form on $\mathcal{M}$, then $f \wedge g$ is a $(k+m)$-form on $\mathcal{M}$.

Remark 7.14. Our construction of the theory of differential form differs from the usual one in an essential way. If we look at Equation (7.57), each $d\chi_j$ is a vector in $\mathbb{R}^n$, however in the standard construction, as carried out in differential form theory, each $d\chi_j$ is a linear functional operating on the tangent spaces of $\mathcal{M}$. That is, if we are evaluating the relevant functions at $x_0 \in \mathcal{M}$, then $d\chi_j$ is a linear map $d\chi_j \colon T_{x_0}\mathcal{M} \to \mathbb{R}$. Equation (7.57) is generally written in the slightly different form

$$f = \sum_{i_1 < \cdots < i_k} \phi_{i_1 \ldots i_k} \, d\chi_{i_1} \cdots d\chi_{i_k},$$

and

$$d\chi_{i_1} \cdots d\chi_{i_k} \colon T_{x_0}\mathcal{M} \times \cdots \times T_{x_0}\mathcal{M} \to \mathbb{R}$$

is now an alternating k-linear functional so that expressions such as $d\chi_i \, d\chi_j$ exhibit the behavior

$$d\chi_i \, d\chi_j = -d\chi_j \, d\chi_i.$$

as though they were vectors with a wedge product between them.

We do not go into any details of the usual development of the theory of differential forms though, as noted above, our construction is "isomorphic" to the usual one. (See, for example, [4] or [25] for the standard construction.) We exhibit a few examples of the fact that our construction produces the "same results" such as the formula for the operation of the exterior derivative and the derivation of the generalized Stokes theorem.

Two fundamental operations of the theory of differential forms are exterior differentiation and integration of forms. We now consider the first of these.

7.5.2 The exterior derivative

Definition 7.9. If f is a $\mathcal{C}^1$ multivector field on a $\mathcal{C}^1$ manifold $\mathcal{M}$, then we define the *left exterior derivative* and *right exterior derivative* of f respectively by

$$\overleftarrow{d}_{\mathcal{M}} f \stackrel{\text{def.}}{=} P_{\mathcal{M}}\left(\overleftarrow{\mathrm{curl}}_{\mathcal{M}}(f)\right) \quad \text{and} \quad \overrightarrow{d}_{\mathcal{M}} f \stackrel{\text{def.}}{=} P_{\mathcal{M}}\left(\overrightarrow{\mathrm{curl}}_{\mathcal{M}}(f)\right). \tag{7.58}$$

A question which immediately suggests itself is "Where did this come from? What is the inspiration for this odd-looking definition?" It looks nothing like the definition of the exterior derivative in the usual construction of differential form theory. However we shall show it gives us the usual computational formula, and we note for the moment that it has the virtue of being independent of any choice of coordinates.

The knowledgeable reader will also note that the usual symbol for exterior derivative of f is df and that there is only a single version of this. We wind up with two versions of this idea because the machinery of geometric calculus is in some instances a little more discriminating than that of differential form theory. As a matter of fact, the exterior derivative of the standard theory is our left exterior derivative:

$$df \qquad = \qquad \overleftarrow{d}_{\mathcal{M}} f.$$
$$\text{(differential form notation)} \qquad \text{(geometric calculus)}$$

Example 7.20. Suppose ϕ is a $\mathcal{C}^1$ real-valued function on $\mathcal{M}$. Let $(\chi_1, \dots, \chi_p)$ be $\mathcal{C}^1$ coordinates on $\mathcal{M}$. We know that the divergence of ϕ must be 0 since ϕ is grade zero and the divergence operation lowers grade, so

$$\overrightarrow{\operatorname{curl}}_{\mathcal{M}}\phi = \overleftarrow{\operatorname{curl}}_{\mathcal{M}}\phi = \overrightarrow{\nabla}_{\mathcal{M}}\phi = \overleftarrow{\nabla}_{\mathcal{M}}\phi = \sum_i \frac{\partial \phi}{\partial \chi_i} d\chi_i.$$

Since $\partial \phi / \partial \chi_i$ is a scalar and $d\chi_i$ is a tangent vector to $\mathcal{M}$, we have

$$\overrightarrow{d}_{\mathcal{M}}\phi = \overleftarrow{d}_{\mathcal{M}}\phi = P_{\mathcal{M}}\left(\overrightarrow{\operatorname{curl}}_{\mathcal{M}}\phi\right) = P_{\mathcal{M}}\left(\overleftarrow{\operatorname{curl}}_{\mathcal{M}}\phi\right) = \sum_i \frac{\partial \phi}{\partial \chi_i} d\chi_i.$$

If in particular we take ϕ to be the coordinate function χ_i that assigns to a point of $\mathcal{M}$ its ith coordinate, then because $\partial \chi_i / \partial \chi_j$ is 1 when $i = j$ and otherwise 0, we have

$$\overrightarrow{d}_{\mathcal{M}}\chi_i = \overleftarrow{d}_{\mathcal{M}}\chi_i = d\chi_i.$$

We now give the rule for calculating the left exterior derivative of a differential form and note that it is the rule that occurs in the standard construction of differential form theory. Since $\overleftarrow{d}_{\mathcal{M}}$ is a linear operator, it suffices to state the rule for a single term of f as described in Equation (7.57).

Proposition 7.27. *Suppose $\mathcal{M}$ is a $\mathcal{C}^2$ p-manifold with coordinates $(\chi_1, \dots, \chi_p)$. If f is the $\mathcal{C}^2$ differential k-form*

$$f = \phi\, d\chi_{i_1} \wedge \cdots \wedge d\chi_{i_k}$$

on $\mathcal{M}$ (where ϕ is a real-valued function), then

$$\overleftarrow{d}_{\mathcal{M}}f = \sum_{j=1}^{p} \frac{\partial \phi}{\partial \chi_j}\, d\chi_j \wedge d\chi_{i_1} \wedge \cdots \wedge d\chi_{i_k}.$$

Example 7.21. Suppose the coordinates on our manifold are (ρ, θ) and we have the differential 1-form $f(\rho, \theta) = \rho \cos(\theta)\, d\rho + \rho \sin(\theta)\, d\theta$. We then compute

$$\overleftarrow{d}_{\mathcal{M}}f = \frac{\partial}{\partial \rho}\left(\rho \cos(\theta)\right) d\rho \wedge d\rho + \frac{\partial}{\partial \theta}\left(\rho \cos(\theta)\right) d\theta \wedge d\rho$$

$$+ \frac{\partial}{\partial \rho}\left(\rho \sin(\theta)\right) d\rho \wedge d\theta + \frac{\partial}{\partial \theta}\left(\rho \sin(\theta)\right) d\theta \wedge d\theta$$

$$= (1 + \rho) \sin(\theta)\, d\rho \wedge d\theta.$$

Before giving the proof of Proposition 7.27, we need an intermediate result. Notice that if $\chi_1, \dots, \chi_p$ are $\mathcal{C}^2$ functions, then the tangent vectors $\partial x / \partial \chi_i$ are $\mathcal{C}^1$ and the reciprocal vectors $d\chi_i$, being constructed algebraically from the different $\partial x / \partial \chi_i$, must also be $\mathcal{C}^1$.

Lemma 7.2. *Given* $\mathcal{M}$ *and coordinates* $(\chi_1, \ldots, \chi_p)$ *as in Proposition 7.27, we have*

$$P_{\mathcal{M}}\left(\overleftarrow{\mathrm{curl}}_{\mathcal{M}}(d\chi_{i_1} \wedge \cdots \wedge d\chi_{i_k})\right) = 0.$$

Proof. We establish the result by induction on k.

We begin with $k = 1$ and see that

$$
\begin{aligned}
P_{\mathcal{M}}\left(\overleftarrow{\mathrm{curl}}_{\mathcal{M}}(d\chi_i)\right) &= P_{\mathcal{M}}\left(\sum_{j=1}^{p} d\chi_j \wedge \frac{\partial(d\chi_i)}{\partial \chi_j}\right) \\
&= \sum_{j=1}^{p} d\chi_j \wedge P_{\mathcal{M}}\left(\frac{\partial(d\chi_i)}{\partial \chi_j}\right)
\end{aligned}
\tag{7.59}
$$

because each $d\chi_j$ is tangent to $\mathcal{M}$ and $P_{\mathcal{M}}$ distributes over $\wedge$.

Next notice that

$$\frac{\partial x}{\partial \chi_k} \cdot d\chi_i = \delta_{ik} \quad \text{(Kronecker's delta)}$$

so that if we differentiate with respect to χ_j, we obtain

$$\frac{\partial^2 x}{\partial \chi_j \partial \chi_k} \cdot d\chi_i + \frac{\partial x}{\partial \chi_k} \cdot \frac{\partial(d\chi_i)}{\partial \chi_j} = 0.
\tag{7.60}$$

Let us set

$$\Gamma^i_{jk} = \frac{\partial^2 x}{\partial \chi_j \partial \chi_k} \cdot d\chi_i.$$

Γ^i_{jk} is called a *Christoffel symbol* and has the property that $\Gamma^i_{jk} = \Gamma^i_{kj}$ since mixed partials commute. Then Equation (7.60) tells us that

$$\frac{\partial(d\chi_i)}{\partial \chi_j} \cdot \frac{\partial x}{\partial \chi_k} = -\Gamma^i_{jk}.$$

Now $\partial(d\chi_i)/\partial\chi_j$ can be decomposed into vectors tangent to $\mathcal{M}$ and orthogonal to $\mathcal{M}$:

$$\frac{\partial(d\chi_i)}{\partial \chi_j} = \left(\sum_{k=1}^{p} \lambda_{ijk}\, d\chi_k\right) + v_{\perp \mathcal{M}}$$

where λ_{ijk} is a scalar. If we dot both sides of this last equation with $\partial x/\partial \chi_k$, we find that

$$\lambda_{ijk} = \frac{\partial(d\chi_i)}{\partial \chi_j} \cdot \frac{\partial x}{\partial \chi_k} = -\Gamma^i_{jk}.$$

Thus

$$P_{\mathcal{M}}\left(\frac{\partial(d\chi_i)}{\partial \chi_j}\right) = -\sum_{k=1}^{p} \Gamma^i_{jk}\, d\chi_k.$$

Plugging this back into (7.59), we see that

$$P_{\mathcal{M}}\left(\overleftarrow{\mathrm{curl}}_{\mathcal{M}}(d\chi_i)\right) = -\sum_{j,k=1}^{p} \Gamma_{jk}^i\, d\chi_j \wedge d\chi_k.$$

Since $d\chi_j \wedge d\chi_k = -d\chi_k \wedge d\chi_j$ and $\Gamma_{jk}^i = \Gamma_{kj}^i$, this last expression must be 0. Thus we have established the $k = 1$ case.

We now suppose the lemma established for r-forms where $r \leq k$ and move to the $k + 1$ case. Suppose we have $d\chi_i \wedge d\chi_I$ where I is the multi-index $(i_1, \ldots, i_k)$ and $d\chi_I$ stands of course for $d\chi_{i_1} \wedge \cdots \wedge d\chi_{i_k}$. We see that

$$
\begin{aligned}
P_{\mathcal{M}}\left(\overleftarrow{\mathrm{curl}}_{\mathcal{M}}(d\chi_i \wedge d\chi_I)\right) &= P_{\mathcal{M}}\left(\sum_{j=1}^{p} d\chi_j \wedge \frac{\partial(d\chi_i \wedge d\chi_I)}{\partial \chi_j}\right) \\
&= P_{\mathcal{M}}\left(\sum_{j=1}^{p}\left(d\chi_j \wedge \frac{\partial(d\chi_i)}{\partial \chi_j} \wedge d\chi_I\right) + \sum_{j=1}^{p}\left(d\chi_j \wedge d\chi_i \wedge \frac{\partial(d\chi_I)}{\partial \chi_j}\right)\right) \\
&= P_{\mathcal{M}}\left(\overleftarrow{\mathrm{curl}}_{\mathcal{M}}(d\chi_i)\right) \wedge d\chi_I - d\chi_i \wedge P_{\mathcal{M}}\left(\overleftarrow{\mathrm{curl}}_{\mathcal{M}}(d\chi_I)\right).
\end{aligned}
$$

By our induction assumption, this is 0 and we are done.   □

Proof of Proposition 7.27. We have $f = \phi\, d\chi_{i_1} \wedge \cdots \wedge d\chi_{i_k}$. Then

$$
\begin{aligned}
P_{\mathcal{M}}\left(\overleftarrow{\mathrm{curl}}_{\mathcal{M}} f\right) &= P_{\mathcal{M}}\left(\sum_{j=1}^{p} d\chi_j \wedge \frac{\partial}{\partial \chi_j}\left(\phi\, d\chi_{i_1} \wedge \cdots \wedge d\chi_{i_k}\right)\right) \\
&= P_{\mathcal{M}}\left(\sum_{j=1}^{p} d\chi_j \wedge \frac{\partial \phi}{\partial \chi_j} d\chi_{i_1} \wedge \cdots \wedge d\chi_{i_k}\right. \\
&\qquad\qquad \left. + \sum_{j=1}^{p} d\chi_j \wedge \phi\, \frac{\partial}{\partial \chi_j}\left(d\chi_{i_1} \wedge \cdots \wedge d\chi_{i_k}\right)\right) \\
&= \sum_{j=1}^{p} \frac{\partial \phi}{\partial \chi_j} d\chi_j \wedge d\chi_{i_1} \wedge \cdots \wedge d\chi_{i_k} + \phi\, P_{\mathcal{M}}\left(\overleftarrow{\mathrm{curl}}(d\chi_{i_1} \wedge \cdots \wedge d\chi_{i_k})\right)
\end{aligned}
$$

By Lemma 7.2 the very last term is 0 which gives us the desired result.   □

We now know how to compute the exterior derivative of a differential form. We state two other properties of the exterior derivative. The first, in the standard notation of differential form theory is usually written as $d^2 f = 0$ or just $d^2 = 0$. The second property does not appear in the standard theory, but we want it so as to convert between the left-hand and right-hand exterior derivatives of geometric calculus. The proofs are simple and are left to the reader.

Proposition 7.28. *Suppose that f is a $\mathcal{C}^2$ differential form on a $\mathcal{C}^2$ manifold $\mathcal{M}$. Then*

1. $\overleftarrow{d}_{\mathcal{M}} \overleftarrow{d}_{\mathcal{M}} f = 0.$

2. $\overrightarrow{d}_{\mathcal{M}}(f^\dagger) = \left(\overleftarrow{d}_{\mathcal{M}} f\right)^\dagger.$

7.5.3 Integrals of differential forms

If we integrate a multivector field over a manifold, $\int_{\mathcal{M}} f$, the result is again a multivector field; indeed, if f has grade k, then so does the integral. However if we integrate a differential form f over a manifold, a process which we represent by exactly the same notation, $\int_{\mathcal{M}} f$, the result is a real number. So we must be doing something different but presumably related.

It turns out that if we integrate a k-form, we want to do this over a k-manifold; that is, the grade of the differential form and the dimension of the manifold should match. On the other hand, it makes perfect sense to construct k-forms for $k = 1, \ldots, p$ on a p-manifold; remember, these are just tangent k-vector fields. Notice that if we have a k-form f on a p-manifold $\mathcal{M}$ and $k < p$, though we cannot integrate f over $\mathcal{M}$, we can still hope to integrate f over k-dimensional submanifolds of $\mathcal{M}$.

This makes it reasonable for us to consider a setting in which $\mathcal{M}$ and $\mathcal{N}$ are $\mathcal{C}^1$ p- and q-manifolds respectively and $\mathcal{M} \subseteq \mathcal{N}$. Given this setting, here is how we shall define integration of a differential form:

Definition 7.10. If f is a differential p-form on $\mathcal{N}$ and the p-dimensional submanifold $\mathcal{M}$ has orientation $w_{\mathcal{M}}$, then we set the integral of f over $\mathcal{M}$ in the sense of differential form theory to the following:

$$\underset{\text{(differential form notation)}}{\int_{\mathcal{M}} f} \quad = \quad \underset{\text{(geometric calculus)}}{\int_{\mathcal{M}} \langle w_{\mathcal{M}}^\dagger f \rangle_0.}$$

We first show that in a simple situation, that of a p-form over a p-cell, the integral of a differential form is nicely tailored to the standard notation of calculus.

Proposition 7.29. *Suppose $\mathcal{M}$ is a $\mathcal{C}^1$ p-cell with orientation $w_{\mathcal{M}}$ and coordinates $(\chi_1, \ldots, \chi_p)$ that agree with the orientation and where $\alpha_i \leq \chi_i \leq \beta_i$. If ϕ is a real-valued function on $\mathcal{M}$ where ϕ is understood to be a function of $\chi_1, \ldots, \chi_p$, then*

$$\underset{\text{(differential form notation)}}{\int_{\mathcal{M}} \phi\, d\chi_1 \wedge \cdots \wedge d\chi_p} \quad = \quad \int_{\alpha_p}^{\beta_p} \cdots \int_{\alpha_1}^{\beta_1} \phi(\chi_1, \ldots, \chi_p)\, d\chi_1 \ldots d\chi_p$$

where the last integral is an iterated integral and hence a real number.

Proof. Every tangent space of $\mathcal{M}$ has dimension p, so $d\chi_1 \wedge \cdots \wedge d\chi_p = \lambda w_{\mathcal{M}}$ for some scalar λ. Thus we must have

$$\left\langle w_{\mathcal{M}}^\dagger (\phi\, d\chi_1 \wedge \cdots \wedge d\chi_p) \right\rangle_0 = w_{\mathcal{M}}^\dagger (\phi\, d\chi_1 \wedge \cdots \wedge d\chi_p).$$

Our desired conclusion then follows from Proposition 6.17. $\qquad \square$

We now consider changes of variables. This is important calculationally and sometimes has a magically simple quality in the theory of differential forms. The general tool for handling it is the idea of the *pullback*, but we will not go into that here since our concern is not to develop the theory of differential forms but merely to indicate how that theory fits into geometric calculus.

We shall, however, look at the following case: Suppose that $\mathcal{M}$ is a submanifold of $\mathcal{N}$ as indicated above, and that the manifolds have, respectively, $\mathcal{C}^1$ coordinates $(\chi_1, \ldots, \chi_p)$ and $(\xi_1, \ldots, \xi_q)$. Let think of how to take a differential p-form f on $\mathcal{N}$ that is written in terms of $\xi_1, \ldots, \xi_q$ and re-express it in terms of $\chi_1, \ldots, \chi_p$ so that one can integrate f over the submanifold $\mathcal{M}$.

Suppose $f = \phi \, d\xi_{i_1} \wedge \cdots \wedge d\xi_{i_p}$, a $\mathcal{C}^1$ p-form on $\mathcal{N}$ and we wish to integrate f over $\mathcal{M}$. Presumably ϕ is given in the form $\phi = \phi(\xi_1, \ldots, \xi_q)$. At points of $\mathcal{M}$, we can hope to solve for ξ_i in terms of $\chi_1, \ldots, \chi_p$ and can write ϕ as a function of the variables on $\mathcal{M}$:

$$\phi = \phi\big(\xi_1(\chi_1, \ldots, \chi_p), \ldots, \xi_q(\chi_1, \ldots, \chi_p)\big).$$

The next question is what do we do with $d\xi_i$?

From Part 5 of Proposition 7.7, we have

$$P_{\mathcal{M}}(d\xi_j) = \sum_{i=1}^{p} \frac{\partial \xi_j}{\partial \chi_i} d\chi_i. \tag{7.61}$$

Now if we really have only one manifold, that is, $\mathcal{M} = \mathcal{N}$, then (7.61) becomes

$$d\xi_j = \sum_{i=1}^{p} \frac{\partial \xi_j}{\partial \chi_i} d\chi_i \tag{7.62}$$

and everything becomes simple.

Example 7.22. Suppose $\mathcal{M}$ is a subset $\mathbb{R}^2$ on which we have both cartesian coordinates (ξ_1, ξ_2) and polar coordinates (ρ, θ) with the relations

$$\xi_1 = \rho \cos(\theta) \quad \text{and} \quad \xi_2 = \rho \sin(\theta).$$

Since (7.62) applies, we have

$$d\xi_1 = \cos(\theta) \, d\rho - \rho \sin(\theta) \, d\theta$$
$$d\xi_2 = \sin(\theta) \, d\rho + \rho \cos(\theta) \, d\theta.$$

Suppose we wish to integrate the 2-form $f = (\xi_2/\xi_1) \, d\xi_1 \wedge d\xi_2$ over $\mathcal{M}$ and for some reason we wish to re-express the integral in terms of ρ and θ. We calculate

$$\frac{\xi_2}{\xi_1} = \tan(\theta),$$

$$d\xi_1 \wedge d\xi_2 = \Big(\cos(\theta) \, d\rho - \rho \sin(\theta) \, d\theta\Big) \wedge \Big(\sin(\theta) \, d\rho + \rho \cos(\theta) \, d\theta\Big)$$
$$= \rho \, d\rho \wedge d\theta.$$

Then we have

$$\int_{\mathcal{M}} \frac{\xi_2}{\xi_1} \, d\xi_1 \wedge d\xi_2 \;=\; \int_{\mathcal{M}} \tan(\theta) \, \rho \, d\rho \wedge d\theta.$$

These integrals may be understood to be in either differential form notation or the notation of geometric calculus. If they are understood to be in differential form notation, then the result will be a real number and will be evaluated using some variation or extension of Proposition 7.29. If, on the other hand, we are thinking of them in terms of geometric calculus, then the result will be a 2-vector.

If, however, $p < q$, then we may have $P_{\mathcal{M}}(d\xi_i) \neq d\xi_i$ so that (7.61) applies rather than (7.62). We therefore examine more closely what is going on and note the following:

First, if we want to treat f as a differential p-form over $\mathcal{M}$, then it should be a tangent p-vector field to $\mathcal{M}$, and in general it is not. This suggests that perhaps we want to look at $P_{\mathcal{M}}(f)$ rather than directly at f.

Second (an exercise for the reader), we have

$$\left\langle w_{\mathcal{M}}^{\dagger} f \right\rangle_0 \;=\; \left\langle w_{\mathcal{M}}^{\dagger} P_{\mathcal{M}}(f) \right\rangle_0, \tag{7.63}$$

so if we integrate over $\mathcal{M}$, we *do* wind up looking at the orthogonal projection of f onto $\mathcal{M}$.

Third, if we look at $d\xi_i$, there is more than one concept behind the symbol. Foremost, we have in mind that $d\xi_1, \ldots, d\xi_q$ are the reciprocal vectors for the tangent vectors $\{\partial x / \partial \xi_i\}_{i=1}^{q}$ that were induced on $\mathcal{N}$ by a chart and the associated coordinates. However we also have

$$d\xi_i \;=\; \overrightarrow{\nabla}_{\mathcal{N}} \xi_i \;=\; \overleftarrow{\nabla}_{\mathcal{N}} \xi_i \;=\; \overrightarrow{d}_{\mathcal{N}} \xi_i \;=\; \overleftarrow{d}_{\mathcal{N}} \xi_i.$$

Notice that if we think of $d\xi_i$ as an exterior derivative and as being associated with $\mathcal{M}$ rather than $\mathcal{N}$, by Parts 1 and 5 of Proposition 7.7, we have

$$\overrightarrow{d}_{\mathcal{M}} \xi_i \;=\; \overleftarrow{d}_{\mathcal{M}} \xi_i \;=\; P_{\mathcal{M}}(d\xi_j) \;=\; \sum_{i=1}^{p} \frac{\partial \xi_j}{\partial \chi_i} \, d\chi_i.$$

These considerations lead us to conclude that if we are integrating a p-form f defined on $\mathcal{N}$ over a submanifold $\mathcal{M}$ of dimension p, then we will not get into any trouble if we convert $d\xi_i$ using the formula (7.62).

Example 7.23. Set $\mathcal{N}$ to $\mathbb{R}^3$ with standard cartesian coordinates (ξ_1, ξ_2, ξ_3), and let us take $\mathcal{M}$ to be a helix centered along the ξ_3-axis and with a chart

$$x(\theta) \;=\; \sum_{i=1}^{3} \xi_i(\theta) e_i \;=\; \cos(\theta) \, e_1 + \sin(\theta) \, e_2 + \frac{\theta}{2\pi} \, e_3$$

where $\alpha \leq \theta \leq \beta$. (See Figure 7.19.) Assume the orientation of $\mathcal{M}$, namely $w_{\mathcal{M}}$, agrees with the parametrization by θ.

Now suppose that we have a 1-form $f = \sum_{i=1}^{3} \phi_i \, d\xi_i$ on $\mathbb{R}^3$ that we wish to integrate along $\mathcal{M}$. We think here of each ϕ_i as being a function of $\mathbb{R}^3$ coordinates, $\phi_i = \phi_i(\xi_1, \xi_2, \xi_3)$. We see that

$$\frac{\partial \xi_1}{\partial \theta} \;=\; -\sin(\theta), \qquad \frac{\partial \xi_2}{\partial \theta} \;=\; \cos(\theta), \qquad \frac{\partial \xi_3}{\partial \theta} \;=\; \frac{1}{2\pi}.$$

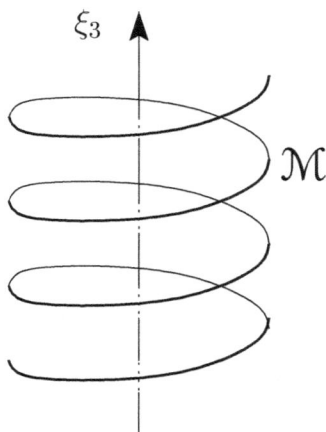

Figure 7.19: Helix in $\mathbb{R}^3$

Since we are preparing to integrate along $\mathcal{M}$, we think of each $d\xi_i$ as being an exterior derivative along $\mathcal{M}$ and use Equation (7.62):

$$d\xi_1 = -\sin(\theta)\,d\theta, \quad d\xi_2 = \cos(\theta)\,d\theta, \quad d\xi_3 = \frac{1}{2\pi}\,d\theta.$$

We then have

$$\underset{\text{(differential form notation)}}{\int_{\mathcal{M}} \sum_{i=1}^{3} \phi_i(\xi_1,\xi_2,\xi_3)\,d\xi_i} = \underset{\text{(geometric calculus)}}{\int_{\mathcal{M}} \left\langle \left(\sum_{i=1}^{3} \phi_i(\xi_1,\xi_2,\xi_3)\,d\xi_i \right) w_{\mathcal{M}} \right\rangle_0}$$

$$= -\int_{\alpha}^{\beta} \phi_1\big(\cos(\theta),\sin(\theta),1/2\pi\big)\,\sin(\theta)\,d\theta$$

$$+ \int_{\alpha}^{\beta} \phi_2\big(\cos(\theta),\sin(\theta),1/2\pi\big)\,\cos(\theta)\,d\theta$$

$$+ \int_{\alpha}^{\beta} \phi_3\big(\cos(\theta),\sin(\theta),1/2\pi\big)\,\frac{1}{2\pi}\,d\theta.$$

7.5.4   The generalized Stokes theorem

We now have some idea of how exterior differentiation and integration of differential forms works. We conclude our discussion by deriving the generalized Stokes theorem.

In differential form theory, this is set down in the equation

$$\int_{\partial\mathcal{M}} f = \int_{\mathcal{M}} df$$

where f is a $(p-1)$-form and $\mathcal{M}$ is a p-manifold. We state and prove it using the language and machinery of geometric calculus and we restrict ourselves, for simplicity, to a p-cell.

There is a peculiar subtlety to this equation. When we use the exterior derivative $d = \overleftarrow{d}_{\mathcal{M}}$, we depend on the construction of the left-hand geometric derivative, but it turns out the orientation we want w to induce on the boundary of the cell is the right-hand version, $\overrightarrow{\partial} w$.

Proposition 7.30 (The generalized Stokes theorem). *Let $\mathcal{M}$ be a $\mathcal{C}^2$ p-cell in $\mathbb{R}^n$ and f a $\mathcal{C}^2$ $(p-1)$-differential form on $\mathcal{M}$. Suppose that $\mathcal{M}$ has orientation w. Then*

$$\int_{\partial\mathcal{M}} \left\langle (\overrightarrow{\partial}w)^\dagger f \right\rangle_0 = \int_{\mathcal{M}} \left\langle w^\dagger (\overleftarrow{d}_{\mathcal{M}}f) \right\rangle_0.$$

Proof. In Equation (6.31) of Theorem 6.1, if we replace f by 1 and g by f and w by $w^\dagger$, then we have the following case of the Fundamental Theorem:

$$\int_{\partial\mathcal{M}} \overleftarrow{\partial}(w^\dagger) f = \int_{\mathcal{M}} w^\dagger (\overleftarrow{\nabla}_{\mathcal{M}}f).$$

We take the grade 0 part of each side of the last equation, move the grade 0 operation inside the integral, and apply Proposition 6.13 to obtain

$$\int_{\partial\mathcal{M}} \left\langle (\overrightarrow{\partial}w)^\dagger f \right\rangle_0 = \int_{\mathcal{M}} \left\langle w^\dagger (\overleftarrow{\nabla}_{\mathcal{M}}f) \right\rangle_0. \tag{7.64}$$

From Equation (7.63), we know that

$$\left\langle w^\dagger (\overleftarrow{\nabla}_{\mathcal{M}}f) \right\rangle_0 = \left\langle w^\dagger P_{\mathcal{M}} (\overleftarrow{\nabla}_{\mathcal{M}}f) \right\rangle_0,$$

and since $\overleftarrow{\nabla} = \overleftarrow{\text{div}} + \overleftarrow{\text{curl}}$, we have

$$\left\langle w^\dagger (\overleftarrow{\nabla}_{\mathcal{M}}f) \right\rangle_0 = \left\langle w^\dagger P_{\mathcal{M}} (\overleftarrow{\text{div}}_{\mathcal{M}}f) \right\rangle_0 + \left\langle w^\dagger P_{\mathcal{M}} (\overleftarrow{\text{curl}}_{\mathcal{M}}f) \right\rangle_0.$$

Now $\overleftarrow{\text{div}}_{\mathcal{M}}f$ must be grade $p-2$ since divergence decreases grade by 1, and because the projection $P_{\mathcal{M}}$ does not change grades, the same must be true for $P_{\mathcal{M}}(\overleftarrow{\text{div}}_{\mathcal{M}}f)$. We see that $w^\dagger P_{\mathcal{M}}(\overleftarrow{\text{div}}_{\mathcal{M}}f)$ is a product of terms of grade p and $p-2$, hence cannot have any terms of grade 0. It follows that

$$\left\langle w^\dagger (\overleftarrow{\nabla}_{\mathcal{M}}f) \right\rangle_0 = \left\langle w^\dagger P_{\mathcal{M}} (\overleftarrow{\text{curl}}_{\mathcal{M}}f) \right\rangle_0 = \left\langle w^\dagger (\overleftarrow{d}_{\mathcal{M}}f) \right\rangle_0.$$

When we plug this result into Equation (7.64), it gives the desired conclusion. $\square$

Exercises 7.5.

1. Compute the left exterior derivatives $\overleftarrow{d}_{\mathcal{M}}f$ of f which is defined in terms of the coordinates $(\chi_1, \dots, \chi_p)$ on a manifold $\mathcal{M}$.

   (a) $f(\chi_1, \chi_2) = \dfrac{\chi_1}{\chi_1^2 + \chi_2^2} d\chi_1 - \dfrac{\chi_2}{\chi_1^2 + \chi_2^2} d\chi_2$

(b) $f(\chi_1, \ldots, \chi_p) = \sum_{i=1}^{p} \phi_i(\chi_i) \, d\chi_i$ where it is to be understood that each ϕ_i is $\mathcal{C}^1$ and is a function of χ_i only, not of any other χ_j.

(c) $f(\chi_1, \ldots, \chi_p) = \sum_{i=1}^{p} \psi_i(\chi_1, \ldots, \chi_p) \, d\chi_1 \wedge \cdots \wedge \widehat{d\chi_i} \wedge \cdots \wedge d\chi_p$ where each ψ_i is $\mathcal{C}^1$.

2. Prove Proposition 7.28.

3. If $f = \phi \, d\chi_{i_1} \wedge \cdots \wedge d\chi_{i_k}$, a $\mathcal{C}^2$ differential form, construct a formula analogous to that of Proposition 7.27 for $\vec{d}_{\mathcal{M}} f$.

4. Show that $\vec{d}_{\mathcal{M}} \, \vec{d}_{\mathcal{M}} = 0$ when applied to $\mathcal{C}^2$ multivector fields on a manifold.

5. If $\mathcal{M}$ is a $\mathcal{C}^1$ p-manifold in $\mathbb{R}^n$ with orientation w and if f is $\mathcal{C}^1$ p-vector field on $\mathcal{M}$, then show that at every point of $\mathcal{M}$ we have

$$\left\langle w^\dagger f \right\rangle_0 = \left\langle w^\dagger P_{\mathcal{M}}(f) \right\rangle_0.$$

7.6   The normal identity

If $\mathcal{M}$ is a manifold in $\mathbb{R}^n$, one sometimes wishes to use calculus-type operations on the identity map $I \colon \mathcal{M} \to \mathcal{M}$. In particular, one wants to calculate directional derivatives. If one extends I to the identity on $\mathbb{R}^n$, this can lead to different results depending on how $\mathcal{M}$ is embedded in $\mathbb{R}^n$. Another way to extend I—a way which we call the *normal identity on* $\mathcal{M}$—leads to results that have a more intrinsic character and are in more accord with the usual results of operations on the identity map in the literature of geometric calculus.

7.6.1   The idea of the normal identity

Suppose that $\mathcal{M}$ is a $\mathcal{C}^r$ p-manifold in $\mathbb{R}^n$ (where $r \geq 2$). Recall that $T_x\mathcal{M}$ is the *tangent space to* $\mathcal{M}$ *at* x. By $N_x\mathcal{M}$, *the normal space to* $\mathcal{M}$ *at* x, we shall mean the orthogonal complement to $T_x\mathcal{M}$. That is, $N_x\mathcal{M}$ is the vector subspace of $\mathbb{R}^n$ consisting of those vectors v such that $v \cdot u = 0$ for all $u \in T_x\mathcal{M}$.

We then introduce the $(n-p)$-dimensional hyperplane $H_x\mathcal{M}$ by

$$H_x\mathcal{M} \overset{\text{def.}}{=} \{x + v : v \in N_x\mathcal{M}\} = x + N_x\mathcal{M}.$$

We think of this as the hyperplane that passes through x and is orthogonal to $\mathcal{M}$ at x. (See Figure 7.20 for the cases in $\mathbb{R}^3$ where $p = 1$ and $p = 2$.)

It is now easy to explain what we want the *normal identity* to be: For $y \in \mathbb{R}^n$, provided y is sufficiently close to $\mathcal{M}$, we set

$$I_{\mathcal{M}}(y) = x \quad \text{when } y \in H_x\mathcal{M}$$

where $I_{\mathcal{M}}$ is our symbol for the normal identity map associated with $\mathcal{M}$. It is clear that for $x \in \mathcal{M}$, we have $I_{\mathcal{M}}(x) = x$.

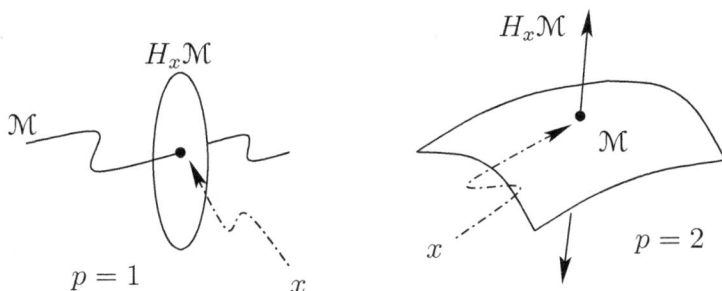

Figure 7.20: $H_x\mathcal{M}$ for 1- and 2-manifolds

Example 7.24. Let $\mathcal{M}$ be the set of points in $\mathbb{R}^3$ of the form $\chi(e_1 + e_2 + e_3)$ where $\chi \in \mathbb{R}$. This amounts to the straight line through the origin that makes a $45°$ angle with each χ_i-axis and passes through the first octant of $\mathbb{R}^3$. Then given $x_0 = (\chi_{01}, \chi_{02}, \chi_{03}) = \sum_{i=1}^{3} \chi_{0i} e_i$ in $\mathbb{R}^3$, we would like to compute $I_{\mathcal{M}}(x_0)$.

Set $e = e_1 e_2 e_3$ and $u = e_1 + e_2 + e_3$. We know that $I_{\mathcal{M}}(x_0) = \chi u$ for some $\chi \in \mathbb{R}$, so our problem is to compute χ from x_0. We need some information about the the orthogonal plane $H_x\mathcal{M}$ to $\mathcal{M}$ that passes through $x \in \mathcal{M}$. Since u is tangent to $\mathcal{M}$, we can compute an orthogonal 2-vector z to $\mathcal{M}$ by dividing u by e:

$$z \overset{\text{def.}}{=} ue^{\dagger} = e_3 e_2 - e_3 e_1 + e_2 e_1.$$

The vector from x_0 to $I_{\mathcal{M}}(x_0)$ must lie in the plane determined by z; that is,

$$\left(I_{\mathcal{M}}(x_0) - x_0\right) \wedge z = 0.$$

Since u is orthogonal to z and $|u|^2 = 3$, we have

$$I_{\mathcal{M}}(x_0) \wedge z = \chi u \wedge z = \chi u u e^{\dagger} = 3\chi e^{\dagger}.$$

We next notice that $e_i \wedge z = e^{\dagger}$ for $i = 1, 2, 3$, and from that fact, we calculate

$$x_0 \wedge z = \sum_{i=1}^{3} \chi_{0i} \, e^{\dagger}.$$

Using the last three equations we can solve for χ and obtain our desired formula:

$$I_{\mathcal{M}}(\chi_{01}, \chi_{02}, \chi_{03}) = \frac{1}{3}(\chi_{01} + \chi_{02} + \chi_{03})(e_1 + e_2 + e_3).$$

The reader may feel that the existence of the normal identity is intuitively obvious, however we wish to give a moderately careful construction and to establish that when $\mathcal{M}$ is $\mathcal{C}^r$, then $I_{\mathcal{M}}$ is at least $\mathcal{C}^{r-1}$.

In the proposition that follows and in its proof, if x_0 is a point in $\mathbb{R}^n$ and ξ is a positive number, then by $B(x_0, \xi)$ we mean the set of $x \in \mathbb{R}^n$ such that $|x - x_0| < \xi$. That is, $B(x_0, \xi)$ is the *open ball* in $\mathbb{R}^n$ centered at x_0 with radius ξ. If $t \in \mathbb{R}^p$ and $t' \in \mathbb{R}^q$, then $B\big((t,t'), \delta\big)$ is the open ball in $\mathbb{R}^{p+q}$ that is centered at $(t,t') \in \mathbb{R}^{p+q}$.

We defer the proof of the proposition to Section A.8 of Appendix A.

Proposition 7.31. *Suppose that $\mathcal{M}$ is a $\mathcal{C}^r$ p-manifold (where $r \geq 2$ and $p \geq 1$) in $\mathbb{R}^n$ and $x_0 \in \mathcal{M}$. Then for every $\xi > 0$ there is an open neighborhood V in $\mathbb{R}^n$ of x_0 and a map $I_{\mathcal{M}} : V \to V$ such that the following hold:*

1. *$V \subseteq B(x_0, \xi)$.*

2. *For all $x \in V \cap \mathcal{M}$ and $y \in V$, $I_{\mathcal{M}}(y) = x$ if and only if $y \in V \cap H_x\mathcal{M}$.*

3. *$I_{\mathcal{M}}$ is $\mathcal{C}^{r-1}$.*

Remark 7.15. We note that the map $I_{\mathcal{M}}$ is *almost* uniquely defined. If both the conditions $y \in V \cap H_x\mathcal{M}$ and $x \in \mathcal{M}$ are satisfied, then $I_{\mathcal{M}}(y) = x$; that is, $I_{\mathcal{M}}$ is unique under those conditions. However if $y \in V$ but $y \notin H_x\mathcal{M}$ for some $x \in \mathcal{M}$—a situation which can occur for y close to the boundary of $\mathcal{M}$—then there may be more than one choice for $I_{\mathcal{M}}$. More is said about this in the proof of the proposition.

7.6.2   Properties

From this point on, we assume $\mathcal{M}$ is a $\mathcal{C}^r$ p-manifold in $\mathbb{R}^n$ with $r \geq 2$ and whenever we use the symbol $I_{\mathcal{M}}$, we mean the normal identity. Unless we say otherwise below, we shall use P_{x_0} for the orthogonal projection map $\mathbb{R}^n \to T_{x_0}\mathcal{M}$.

We must not confuse $I_{\mathcal{M}}$ with an orthogonal projection onto a vector subspace. The connection with subspace orthogonal projections is this:

Corollary 7.7. *Whenever $x_0 \in \mathcal{M}$, then $I_{\mathcal{M}}(y) = x_0$ implies $P_{x_0}(y) = P_{x_0}(x_0)$.*

Proof. By Proposition 7.31, if $I_{\mathcal{M}}(y) = x_0$, it follows that $y \in H_{x_0}\mathcal{M}$. From this we see that we can write $y = x_0 + v$ where v is orthogonal to $T_{x_0}\mathcal{M}$. Therefore

$$P_{x_0}(y) \; = \; P_{x_0}(x_0) + P_{x_0}(v) \; = \; P_{x_0}(x_0). \qquad \square$$

We next note that directional derivatives of the normal identity are always nicely related to the manifold.

Corollary 7.8. *If $x_0 \in \mathcal{M}$ and $a \in \mathbb{R}^n$, then*

$$\partial_a I_{\mathcal{M}}(x_0) \; = \; P_{x_0} a.$$

Proof. We treat x_0 as an interior point of $\mathcal{M}$.

Define a function h from an open neighborhood of 0 in $\mathbb{R}$ into $\mathcal{M}$ by $h(\lambda) = I_{\mathcal{M}}(x_0 + \lambda\, a)$. Since $I_{\mathcal{M}}$ is $\mathcal{C}^{r-1}$, we see that h defines a $\mathcal{C}^{r-1}$ path in $\mathcal{M}$ passing through x_0 at $\lambda = 0$. Then

$$\partial_a I_{\mathcal{M}}(x_0) \; = \; \lim_{\lambda \to 0} \frac{1}{\lambda}\Big(I_{\mathcal{M}}(x_0 + \lambda\, a) - I_{\mathcal{M}}(x_0)\Big) \; = \; h'(0).$$

Since $h(\lambda) = I_{\mathcal{M}}(x_0 + \lambda\, a)$, by Corollary 7.7, $P_{h(\lambda)}\big(h(\lambda)\big) = P_{h(\lambda)}(x_0 + \lambda\, a)$. This amounts to

$$P_{h(\lambda)}\left(\frac{h(\lambda) - h(0)}{\lambda} - a\right) = 0. \tag{7.65}$$

It is straightforward to construct an orthonormal basis $\{u_i(x)\}_{i=1}^{p}$ for $T_x\mathcal{M}$ for all x in some neighborhood in $\mathcal{M}$ of x_0 such that $x \mapsto u_i(x)$ is $\mathcal{C}^{r-1}$. For such x, the orthogonal projection map is given by

$$P_x(v) = \sum_{i=1}^{p} \big(v \cdot u_i(x)\big) u_i(x).$$

Then Equation (7.65) becomes

$$\sum_{i=1}^{p} \left[u_i\big(h(\lambda)\big) \cdot \left(\frac{h(\lambda) - h(0)}{\lambda} - a\right) \right] u_i\big(h(\lambda)\big) = 0.$$

Using the facts that all the functions involved are $\mathcal{C}^{r-1}$ and $h'(0)$ is a tangent vector to $\mathcal{M}$, we let $\lambda \to 0$ and obtain $h'(0) = P_{x_0}(a)$. This gives us the desired result. $\qquad\square$

7.6.3 Applications

If f is a multivector field on $\mathcal{M}$, to say f is $\mathcal{C}^r$ means that at every point x_0 of $\mathcal{M}$ there must exist an extension of f to an open neighborhood of x_0 in $\mathbb{R}^n$ such that the extended f is $\mathcal{C}^r$ on that open neighborhood.

Suppose we use this f to form the new function $f \circ I_{\mathcal{M}}$. This function will be $\mathcal{C}^{r-1}$ on $\mathcal{M}$ since $I_{\mathcal{M}}$ is $\mathcal{C}^{r-1}$. We cannot expect f and $f \circ I_{\mathcal{M}}$ to have the same values at points not on $\mathcal{M}$, however $f(x_0) = \big(f \circ I_{\mathcal{M}}\big)(x_0)$ whenever $x_0 \in \mathcal{M}$. This has the following interesting consequences:

1. The values of $f \circ I_{\mathcal{M}}$ are uniquely determined by the values of f at points on $\mathcal{M}$.

2. $\partial_v f = \partial_v(f \circ I_{\mathcal{M}})$ whenever v is a tangent vector to $\mathcal{M}$.

3. $\vec{\nabla}_{\mathcal{M}} f = \vec{\nabla}_{\mathcal{M}}(f \circ I_{\mathcal{M}})$ and $\overleftarrow{\nabla}_{\mathcal{M}} f = \overleftarrow{\nabla}_{\mathcal{M}}(f \circ I_{\mathcal{M}})$.

These considerations suggest the following idea:

Definition 7.11. Given a manifold $\mathcal{M}$ in $\mathbb{R}^n$ and a function f whose domain lies in $\mathbb{R}^n$, we say f is *normally extended from* $\mathcal{M}$ provided $f = f \circ I_{\mathcal{M}}$ in neighborhoods of all points of $\mathcal{M}$ at which f is defined.

Example 7.25. Suppose $\mathcal{M}$ is the plane $\chi_3 = \beta$ in $\mathbb{R}^3$. The function $\phi(\chi_1, \chi_2, \chi_3) = \chi_1^2 + \chi_2^2 + \chi_3^2$ is $\mathcal{C}^\infty$ on all of $\mathbb{R}^3$, hence on $\mathcal{M}$. The normal identity for $\mathcal{M}$ is clearly

$$I_{\mathcal{M}}(\chi_1, \chi_2, \chi_3) = (\chi_1, \chi_2, \beta).$$

Then

$$\big(\phi \circ I_{\mathcal{M}}\big)(\chi_1, \chi_2, \chi_3) = \chi_1^2 + \chi_2^2 + \beta^2.$$

The functions ϕ and $\phi \circ I_{\mathcal{M}}$ agree on $\mathcal{M}$, have the same directional derivatives (with respect to tangent vectors), and the same geometric derivatives.

We see then that we can always hope to replace f on a manifold by a function that is normally extended. The price we pay is moving from a $\mathcal{C}^r$ to a $\mathcal{C}^{r-1}$ function. However the normally extended function can be more well-behaved. Here is the principal way that is true:

Corollary 7.9. *If f is a normally extended $\mathcal{C}^1$ function on a $\mathcal{C}^2$ manifold $\mathcal{M}$, x_0 is a point on $\mathcal{M}$, a is a vector in $\mathbb{R}^n$, and $b = P_{x_0}a$, then*

$$\partial_a f(x_0) \;=\; \partial_b f(x_0).$$

Proof. We know that the normal identity $I_{\mathcal{M}}$ must be $\mathcal{C}^1$, so we can apply the chain rule to the composition of $\mathcal{C}^1$ functions $f \circ I_{\mathcal{M}}$. It then follows from Corollary 7.8 that

$$\left[f'(x_0)\right]a \;=\; \left[f'(x_0)\,I'_{\mathcal{M}}(x_0)\right]a \;=\; \left[f'(x_0)\right]b. \qquad \square$$

Now let us return to the setting in which $\mathcal{M}$ and $\mathcal{N}$ are $\mathcal{C}^2$ p- and q-manifolds respectively in $\mathbb{R}^n$ such that

1. $\mathcal{M} \subseteq \mathcal{N}$,

2. x_0 is a point in $\mathcal{M}$,

3. $\mathcal{M}$ has $\mathcal{C}^2$ coordinates $(\chi_1, \dots, \chi_p)$ and $\mathcal{N}$ has $\mathcal{C}^2$ coordinates $(\xi_1, \dots, \xi_q)$ associated with charts that both cover x_0.

By $I_{\mathcal{M}}$ and $I_{\mathcal{N}}$ we mean the normal identities of the indicated manifolds.

We wish to now add to the results that we obtained in Proposition 7.7. Recall that in that proposition we considered how expressions changed when one switched from partial derivatives in terms of one coordinate system to those in terms of another coordinate system.

Proposition 7.32. *If a $\mathcal{C}^1$ multivector field f is normally extended from $\mathcal{M}$, then $\vec{\nabla}_{\mathcal{N}} f = \vec{\nabla}_{\mathcal{M}} f$ on $\mathcal{M}$.*

Proof. We know that the tangent spaces $T_{x_0}\mathcal{M}$ and $T_{x_0}\mathcal{N}$ have dimensions p and q respectively and $T_{x_0}\mathcal{M} \subseteq T_{x_0}\mathcal{N}$. At x_0, let us set

$$u_i \;=\; \frac{\partial I_{\mathcal{M}}}{\partial \chi_i}(x_0) \quad \text{for } i = 1, \dots, p$$

and let us take $u_{p+1}, u_{p+2}, \dots, u_q$ to be orthonormal vectors that are in $T_{x_0}\mathcal{N}$ and orthogonal to $T_{x_0}\mathcal{M}$. Let $\{m_i\}_{i=1}^q$ be the reciprocal frame for $\{u_i\}_{i=1}^q$. It is easily seen that at x_0,

$$m_i \;=\; \begin{cases} d\chi_i & \text{for } i = 1, \dots, p, \\ u_i & \text{for } i = p+1, \dots, q. \end{cases}$$

The geometric derivative of f over $\mathcal{N}$ at x_0 is then

$$\vec{\nabla}_{\mathcal{N}} f \;=\; \sum_{i=1}^{q} (\partial_{u_i} f)\, m_i.$$

The orthogonal projection of u_i to $T_{x_0}\mathcal{M}$ is 0 for $i = p+1,\ldots,q$, so by Corollary 7.9, $\partial_{u_i} f = 0$ for those values of i. Thus at x_0 we have

$$\vec{\nabla}_{\mathcal{N}} f = \sum_{i=1}^{p} (\partial_{u_i} f)\, m_i = \sum_{i=1}^{p} \frac{\partial f}{\partial \chi_i}\, d\chi_i.$$

But this is $\vec{\nabla}_{\mathcal{M}} f$, so the result is established. □

Proposition 7.33. *If the coordinate function χ_i is normally extended from $\mathcal{M}$, then*

$$\frac{\partial \chi_i}{\partial \xi_j} = d\chi_i \cdot \frac{\partial I_{\mathcal{N}}}{\partial \xi_j}.$$

Proof. Note that

$$\vec{\nabla}_{\mathcal{N}} \chi_i = \sum_{j=1}^{q} \frac{\partial \chi_i}{\partial \xi_j}\, d\xi_j.$$

We know that $\{\partial I_{\mathcal{N}}/\partial \xi_j\}_{j=1}^{q}$, if evaluated at a point of $\mathcal{N}$, is a basis for the tangent space to $\mathcal{N}$ at that point and that $\{d\xi_j\}_{j=1}^{q}$ is the reciprocal basis. So if we dot the last equation by $\partial I_{\mathcal{N}}/\partial \xi_j$, then

$$\left(\vec{\nabla}_{\mathcal{N}} \chi_i\right) \cdot \frac{\partial I_{\mathcal{N}}}{\partial \xi_j} = \frac{\partial \chi_i}{\partial \xi_j}.$$

By Proposition 7.32, we have $\vec{\nabla}_{\mathcal{N}} \chi_i = \vec{\nabla}_{\mathcal{M}} \chi_i$. By Corollary 6.1, we have $d\chi_i = \vec{\nabla}_{\mathcal{M}} \chi_i$. This gives us the desired formula. □

The following formula is exactly what one expects based on an introductory course in multivariable calculus.

Proposition 7.34. *If f is a $\mathcal{C}^1$ multivector field on $\mathcal{M}$ that is normally extended from $\mathcal{M}$ and the coordinate functions χ_i are also normally extended from $\mathcal{M}$, then at all points of $\mathcal{M}$ we have*

$$\frac{\partial f}{\partial \xi_j} = \sum_{i=1}^{p} \frac{\partial \chi_i}{\partial \xi_j} \frac{\partial f}{\partial \chi_i}.$$

Proof. Let ϕ be a real-valued $\mathcal{C}^1$ function defined on $\mathcal{M}$ that is normally extended from $\mathcal{M}$. By Proposition 7.32, $\vec{\nabla}_{\mathcal{N}} \phi = \vec{\nabla}_{\mathcal{M}} \phi$. Thus

$$\sum_{j=1}^{q} \frac{\partial \phi}{\partial \xi_j}\, d\xi_j = \sum_{i=1}^{p} \frac{\partial \phi}{\partial \chi_i}\, d\chi_i$$

on $\mathcal{M}$. We dot this last equation with $\partial I_{\mathcal{N}}/\partial \xi_j$ and obtain

$$\frac{\partial \phi}{\partial \xi_j} = \sum_{i=1}^{p} \frac{\partial \phi}{\partial \chi_i} \left(d\chi_i \cdot \frac{\partial I_{\mathcal{N}}}{\partial \xi_j} \right)$$

$$= \sum_{i=1}^{p} \frac{\partial \phi}{\partial \chi_i} \frac{\partial \chi_i}{\partial \xi_j}$$

$$(7.66)$$

where the second step is justified by Proposition 7.33. Since f can be written as a sum of terms of the form $\phi\, e_I$ (where $e_1, \ldots, e_n$ is the standard basis for $\mathbb{R}^n$ and I is an ordered multi-index) and since

$$\frac{\partial(\phi\, e_I)}{\partial \xi_j} = \frac{\partial \phi}{\partial \xi_j}\, e_I,$$

we see that (7.66) gives us the desired result.   □

One final comment concerns a matter of notation from the literature of geometric calculus.

Given a manifold $\mathcal{M}$, the symbol x is used in geometric calculus, in works such as [22] and [42], to denote the identity function on $\mathcal{M}$ or an arbitrary point of $\mathcal{M}$. Its behavior is like that of our concept of the normal identity, and this suggests the use of x as an alternate symbol for $I_{\mathcal{M}}$.

A drawback of this practice is that we now have three ways in which we use this small and innocent-looking symbol:

1. As a chart, $x\colon U \to \mathbb{R}^n$ on $\mathcal{M}$.

2. As a point on the manifold, $x \in \mathcal{M}$.

3. And now as a particular map, the normal identity, $x = I_{\mathcal{M}}\colon V \to V$ acting on neighborhoods of points of $\mathcal{M}$.

There is a possibility of confusion. However in practice, the particular meaning of any given x should be clear, and the resulting applications have an appealing quality of *naturalness*. For example, it seems reasonably sensible to write

$$f = f \circ I_{\mathcal{M}} = f(x).$$

Exercises 7.6.

1. Let $\mathcal{M}$ be the unit circle $\chi_1^2 + \chi_2^2 = 1$ and $\chi_3 = 0$ in the $\chi_1 \chi_2$-plane in $\mathbb{R}^3$. Construct a normal identity $I_{\mathcal{M}}\colon U \to \mathcal{M}$ where U is the set of points (χ_1, χ_2, χ_3) not lying on the χ_3-axis, that is, the set of points for which either χ_1 or χ_2 is nonzero.

2. Suppose f is a normally extended $\mathcal{C}^r$ function on a $\mathcal{C}^r$ manifold $\mathcal{M}$ and the vector a is orthogonal to $\mathcal{M}$ at the point $x_0 \in \mathcal{M}$. Show that $\partial_a f(x_0) = 0$.

3. Prove a version of Proposition 7.32 which is good for left-hand geometric derivatives.

4. Recall Definition 7.11 of what it means for a function to be normally extended from a manifold. Let $\mathcal{M}$ be the unit circle $\chi_1^2 + \chi_2^2 = 1$ in $\mathbb{R}^2$. Are the following functions normally extended from $\mathcal{M}$?

    (a) $\phi(\chi_1, \chi_2) = \dfrac{\chi_2}{\chi_1}$.

    (b) $\psi(\chi_1, \chi_2) = \chi_1^2 + \chi_2^2$.

Appendices

Appendix A

Proofs

A.1 Manifolds

Proposition 2.7. *Suppose ϕ is a real-valued $\mathcal{C}^1$ function defined on an open subset of $\mathbb{R}^n$. Let x_0 be a point in the domain of ϕ and v be a vector in $\mathbb{R}^n$. Let $c: J \to \mathbb{R}^n$ be a $\mathcal{C}^1$ arc in $\mathbb{R}^n$ such that $c(\tau_0) = x_0$ and $c'(\tau_0) = v$. Then (2.7) and (2.8) both give the same result.*

Proof. Since ϕ and c are $\mathcal{C}^1$, by Definition 2.6, we can write

$$\phi(x) - \phi(x_0) = \left[\phi'(x_0)\right](x - x_0) + g_0(x - x_0),$$
$$c(\tau) - x_0 = c'(\tau_0)(\tau - \tau_0) + g_1(\tau - \tau_0) = v(\tau - \tau_0) + g_1(\tau - \tau_0),$$

where x is sufficiently close to x_0, τ is sufficiently close to τ_0,

$$\lim_{x \to x_0} \frac{g_0(x - x_0)}{|x - x_0|} = 0, \quad \text{and} \quad \lim_{\tau \to \tau_0} \frac{g_1(\tau - \tau_0)}{\tau - \tau_0} = 0.$$

Then

$$\phi(c(\tau)) - \phi(x_0) = \left[\phi'(x_0)\right]\left(c(\tau) - x_0\right) + g_0\left(c(\tau) - x_0\right)$$
$$= \left[\phi'(x_0)\right]\left(v(\tau - \tau_0) + g_1(\tau - \tau_0)\right) + g_0\left(c(\tau) - x_0\right).$$

Thus

$$\frac{\phi(c(\tau)) - \phi(x_0)}{\tau - \tau_0} = \left[\phi'(x_0)\right] v + \left[\phi'(x_0)\right] \frac{g_1(\tau - \tau_0)}{\tau - \tau_0} + \frac{g_0\left(c(\tau) - x_0\right)}{\tau - \tau_0} \tag{A.1}$$

where $\tau \neq \tau_0$. We know that $g_1(\tau - \tau_0)/(\tau - \tau_0) \to 0$ as $\tau \to \tau_0$. Next we have

$$\frac{g_0\left(c(\tau) - x_0\right)}{\tau - \tau_0} = \frac{g_0\left(c(\tau) - x_0\right)}{|c(\tau) - x_0|} \frac{|c(\tau) - x_0|}{\tau - \tau_0}$$

255

$$= \pm \frac{g_0\big(c(\tau)-x_0\big)}{|c(\tau)-x_0|} \left| \frac{c(\tau)-x_0}{\tau-\tau_0} \right| \to 0 \quad \text{as } \tau \to \tau_0.$$

(We have used here the fact that c is a arc so that $c(\tau) \neq c(\tau_0) = x_0$ if $\tau \neq \tau_0$.) We then see from Equation (A.1) and Proposition 2.3 that

$$\left[\phi'(x_0)\right] v = \partial_v \phi(x_0) = \lim_{\tau \to \tau_0} \frac{\phi(c(\tau)) - \phi(x_0)}{\tau - \tau_0}$$

as promised. $\square$

A.2   Simple k-vectors

Proposition 3.3.   *Suppose $a_1, \ldots, a_k \in \mathbb{R}^n$. Then the following are equivalent:*

1. *$vol(a_1, \ldots, a_k) > 0$.*

2. *$a_1 \wedge \cdots \wedge a_k \neq 0$.*

3. *$a_1, \ldots, a_k$ are linearly independent.*

Proof. We know, of course, that $a_1 \wedge \cdots \wedge a_k = 0$ if and only if $a_1, \ldots, a_k$ are linearly dependent; so the equivalence of the second and third parts is trivial.

The only part that requires a moment of thought is the first one. Let U be a k-dimensional vector subspace of $\mathbb{R}^n$ containing $a_1, \ldots, a_k$. Choose an orthonormal basis $\{u_i\}_{i=1}^k$ for U. Next construct two $k \times k$ matrices: The first is $A = (a_i \cdot a_j)_{i,j=1}^k$, the second is $B = (a_1 \cdots a_k)$ where each a_i is written as a column vector in terms of its expansion in the basis $\{u_i\}_{i=1}^k$. These are all square matrices and satisfy $A = B^T B$ where B^T is the transpose of B. It then easily follows that

$$\big(\mathrm{vol}(a_1, \ldots, a_k)\big)^2 = \det(A) = \big(\det(B)\big)^2.$$

We know that $\det(B) \neq 0$ if and only if $a_1, \ldots, a_k$ are linearly independent. This establishes the result. $\square$

Proposition 3.4.   *Let $\{a_1, \ldots, a_k\}$ and $\{b_1, \ldots, b_k\}$ be sets of linearly independent vectors in $\mathbb{R}^n$ that span the same vector subspace V. Let A and B be the $k \times k$ matrices $A = (a_1, \ldots, a_k)$ and $B = (b_1, \ldots, b_k)$ in terms of some orthonormal basis of V. Let $f : V \to V$ be the unique linear transformation satisfying $f(a_i) = b_i$ for all i. Then the following are equivalent:*

1. *$a_1 \wedge \cdots \wedge a_k = b_1 \wedge \cdots \wedge b_k$.*

2. *$\det A = \det B$.*

3. *$\det f = 1$.*

Proof. Let $\{u_i\}_{i=1}^k$ be the orthonormal basis of V with respect to which we calculate A and B. Take F to be the $k \times k$ matrix of f calculated with respect to $\{u_i\}_{i=1}^k$. Since $b_i = f(a_i)$, we have the matrix equation $B = FA$.

Because we have calculated our matrices with respect to an orthonormal basis, we must have

$$\begin{aligned}
A^T A &= (a_i \cdot a_j)_{k \times k}, \\
B^T B &= (b_i \cdot b_j)_{k \times k}, \\
A^T B &= (a_i \cdot b_j)_{k \times k}
\end{aligned}$$

where A^T and B^T are the transposes of A and B. It follows that

$$\begin{aligned}
\mathrm{vol}(a_1, \ldots, a_k)^2 &= \det(a_i \cdot a_j) = \det(A^T A) = (\det A)^2, \\
\mathrm{vol}(b_1, \ldots, b_k)^2 &= \det(b_i \cdot b_j) = \det(B^T B) = (\det B)^2, \\
\det(a_i \cdot b_j) &= \det(A^T B) = (\det A)(\det B).
\end{aligned}$$

Now we prove our equivalences.

Suppose that $a_1 \wedge \cdots \wedge a_k = b_1 \wedge \cdots \wedge b_k$. We have $\mathrm{vol}(a_1, \ldots, a_k) = \mathrm{vol}(b_1, \ldots, b_k)$ and $\det(a_i \cdot b_j) > 0$. This forces $(\det A)(\det B) > 0$ and $(\det A)^2 = (\det B)^2$. Thus $\det A = \det B$.

Now suppose that $\det A = \det B$. We know these determinants are nonzero since $\{a_i\}_{i=1}^k$ is a linearly independent set. Then

$$\mathrm{vol}(a_1, \ldots, a_k) = \sqrt{(\det A)^2} = \sqrt{(\det B)^2} = \mathrm{vol}(b_1, \ldots, b_k),$$

and $\det(a_i \cdot b_j) = (\det A)^2 > 0$. These conditions amount to $a_1 \wedge \cdots \wedge a_k = b_1 \wedge \cdots \wedge b_k$.

Finally, since the condition $f(a_i) = b_i$ amounts to the matrix equation $B = FA$, we have

$$\det B = (\det F)(\det A) = (\det f)(\det A).$$

We know that both $\det A$ and $\det B$ are nonzero. Then $\det f = 1$ if and only if $\det A = \det B$. $\qquad\square$

Proposition 3.5 *Let $a_1, \ldots, a_k$ be vectors in $\mathbb{R}^n$ and λ be a scalar. Set*

$$\begin{aligned}
a &= a_1 \wedge \cdots \wedge a_i \wedge \cdots \wedge a_k, \\
b &= a_1 \wedge \cdots \wedge \lambda a_i \wedge \cdots \wedge a_k.
\end{aligned}$$

Then the following hold:

1. *If $\lambda = 0$, then $b = 0$.*

2. *$\mathrm{vol}(b) = |\lambda| \, \mathrm{vol}(a)$.*

3. *If $\lambda \neq 0$, then $a \neq 0$ if and only if $b \neq 0$.*

4. Suppose $a, b \neq 0$. Then a and b have the same orientation if and only if $\lambda > 0$. If $\lambda < 0$, then they have opposite orientations.

Proof. Suppose $\lambda = 0$. Then $\lambda a_i = 0$, and $\{a_1, \ldots, 0, \ldots, a_k\}$ is a linearly dependent set. Thus $b = 0$.

For the remaining items, we check only the case $k = 2$ and attach λ to a_1. The general proofs work the same way.

To check the statement about the volume, we calculate

$$
\begin{aligned}
\mathrm{vol}(\lambda a_1 \wedge a_2) &= \sqrt{\det \begin{pmatrix} (\lambda a_1) \cdot (\lambda a_1) & (\lambda a_1) \cdot a_2 \\ a_2 \cdot (\lambda a_1) & a_2 \cdot a_2 \end{pmatrix}} \\
&= \sqrt{\lambda^2 \det \begin{pmatrix} a_1 \cdot a_1 & a_1 \cdot a_2 \\ a_2 \cdot a_1 & a_2 \cdot a_2 \end{pmatrix}} \\
&= |\lambda| \, \mathrm{vol}(a_1 \wedge a_2).
\end{aligned}
$$

That is, $\mathrm{vol}(b) = |\lambda| \, \mathrm{vol}(a)$.

We see from this last statement, that if we know $\lambda \neq 0$, then $\mathrm{vol}(a) > 0$ if and only if $\mathrm{vol}(b) > 0$. It follows from Proposition 3.3 that in this case, $a \neq 0$ if and only if $b \neq 0$.

Finally, suppose that $a, b \neq 0$. We must have $\lambda \neq 0$. We then easily check that

$$
\mathrm{span}\{\lambda a_1, a_2\} = \mathrm{span}\{a_1, a_2\}
$$

so that $a_1 \wedge a_2$ and $\lambda a_1 \wedge a_2$ have comparable orientations. By definition, these will have the same orientation if and only if the determinant of

$$
C = \begin{pmatrix} a_1 \cdot (\lambda a_1) & a_1 \cdot a_2 \\ a_2 \cdot (\lambda a_1) & a_2 \cdot a_2 \end{pmatrix}
$$

is positive. We easily check that $\det C = \lambda \, \mathrm{vol}(a_1 \wedge a_2)$. Since $a \neq 0$, it has positive volume, so we see that $\det C > 0$ precisely when $\lambda > 0$. □

Proposition A.1. *Suppose that $a_1, \ldots, a_k$ are vectors in $\mathbb{R}^n$ and we set $a = a_1 \wedge \cdots \wedge a_k$. Now suppose that we switch the order of a_i and a_j (where $i < j$) to obtain a new simple k-vector b. Thus*

$$
\begin{aligned}
a &= a_1 \wedge \cdots \wedge a_i \wedge \cdots \wedge a_j \wedge \cdots \wedge a_k, \\
b &= a_1 \wedge \cdots \wedge a_j \wedge \cdots \wedge a_i \wedge \cdots \wedge a_k.
\end{aligned}
$$

Then the following hold:

1. $\mathrm{vol}(b) = \mathrm{vol}(a)$.

2. If $a, b \neq 0$, then they have opposite orientations.

Proof. We leave to the reader the question of equality of volumes.

For the second item, we suppose $a, b \neq 0$ and check only the case $k = 2$. Thus $a = a_1 \wedge a_2$ and $b = a_2 \wedge a_1$. To determine how their orientations are related, we must check the sign of $\det A$ where

$$A = \begin{pmatrix} a_1 \cdot a_2 & a_1 \cdot a_1 \\ a_2 \cdot a_2 & a_2 \cdot a_1 \end{pmatrix}.$$

But

$$\det A = -\det \begin{pmatrix} a_1 \cdot a_1 & a_1 \cdot a_2 \\ a_2 \cdot a_1 & a_2 \cdot a_2 \end{pmatrix} = -\mathrm{vol}(a_1 \wedge a_2) < 0.$$

That is, a and b have opposite orientations. $\qquad\square$

Proposition 3.8. *Let $a_1, \ldots, a_k$ and $b_1, \ldots, b_k$ be vectors in $\mathbb{R}^n$ such that $a_i = \sum_{j=1}^{k} \gamma_{ij} b_j$ for all i. Set $C = (\gamma_{ij})_{k \times k}$, a $k \times k$ matrix. Then*

$$a_1 \wedge \cdots \wedge a_k = \det(C) \, (b_1 \wedge \cdots \wedge b_k).$$

Proof. We must be able to find a k-dimensional vector subspace U of $\mathbb{R}^n$ that contains all a_i and b_i. Let $\{u_i\}_{i=1}^{k}$ be an orthonormal basis for U. We set $a = a_1 \wedge \cdots \wedge a_k$ and $b = b_1 \wedge \cdots \wedge b_k$ and $u = u_1 \wedge \cdots \wedge u_k$.

We expand the a_i vectors in terms of the orthonormal basis: $a_i = \sum_{i=1}^{k} \alpha_{ij} u_j$. Set $A = (\alpha_{ij})_{k \times k}$. We will first show that $a = \det(A) u$.

Since $\{u_i\}_{i=1}^{k}$ is an orthonormal basis, we compute

$$\mathrm{vol}(u) = \sqrt{\det(u_i \cdot u_j)} = 1.$$

Next,

$$\mathrm{vol}(a) = \sqrt{\det(a_i \cdot a_j)} = \sqrt{\det(A^T A)} = |\det(A)|.$$

Then

$$\mathrm{vol}(a) = |\det(A)| \, \mathrm{vol}(u) = \mathrm{vol}\big(\det(A) \, u\big).$$

We next check equality of orientations.

We need to know $\det(a_i \cdot u_j)$ to compare orientations of a and u, and

$$\det(a_i \cdot u_j) = \det(\alpha_{ij}) = \det(A).$$

If $\det(A) = 0$, then $\{a_1, \ldots, a_k\}$ must be linearly dependent since they are the column vectors of $(a_i \cdot u_j)_{k \times k}$. Then $a = \det(A) u = 0$ trivially. So suppose $\det(A) > 0$. Then the orientations of $\det(A) u$ and u and a must all agree. On the other hand, if $\det(A) < 0$, then $\det(A) u$ and a must have the opposite orientation to that of u, so their orientations agree. It then follows that for all cases,

$$a_1 \wedge \cdots \wedge a_k = \det(A) \, (u_1 \wedge \cdots \wedge u_k).$$

We now bring $b = b_1 \wedge \cdots \wedge b_k$ back into the picture.

If $\{b_1, \ldots, b_k\}$ is a linearly dependent set, then the same must be true for $\{a_1, \ldots, a_k\}$, and we have $a = b = 0$ trivially. So let us suppose that $b \neq 0$.

We expand thus, $b_i = \sum_{j=1}^{k} \beta_{ij} u_j$, and we set $B = (\beta_{ij})_{k \times k}$. We know that we must have $b = \det(B) u$ and also that $\det(B) \neq 0$ since $b \neq 0$. We know that we can write $a_i = \sum_{j=1}^{k} \gamma_{ij} b_j$ for appropriate scalars γ_{ij}. It is not hard to see that $A B^{-1} = (\gamma_{ij})_{k \times k}$. Finally, we see that

$$a = \frac{\det(A)}{\det(B)} \det(B) u = \det(A) \det(B^{-1}) b$$
$$= \det(A B^{-1}) b.$$

This is the desired result. $\square$

Proposition A.2. *The dot product of simple k-vectors is well-defined.*

Proof. Let $a = a_1 \wedge \cdots \wedge a_k$. It suffices to show that if $b_k \wedge \cdots \wedge b_1 = c_k \wedge \cdots \wedge c_1 \neq 0$, then $\det(a_i \cdot b_j) = \det(a_i \cdot c_j)$.

Let V be the k-dimensional vector subspace of $\mathbb{R}^n$ spanned by both $\{b_1, \ldots, b_k\}$ and $\{c_1, \ldots, c_k\}$. Let $\{v_1, \ldots, v_k\}$ be an orthonormal basis for V and extend it to an orthonormal basis $\{v_1, \ldots, v_n\}$ for $\mathbb{R}^n$. For $x \in \mathbb{R}^n$ we can write $x = \sum_{i=1}^{n} \chi_i v_i$ and then define the *orthogonal projection* of x onto V by $x' = \sum_{i=1}^{k} \chi_i v_i$. Notice that if $y \in V$, then $x \cdot y = x' \cdot y$.

Let A', B, and C be the $k \times k$ matrices $A' = (a_1', \ldots, a_k')$, $B = (b_1, \ldots, b_k)$, and $C = (c_1, \ldots, c_k)$ computed with respect to the orthonormal basis for V, where a_i' is the orthogonal projection of a_i onto V. By Proposition 3.4 we know that $\det B = \det C$. Then

$$\det(a_i \cdot b_j) = \det(a_i' \cdot b_j) = \det(A'^{T} B) = (\det A')(\det B)$$
$$= (\det A')(\det C) = \det(a_i' \cdot c_j) = \det(a_i \cdot c_j). \qquad \square$$

Proposition 3.11. *If a and b are simple k-vectors in $\mathbb{R}^n$, then $a = b$ if and only if $a \cdot c = b \cdot c$ for all simple k-vectors c in $\mathbb{R}^n$.*

Proof. If $a = b$, then $a \cdot c = b \cdot c$ trivially.

So let us suppose that $a \cdot c = b \cdot c$ for all simple k-vectors c in $\mathbb{R}^n$. We want to show that $a = b$.

First, we check equality of volume:

$$\mathrm{vol}(a)^2 = a \cdot a^{\dagger} = b \cdot a^{\dagger}$$
$$= a \cdot b^{\dagger} = b \cdot b^{\dagger} = \mathrm{vol}(b)^2$$

where we have made use of some of the properties listed in Proposition 3.10. Thus $\mathrm{vol}(a) = \mathrm{vol}(b)$.

Now we need to check equality of orientation.

If $a = 0$, then $\text{vol}(b) = \text{vol}(a) = 0$ so that $b = 0$ as well. Thus $a = b$ trivially.

From this point on, we assume that a (and hence, implicitly, also b) to not be zero. If a and b lie in some common k-dimensional subspace, since we know on the basis of previous calculations that $a \cdot b^\dagger = \text{vol}(a)^2 > 0$, we see that a and b have the same orientation. Thus $a = b$.

So now we assume that a and b do not lie in any common k-dimensional vector subspace. We can write $a = a_1 \wedge \cdots \wedge a_k$ where each a_i is a vector in $\mathbb{R}^n$. Let us take U to be the span of $\{a_i\}_{i=1}^k$. Now extend $\{a_i\}_{i=1}^k$, a basis of U, to $\{a_i\}_{i=1}^n$, a basis for all of $\mathbb{R}^n$, in such a way that $a_{k+1}, \ldots, a_n$ are orthogonal to U. Next we write $b = b_1 \wedge \cdots \wedge b_k$ and let V be the k-dimensional subspace spanned by $\{b_i\}_{i=1}^k$. Let $\{w_i\}_{i=1}^k$ be an orthonormal basis for V. Since b cannot lie in U, we may, without loss of generality, suppose that $a_{k+1} \cdot w_1 \neq 0$.

We now construct the simple k-vector

$$c = a_{k+1} \wedge w_2 \wedge \cdots \wedge w_k.$$

By Proposition 3.8, we can write

$$b = b_1 \wedge \cdots \wedge b_k = \lambda\, w_1 \wedge \cdots \wedge w_k$$

for some scalar λ, and we must have $\lambda \neq 0$ since $b \neq 0$. Then

$$
\begin{aligned}
b \cdot c^\dagger &= \lambda\, (w_1 \wedge \cdots \wedge w_k) \cdot (w_k \wedge \cdots \wedge a_{k+1}) \\
&= \lambda\, w_1 \cdot a_{k+1} \neq 0.
\end{aligned}
$$

However

$$
a \cdot c^\dagger = \det \begin{pmatrix} a_1 \cdot a_{k+1} & a_1 \cdot w_2 & \cdots & a_1 \cdot w_k \\ \cdots & & & \\ a_k \cdot a_{k+1} & a_k \cdot w_2 & \cdots & a_k \cdot w_k \end{pmatrix} = 0
$$

since a_{k+1} is orthogonal to $a_1, \ldots, a_k$. We have $a \cdot c^\dagger \neq b \cdot c^\dagger$, a contradiction. So a and b must lie in some common k-dimensional subspace, and this forces $a = b$. $\qquad\square$

Proposition A.3 (The operator $\wedge^k f$ is well-defined). *Suppose that $f : V \to \mathbb{R}^n$ is a linear transformation defined on a k-dimensional vector subspace V of $\mathbb{R}^m$. If $a_1, \ldots, a_k$ and $b_1, \ldots, b_k$ are vectors in V and $a_1 \wedge \cdots \wedge a_k = b_1 \wedge \cdots \wedge b_k$, then $f(a_1) \wedge \cdots \wedge f(a_k) = f(b_1) \wedge \cdots \wedge f(b_k)$.*

Proof. The case where $a_1 \wedge \cdots \wedge a_k = 0$ is trivial. We consider only the case where $a_1 \wedge \cdots \wedge a_k \neq 0$.

There must exist a k-dimensional vector subspace W of $\mathbb{R}^n$ such that $f(a_i)$ and $f(b_i)$ belong to W for all i. Since $a_1 \wedge \cdots \wedge a_k \neq 0$, we see that $\{a_1, \ldots, a_k\}$ must be a basis for V. It follows that we can expand each b_i in terms of $a_1, \ldots, a_k$ thus:

$$b_i = \sum_{j=1}^k \gamma_{ij} a_j. \tag{A.2}$$

We set $C = (\gamma_{ij})_{k \times k}$. By Proposition 3.8, we have

$$b_1 \wedge \cdots \wedge b_k = \det(C) \, a_1 \wedge \cdots \wedge a_k.$$

Since $a_1 \wedge \cdots \wedge a_k = b_1 \wedge \cdots \wedge b_k$, it follows that $\det(C) = 1$. Applying f to Equation (A.2) gives us

$$f(b_i) = \sum_{j=1}^{k} \gamma_{ij} f(a_i)$$

for all i. Appealing again to Proposition 3.8, we have

$$\begin{aligned} f(b_1) \wedge \cdots \wedge f(b_k) &= \det(C) \, f(a_1) \wedge \cdots \wedge f(a_k) \\ &= f(a_1) \wedge \cdots \wedge f(a_k). \end{aligned} \qquad \square$$

Proposition A.4. *The wedge product of simple p- and q-vectors is well-defined.*

Proof. Let $a = a_1 \wedge \cdots \wedge a_p$ and $b = b_1 \wedge \cdots \wedge b_q$ be simple p- and q-vectors in $\mathbb{R}^n$. It will suffice to suppose we can also write $a = a_1' \wedge \cdots \wedge a_p'$ and to show that

$$a_1 \wedge \cdots \wedge a_p \wedge b_1 \wedge \cdots \wedge b_q = a_1' \wedge \cdots \wedge a_p' \wedge b_1 \wedge \cdots \wedge b_q. \qquad (A.3)$$

One can show that $\{a_1, \ldots, a_p, b_1, \ldots, b_q\}$ is a linearly dependent set if and only if the same is true of $\{a_1', \ldots, a_p', b_1, \ldots, b_q\}$; this requires working through several cases which we leave to the reader. In this instance, (A.3) is trivially true; both sides are zero.

Assume $\{a_1, \ldots, a_p, b_1, \ldots, b_q\}$ and $\{a_1', \ldots, a_p', b_1, \ldots, b_q\}$ are both independent. Let

$$V \stackrel{\text{def.}}{=} \mathrm{span}\{a_1, \ldots, a_p\} = \mathrm{span}\{a_1', \ldots, a_p'\}$$

and

$$W \stackrel{\text{def.}}{=} \mathrm{span}\{b_1, \ldots, b_q\}.$$

One can show V and W have in common only the element 0. Let $V \oplus W$ be the smallest vector space in $\mathbb{R}^n$ containing both V and W; it can be seen that $z \in V \oplus W$ if and only if $z = x + y$ where $x \in V$ and $y \in W$.

Define a linear transformation $f : V \to V$ by $f(a_i) = a_i'$ for all i and let F be the $p \times p$ matrix of f written in terms of the basis $\{a_1, \ldots, a_p\}$. We know by Proposition 3.4 that $\det(f) = \det(F) = 1$. Next define a linear transformation $f' : V \oplus W \to V \oplus W$ by $f'(a_i) = a_i'$ and $f'(b_j) = b_j$ for all i, j and let F' be the $(p+q) \times (p+q)$ matrix of f' written in terms of the basis $\{a_1, \ldots, a_p, b_1, \ldots, b_q\}$. Then F and F' are related thus:

$$F' = \begin{pmatrix} F & 0 \\ 0 & I \end{pmatrix}$$

where I is the $q \times q$ identity matrix. We see that $\det(F') = \det(F) = 1$. It then follows from Proposition 3.4 that Equation (A.3) holds; this is our desired conclusion. $\qquad \square$

A.3 The wedge product for arbitrary k-vectors

Let $v = \{v_i\}_{i=1}^n$ be a fixed, orthonormal basis for $\mathbb{R}^n$. Recall that if $I = (i_1, \ldots, i_k)$, a multi-index, then $v_I = v_{i_1} \wedge \cdots \wedge v_{i_k}$. Let $\mathcal{O}_k$ be the set of ordered multi-indices of length k; that is, if $I \in \mathcal{O}_k$, then $i_1 < \cdots < i_k$. Let a and b be p- and q-vectors respectively in $\mathbb{R}^n$ and write their unique expansions in terms of the basis:

$$a = \sum_{I \in \mathcal{O}_p} \alpha_I v_I \quad \text{and} \quad b = \sum_{J \in \mathcal{O}_q} \beta_J v_J.$$

We want to define the wedge product of a and b, and ultimately we will write it as $a \wedge b$. But let us keep in mind that when we write down symbolism such as $c_1 \wedge \cdots \wedge c_k$, we are not thinking of $\wedge$ as a binary operation; rather it is only part of the symbol for a simple k-vector. The wedge product we are about to introduce will depend on our choice of the basis v, so to distinguish it from our previous use of $\wedge$, let us use $\wedge_v$ to denote our new binary operation. Of course we will have to show at some point that we can identify $\wedge_v$ with $\wedge$ as it appears in $c_1 \wedge \cdots \wedge c_k$.

Using the unique expansions of a and b and the fact that we have at least defined the wedge product of simple k-vectors, we define the binary operation $\wedge_v : \Lambda^p \mathbb{R}^n \times \Lambda^q \mathbb{R}^n \to \Lambda^{p+q} \mathbb{R}^n$ by

$$a \wedge_v b \overset{\text{def.}}{=} \sum_{I,J} \alpha_I \beta_J (v_I \wedge v_J).$$

If a is a scalar, $a = \lambda$, then we set

$$\lambda \wedge_v b \overset{\text{def.}}{=} \lambda b = \sum_J \lambda \beta_J v_J.$$

A similar remark applies if b is a scalar.

The following is then trivial:

Proposition A.5. *Let a be a p-vector, b and c be q-vectors, and λ a scalar. Then*

1. $1 \wedge_v a = a \wedge_v 1 = a.$

2. $a \wedge_v (b+c) = a \wedge_v b + a \wedge_v c.$

3. $(b+c) \wedge_v a = b \wedge_v a + c \wedge_v a.$

4. $\lambda (a \wedge_v b) = (\lambda a) \wedge_v b = a \wedge_v (\lambda b).$

We now develop a crucial property of basis k-vectors:

Proposition A.6. *If I and J are multi-indices, not necessarily ordered, then*

$$v_I \wedge_v v_J = v_I \wedge v_J. \tag{A.4}$$

Proof. **Case 1.** If I and J are ordered, then (A.4) is a trivial consequence of the definition of $\wedge_v$.

Case 2. Suppose that I has a repeated index. That is, if $I = (i_1, \ldots, i_p)$, then there exist $s \neq t$ such that $i_s = i_t$. Then $v_I = 0$ and $v_I \wedge v_J = 0$. On the other hand,

$v_I \wedge_v v_J = 0 \wedge_v v_J$, and this last expression must be 0 by the definition of $\wedge_v$. Thus (A.4) holds. We have the same result if J has a repeated index.

Case 3. Suppose that neither I nor J has a repeated index but that there is an index that is common to both I and J. Then $v_I \wedge v_J = 0$. Let I' and J' be the ordered multi-indices that are the ordered rearrangements of I and J respectively. There exist r_I and r_J such that $v_I = (-1)^{r_I} v_{I'}$ and $v_J = (-1)^{r_J} v_{J'}$. Appealing to the definition of $\wedge_v$, we can write

$$v_I \wedge_v v_J = (-1)^{r_I+r_J} v_{I'} \wedge_v v_{J'} = (-1)^{r_I+r_J} v_{I'} \wedge v_{J'}.$$

But $v_{I'} \wedge v_{J'} = 0$ since there is a vector "factor" that is common to both. Therefore (A.4) holds once again.

Case 4. Suppose that I and J have no repeated indices and no index common to both of them, but they are not necessarily ordered. Let I' and J' be as in the last case. Then

$$\begin{aligned}
v_I \wedge_v v_J &= \big((-1)^{r_I} v_{I'}\big) \wedge_v \big((-1)^{r_J} v_{J'}\big) \\
&= (-1)^{r_I+r_J} v_{I'} \wedge v_{J'} \\
&= v_I \wedge v_J.
\end{aligned}$$

Hence (A.4) holds in all cases.   □

Corollary A.1. *Given* $v_{i_1}, \ldots, v_{i_p}$ *with any indices whatsoever, we have*

$$v_{i_1} \wedge_v \cdots \wedge_v v_{i_p} = v_{i_1} \wedge \cdots \wedge v_{i_p}$$

with no necessity to insert parentheses to indicate the order of multiplication.

Proof. We show how this goes in the case $p = 3$.

$$\begin{aligned}
v_{i_1} \wedge_v (v_{i_2} \wedge_v v_{i_3}) &= v_{i_1} \wedge_v (v_{i_2} \wedge v_{i_3}) \\
&= v_{i_1} \wedge (v_{i_2} \wedge v_{i_3}) \\
&= (v_{i_1} \wedge v_{i_2}) \wedge v_{i_3} = \cdots \\
&= (v_{i_1} \wedge_v v_{i_2}) \wedge_v v_{i_3}.
\end{aligned}$$
                                                                                                   □

Corollary A.2. *For any multi-indices I, J, and K,*

$$v_I \wedge_v (v_J \wedge_v v_K) = (v_I \wedge_v v_J) \wedge_v v_K.$$

Showing that $\wedge_v$ is associative for basis elements immediately gives us the following:

Corollary A.3. *For arbitrary p-, q-, and r-vectors a, b, and c respectively,*

$$a \wedge_v (b \wedge_v c) = (a \wedge_v b) \wedge_v c.$$

To show that $\wedge_v$ is an extension of $\wedge$ as it appears in the notation for simple k-vectors, we must show that for arbitrary vectors $a_1, \ldots, a_k$, we have

$$a_1 \wedge \cdots \wedge a_k = a_1 \wedge_v \cdots \wedge_v a_k.$$

We accomplish this by the following result:

Proposition A.7. *Let $a_1, \ldots, a_k$ be vectors in $\mathbb{R}^n$ and expand them in terms of $\{v_i\}_{i=1}^n$:*

$$a_i = \sum_{j=1}^n \alpha_{ij} v_j.$$

Then

$$a_1 \wedge_v \cdots \wedge_v a_k$$

$$= \sum_{i_1 < \cdots < i_k} \det \begin{pmatrix} \alpha_{1i_1} & \cdots & \alpha_{1i_k} \\ \cdots & & \\ \alpha_{ki_1} & \cdots & \alpha_{ki_k} \end{pmatrix} v_{i_1} \wedge \cdots \wedge v_{i_k} \qquad (A.5)$$

$$= a_1 \wedge \cdots \wedge a_k.$$

Proof. Using the linearity and distributivity of $\wedge_v$, we can write

$$a_1 \wedge_v \cdots \wedge_v a_k = \sum_{i_1, \ldots, i_k} \alpha_{1i_1} \cdots \alpha_{ki_k} (v_{i_1} \wedge_v \cdots \wedge_v v_{i_k})$$

where the sum is over all multi-indices $I = (i_1, \ldots, i_k)$ regardless of order or whether or not there are repetitions of indices. Since $v_{i_1} \wedge_v \cdots \wedge_v v_{i_k} = v_{i_1} \wedge \cdots \wedge v_{i_k}$, if there are repetitions in $I = (i_1, \ldots, i_k)$, then $v_{i_1} \wedge_v \cdots \wedge_v v_{i_k} = 0$. We may therefore assume each I has no repeated indices.

Let us introduce the symbolism $J \sim I$ if $J = (j_1, \ldots, j_k)$ is a permutation of $I = (i_1, \ldots, i_k)$. If I is an ordered multi-index and $J \sim I$, then by $\text{sgn}(J)$ we mean the sign of the permutation that transforms $J = (j_1, \ldots, j_k)$ into $I = (i_1, \ldots, i_k)$. Since $v_i \wedge_v v_j = v_i \wedge v_j$, we know that $v_i \wedge_v v_j = -v_j \wedge_v v_i$. It follows that we may write $v_{j_1} \wedge_v \cdots \wedge_v v_{j_k} = \text{sgn}(J) v_{i_1} \wedge_v \cdots \wedge_v v_{i_k}$ provided we know that $J \sim I$ and I is ordered.

Recall that $\mathcal{O}_k$ is the set of ordered multi-indices of length k. Returning to $a_1 \wedge_v \cdots \wedge_v a_k$, we see that we may write

$$a_1 \wedge_v \cdots \wedge_v a_k = \sum_{I \in \mathcal{O}_k} \sum_{J \sim I} \text{sgn}(J) \alpha_{1j_1} \cdots \alpha_{kj_k} (v_{i_1} \wedge_v \cdots \wedge_v v_{i_k}).$$

Now

$$\sum_{J \sim I} \text{sgn}(J) \alpha_{1j_1} \cdots \alpha_{kj_k} = \det \begin{pmatrix} \alpha_{1i_1} & \cdots & \alpha_{1i_k} \\ \cdots & & \\ \alpha_{ki_1} & \cdots & \alpha_{ki_k} \end{pmatrix}$$

and $v_{i_1} \wedge_v \cdots \wedge_v v_{i_k} = v_{i_1} \wedge \cdots \wedge v_{i_k}$, so this gives us the first equality in Equation (A.5).

To obtain the second equality, we first note that $a_1 \wedge \cdots \wedge a_k$, being a k-vector, must have a unique expansion

$$a_1 \wedge \cdots \wedge a_k = \sum_{I \in \mathcal{O}_k} \lambda_I v_I \quad \text{where } \lambda_I \in \mathbb{R}.$$

We also know that $a_1 \wedge \cdots \wedge a_k$ acts as a linear functional on the set of simple k-vectors. Choose $J \in \mathcal{O}_k$. Then

$$(a_1 \wedge \cdots \wedge a_k) \bullet v_J^\dagger = \sum_{I \in \mathcal{O}_k} \lambda_I (v_I \bullet v_J^\dagger).$$

Now here, we use for the first time the fact that $v = \{v_i\}_{i=1}^n$ is a basis that is orthonormal: By Proposition 4.3, we know that $v_I \cdot v_J^\dagger$ is 1 if $I = J$ and 0 otherwise. Thus

$$\lambda_J = (a_1 \wedge \cdots \wedge a_k) \cdot v_J^\dagger.$$

We may suppose that $J = (j_1, \ldots, j_k)$ where $j_1 < \cdots < j_k$. We know that $a_p \cdot v_{j_q} = \alpha_{p j_q}$. Therefore,

$$\lambda_J = \det(a_p \cdot v_{j_q})_{p,q=1}^k = \det \begin{pmatrix} \alpha_{1 j_1} & \cdots & \alpha_{1 j_k} \\ \cdots & & \\ \alpha_{k j_1} & \cdots & \alpha_{k j_k} \end{pmatrix}.$$

This gives us the second equality of Equation (A.5). $\square$

We thus know that $\wedge$, defined for simple k-vectors, can be extended to a binary operation

$$\wedge : \Lambda^p \mathbb{R}^n \times \Lambda^q \mathbb{R}^n \to \Lambda^{p+q} \mathbb{R}^n, \quad 0 \le p, q \le n,$$

that satisfies all the linearity and distributivity properties of $\wedge_v$.

Notice that if we return to the definition of $\wedge_v$ and use a different orthonormal basis $w = \{w_i\}_{i=1}^n$ to define an extension $\wedge_w$ of $\wedge$, this must be the same $\wedge_v$. This follows from the fact that

$$a_1 \wedge_v \cdots \wedge_v a_k = a_1 \wedge \cdots \wedge a_k = a_1 \wedge_w \cdots \wedge_w a_k$$

for all vectors a_i.

A.4   Integrals over manifolds

Proposition 4.18. *Suppose that ϕ is an integrable, real-valued function defined on the k-dimensional $\mathbb{C}^1$ manifold $\mathcal{M}$. Then the integral of ϕ as defined by (4.11) is independent of our choice of parametrization.*

Proof. Suppose that $\mathcal{M}$ has two $\mathbb{C}^1$ parametrizations, $x : U \to \mathcal{M}$ and $y : V \to \mathcal{M}$. U and V are, of course, subsets of $\mathbb{R}^k$, and we will suppose that x assigns coordinates $(\rho_1, \ldots, \rho_k)$ to points of $\mathcal{M}$ while y assigns coordinates $(\sigma_1, \ldots, \sigma_k)$. To show that $\int_{\mathcal{M}} \phi$ is independent of our choice of parametrization, we must show that

$$\int_U (\phi \circ x) \left| \frac{\partial x}{\partial \rho_1} \wedge \cdots \wedge \frac{\partial x}{\partial \rho_k} \right| = \int_V (\phi \circ y) \left| \frac{\partial y}{\partial \sigma_1} \wedge \cdots \wedge \frac{\partial y}{\partial \sigma_k} \right|.$$

We have the setup

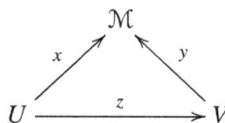

where z is the one-to-one $\mathcal{C}^1$ function that satisfies $y \circ z = x$. Let $p \in \mathcal{M}$ and let r and s be the points in U and V respectively such that $p = x(r) = y(s)$. We know that

$$\frac{\partial x}{\partial \rho_i}(r) = x'(r)e_i,$$

$$\frac{\partial y}{\partial \sigma_i}(s) = y'(s)e_i,$$

$$\frac{\partial z}{\partial \rho_i}(r) = z'(r)e_i.$$

We must be able to write z in the form

$$z(r) = \sum_{j=1}^{k} \zeta_j(r) e_j$$

where each ζ_j is a real-valued, $\mathcal{C}^1$ function. Thus

$$\frac{\partial z}{\partial \rho_i}(r) = \sum_{j=1}^{k} \frac{\partial \zeta_j}{\partial \rho_i}(r) e_j.$$

By the fact that $y \circ z = x$ and the chain rule, we know that $y'(s) \circ z'(r) = x'(r)$. We then calculate that

$$\frac{\partial x}{\partial \rho_i}(r) = x'(r) e_i = \left(y'(s) \circ z'(r)\right) e_i$$

$$= \sum_{j=1}^{k} \frac{\partial \zeta_j}{\partial \rho_i}(r) \frac{\partial y}{\partial \sigma_j}(s).$$

Applying Proposition 3.8 to this last result, we see that we have

$$\frac{\partial x}{\partial \rho_1}(r) \wedge \cdots \wedge \frac{\partial x}{\partial \rho_k}(r) = \det\left(\frac{\partial \zeta_j}{\partial \rho_i}(r)\right)_{k \times k} \left(\frac{\partial y}{\partial \sigma_1}(s) \wedge \cdots \wedge \frac{\partial y}{\partial \sigma_k}(s)\right).$$

But the determinant in this last equation is the Jacobian of z, that is,

$$\det\left(z'(r)\right) = \det\left(\frac{\partial \zeta_j}{\partial \rho_i}(r)\right)_{k \times k}.$$

From this we can deduce that

$$\left|\frac{\partial x}{\partial \rho_1}(r) \wedge \cdots \wedge \frac{\partial x}{\partial \rho_k}(r)\right| = \left|\det\left(z'(r)\right)\right| \left|\frac{\partial y}{\partial \sigma_1}(s) \wedge \cdots \wedge \frac{\partial y}{\partial \sigma_k}(s)\right|.$$

Finally, appealing to the change-of-variables formula, we see that

$$\int_V (\phi \circ y) \left|\frac{\partial y}{\partial \sigma_1} \wedge \cdots \wedge \frac{\partial y}{\partial \sigma_k}\right|$$

$$= \int_U (\phi \circ y \circ z)(r) \left|\left[\left(\frac{\partial y}{\partial \sigma_1} \wedge \cdots \wedge \frac{\partial y}{\partial \sigma_k}\right) \circ z\right](r)\right| \left|\det\left(z'(r)\right)\right| dr$$

$$= \int_U (\phi \circ x) \left|\frac{\partial x}{\partial \rho_1} \wedge \cdots \wedge \frac{\partial x}{\partial \rho_k}\right|.$$

This is the desired result. $\qquad\square$

A.5   Geometric algebra

Proposition 5.4. *The geometric product is associative.*

Proof. It suffices to show that if I, J, and K are ordered multi-indices, then $v_I(v_Jv_K) = (v_Iv_J)v_K$.

Concatenation is trivially associative; that is, $I(JK) = (IJ)K$. We associate roots of certain multi-indices and the parities of the corresponding proper transformations thus:

$$R \overset{\sigma}{\sim} IJK,$$

$$R_{IJ} \overset{\sigma_{IJ}}{\sim} IJ,$$

$$R_{JK} \overset{\sigma_{JK}}{\sim} JK.$$

By the definition of geometric product, $v_Jv_K = \sigma_{JK}v_{R_{JK}}$. Thus we have $v_I(v_Jv_K) = \sigma_{JK}v_Iv_{R_{JK}}$. Since $I \overset{+1}{\sim} I$ and $JK \overset{\sigma_{JK}}{\sim} R_{JK}$, by Lemma 5.4, we have $IJK \overset{\sigma_{JK}}{\sim} IR_{JK}$. From $R \overset{\sigma}{\sim} IJK \overset{\sigma_{JK}}{\sim} IR_{JK}$ and Lemma 5.2 we conclude that $R \overset{\sigma\sigma_{JK}}{\sim} IR_{JK}$. Because R is ordered, it must be the root of IR_{JK}, therefore $v_Iv_{R_{JK}} = \sigma\sigma_{JK}v_R$. Hence

$$\begin{aligned}
v_I(v_Jv_K) &= \sigma_{JK}v_Iv_{R_{JK}} \\
&= \sigma_{JK}\sigma\sigma_{JK}v_R \\
&= \sigma v_R.
\end{aligned}$$

By a similar analysis involving R_{IJ} and σ_{IJ}, we can show that $(v_Iv_J)v_K = \sigma v_R$. This establishes the result.   □

Proposition 5.5. *Let $a, b_1, \ldots, b_p \in \mathbb{R}^n$. Then*

$$\begin{aligned}
a(b_1 \wedge \cdots \wedge b_p) &= \sum_{r=1}^{p}(-1)^{r-1}(a \cdot b_r)(b_1 \wedge \cdots \wedge \widehat{b_r} \wedge \cdots \wedge b_p) \\
&\quad + a \wedge b_1 \wedge \cdots \wedge b_p \quad and \\
(b_1 \wedge \cdots \wedge b_p)a &= \sum_{r=1}^{p}(-1)^{p-r}(a \cdot b_r)(b_1 \wedge \cdots \wedge \widehat{b_r} \wedge \cdots \wedge b_p) \\
&\quad + b_1 \wedge \cdots \wedge b_p \wedge a.
\end{aligned}$$

Proof. We prove only the first identity. The other one is established similarly.

Let us compute the geometric product $v_j(v_{k_1} \wedge \cdots \wedge v_{k_p})$ where $k_1 < \cdots < k_p$. Set $J = (j)$ and $K = (k_1, \ldots, k_p)$.

Case 1. Suppose that j occurs in K, say, $j = k_r$. Then the root of JK is $R = (k_1, \ldots, \widehat{k_r}, \ldots, k_p)$ and $JK \overset{(-1)^{r-1}}{\sim} R$. So

$$v_Jv_K = (-1)^{r-1}(v_{k_1} \wedge \cdots \wedge \widehat{v_{k_r}} \wedge \cdots \wedge v_{k_p}).$$

Case 2. Suppose that j does not occur in K. Let $R = (i_0, i_1, \ldots, i_p)$ be the rearrangement of JK into an ordered sequence with $j = i_r$. R is the root of JK. So

$$v_J v_K = (-1)^r v_R = v_j \wedge v_{k_1} \wedge \cdots \wedge v_{k_p}.$$

Combining the results of these two cases gives us

$$v_j(v_{k_1} \wedge \cdots \wedge v_{k_p}) = \sum_{r=1}^{p} (-1)^{r-1} (v_j \cdot v_{k_r})(v_{k_1} \wedge \cdots \wedge \widehat{v_{k_r}} \wedge \cdots \wedge v_{k_p})$$
$$+ v_j \wedge v_{k_1} \wedge \cdots \wedge v_{k_p}.$$

Using linearity, we can extend this to

$$a(b_1 \wedge \cdots \wedge b_p) = \sum_{r=1}^{p} (-1)^{r-1} (a \cdot b_r)(b_1 \wedge \cdots \wedge \widehat{b_r} \wedge \cdots \wedge b_p)$$
$$+ a \wedge b_1 \wedge \cdots \wedge b_p$$

where $a, b_1, \ldots, b_p$ are vectors in $\mathbb{R}^n$.   □

Corollary 5.4. *The definition of the geometric product is independent of our choice of the orthonormal basis $v = \{v_i\}_{i=1}^n$.*

Proof. Choose two new orthonormal bases $\{u_i\}_{i=1}^n$ and $\{w_i\}_{i=1}^n$ for $\mathbb{R}^n$. Let us pick $a, b \in \mathbb{G}^n$ and write out their unique expansions in terms of $\{u_i\}_{i=1}^n$:

$$a = \sum_I \alpha_I u_I \quad \text{and} \quad b = \sum_J \beta_J u_J$$

where it is understood that the summations are over all possible ordered multi-indices. These expansions do not depend on any use of the geometric product. The geometric product is then

$$ab = \sum_{I,J} \alpha_I \beta_J (u_I u_J).$$

Suppose we show the dependence of the geometric product on the orthonormal basis v by writing $a *_v b$ instead of ab. Now let us go back and reconstruct the geometric product, only this time we use $\{w_i\}_{i=1}^n$ instead of $\{v_i\}_{i=1}^n$, and we denote the new geometric product by $a *_w b$. Then to show independence of the choice of orthonormal basis, we need to show that $a *_v b = a *_w b$. This reduces to showing that $u_I *_v u_J = u_I *_w u_J$.

We know from Corollaries 5.2 and 5.3 that

$$u_i *_v u_i = u_i *_w u_i = 1,$$
$$u_i *_v u_j = -u_j *_v u_i \quad \text{and} \quad u_i *_w u_j = -u_j *_w u_i \quad \text{when } i \neq j,$$
$$u_{i_1} *_v \cdots *_v u_{i_p} = u_{i_1} *_w \cdots *_w u_{i_p} = u_{i_1} \wedge \cdots \wedge u_{i_p}.$$

Using these facts and the associativity of both $*_v$ and $*_w$, it is obvious that when we compute them, we will have $u_I *_v u_J = u_I *_w u_J$.   □

Proposition 5.21. *Let V be a nontrivial p-dimensional vector subspace of $\mathbb{R}^n$ and v a p-blade parallel to V. Suppose that $P : \mathbb{R}^n \to V$ is the (orthogonal) projection onto V, and $Q : \mathbb{R}^n \to V_\perp$ is the (orthogonal) projection onto its orthogonal complement. Let $\underline{P}$ and $\underline{Q}$ be the corresponding outermorphisms of $\mathbb{G}^n$ onto $\mathbb{G}(V)$ and $\mathbb{G}(V_\perp)$ respectively. Then for all $a \in \mathbb{G}^n$, we have*

$$\underline{P}(a) = (a \cdot_L v)v^{-1} \quad and \quad \underline{Q}(a) = (a \wedge v)v^{-1}.$$

Proof. It is sufficient to check the proposition on basis elements constructed from an orthonormal basis $\{u_i\}_{i=1}^n$ for $\mathbb{R}^n$. We may, without loss of generality, suppose that $v = \lambda (u_1 \cdots u_p)$ for some nonzero scalar λ, and since v^{-1} has the form $(1/\lambda)(u_p \cdots u_1)$, we may as well assume that $\lambda = 1$.

We dispose first of the trivial case where $a = \alpha$, a scalar. We know that $\underline{P}$ and $\underline{Q}$ must be the identity on α by definition of an outermorphism. And by the definitions of $\cdot_L$ and $\wedge$, we have

$$(\alpha \cdot_L v)v^{-1} = \alpha vv^{-1} = \alpha$$

and

$$(\alpha \wedge v)v^{-1} = \alpha vv^{-1} = \alpha.$$

Now suppose that $a = u_I$ where I is the multi-index $(i_1, \ldots, i_q)$. That is, we suppose that $a = u_{i_1} \cdots u_{i_q}$ where $i_1 < \cdots < i_q$. There are three cases:

Case 1. Suppose that $i_q \leq p$. That is, each of $i_1, \ldots, i_q$ belongs to the set $\{1, \ldots, p\}$. We must be able to write $v = (-1)^r u_I^\dagger u_J$ for some uniquely determined r and multi-index J where I and J have no elements in common. Of course $v^{-1} = (-1)^r u_J^\dagger u_I$. We see that $\underline{P}(u_I) = u_I$ and $\underline{Q}(u_I) = 0$. Then

$$(u_I \cdot_L v)v^{-1} = (-1)^r \big(u_I \cdot_L (u_I^\dagger u_J)\big)(-1)^r u_J^\dagger u_I = u_I.$$

On the other hand, $(u_I \wedge v)v^{-1} = 0$ since u_{i_1} occurs in both u_I and v. This disposes of the first case.

Case 2. Suppose that $i_1 \leq p$ and $i_q > p$. Then we must be able to write $u_I = u_{I_0} u_{I_1}$ where I_0 and I_1 are multi-indices having no element in common such that each element of I_0 is less than or equal to p while each element of I_1 is greater than p. We see that $\underline{P}(u_{I_1}) = 0$ and $\underline{Q}(u_{I_0}) = 0$, so $\underline{P}(u_I) = \underline{Q}(u_I) = 0$. Now $u_I \cdot_L v = 0$ since u_{I_1} must have a factor u_{i_r} that does not occur in v. And $u_I \wedge v = 0$ since u_{I_0} has at least one factor u_{i_r} that occurs in v. Thus

$$(u_I \cdot_L v)v^{-1} = (u_I \wedge v)v^{-1} = 0.$$

This establishes the result for the second case.

Case 3. Suppose finally that $i_1 > p$, that is, every element of I is greater than p. We see that $\underline{P}(u_I) = 0$ and $\underline{Q}(u_I) = u_I$. Now $u_I \cdot_L v = 0$ since u_I has 1-vector factors that do not occur in v; so $(u_I \cdot_L v) v^{-1} = 0$. However, since every 1-vector factor of u_I is orthogonal to every 1-vector factor of v, we have $u_I \wedge v = u_I v$. Therefore $(u_I \wedge v) v^{-1} = u_I v v^{-1} = u_I$. This completes the proof. $\square$

Proposition 5.24. *The blades a and b are orthogonal if and only if $ab = a \wedge b$.*

Proof. Assume that a and b are blades in $\mathbb{R}^n$ and that the associated linear subspaces are A and B respectively. Let $\{u_1, \ldots, u_p\}$ and $\{v_1, \ldots, v_q\}$ be orthonormal bases for A and B respectively. We can then write $a = \alpha (u_1 \cdots u_p)$ and $b = \beta (v_1 \cdots v_q)$ where $\alpha, \beta \neq 0$.

Suppose that a and b are orthogonal. Then u_i and and v_j are orthogonal for all i, j. It follows that

$$
\begin{aligned}
a \wedge b &= \alpha\beta (u_1 \cdots u_p) \wedge (v_1 \cdots v_q) \\
&= \alpha\beta (u_1 \cdots u_p v_1 \cdots v_q) = ab.
\end{aligned}
$$

Suppose on the other hand that $ab = a \wedge b$. Let us set $u = u_1 \cdots u_p$ and $v = v_1 \cdots v_q$. Since $\alpha, \beta \neq 0$, we see that $uv = u \wedge v \neq 0$ since the geometric product of nonzero blades must be nonzero.

Let us extend $\{u_1, \ldots, u_p\}$ to $\{u_1, \ldots, u_n\}$, an orthonormal basis for $\mathbb{R}^n$. Recall that for $x = \sum_{i=1}^n \chi_i u_i$ in $\mathbb{R}^n$, the projection of x on A and the perpendicular part of x with respect to A are given by

$$
x\|a = \sum_{i \leq p} \chi_i u_i \quad \text{and} \quad x \perp a = \sum_{i > p} \chi_i u_i.
$$

Now $u \wedge (v_i\|u) = 0$ since $v_i\|u$ is a linear combination of $u_1, \ldots, u_p$. So if we replace every v_i in $u \wedge v$ by $(v_i\|u) + (v_i \perp u)$ and multiply out the result, we see that $uv = u \wedge v = u \wedge w$ where $w = (v_1 \perp u) \wedge \cdots \wedge (v_q \perp u)$. Because $uv \neq 0$, we have $w \neq 0$ and the vectors $\{v_i \perp u\}_{i=1}^q$ are linearly independent. Let

$$
W = \text{span}\{v_1 \perp u, \ldots, v_q \perp u\}.
$$

Of course, W is the linear subspace determined by the blade w. Clearly, if $y \in W$, then $y \cdot u_i = 0$ for $i = 1, \ldots, p$. Next let $\{w_1, \ldots, w_q\}$ be an orthonormal basis for W. We must have $w = \gamma(w_1 \cdots w_q) = \gamma(w_1 \wedge \cdots \wedge w_q)$ for some $\gamma \neq 0$. We can then write

$$
uv = u \wedge w = \gamma u_1 \wedge \cdots \wedge u_p \wedge w_1 \wedge \cdots \wedge w_q.
$$

Because the vectors $u_1, \ldots, u_p, w_1, \ldots, w_q$ are all orthogonal, we have

$$
uv = \gamma u_1 \cdots u_p w_1 \cdots w_q = uw.
$$

Dividing by u, we obtain $v = w$. This means that the blades v and w determine the same subspace of $\mathbb{R}^n$, that is, $B = W$. Therefore all the vectors of B are orthogonal to those of A. Thus A and B (and hence a and b) are orthogonal. $\square$

Before we can prove Proposition 5.25, we need a lemma:

Lemma A.1. *Let a and b be blades in $\mathbb{R}^n$ and suppose that they determine the spaces A and B respectively. Then $a \bullet_L b = 0$ if and only if there exists a nonzero $x_0 \in A$ such that $x_0 \bullet y = 0$ for all $y \in B$.*

Proof. Let us take A and B to have dimensions p and q respectively.

Assume there exists nonzero $x_0 \in A$ such that $x_0 \bullet y = 0$ for all $y \in B$.

We may, without loss of generality, assume that $|x_0| = 1$. We can then construct an orthonormal basis $\{u_1, \ldots, u_n\}$ for $\mathbb{R}^n$ such that $u_1 = x_0$ and $\{u_i\}_{i=1}^p$ is a basis for A. Let us set $u = u_1 \cdots u_p$. This is a blade that determines A, so we can write $a = \alpha u$ for some nonzero scalar α.

We now calculate $a \bullet_L b$. We know that a is a p-blade and b a q-blade. We can write b in the form

$$b = \sum \{\beta_I u_I : I \in \mathcal{O}(q) \text{ and } u_I \in \Lambda^q B\}.$$

Then

$$a \bullet_L b = \sum_I \alpha \beta_I \langle u u_I \rangle_{q-p}.$$

Now u_1 cannot be a factor of any $u_I = u_{i_1} \cdots u_{i_q}$ in the expansion of b because u_1 is orthogonal to B. So $u u_I$ cannot have grade $q - p$; thus $\langle u u_I \rangle_{q-p} = 0$. Therefore $a \bullet_L b = 0$.

Now for the other direction in the proof. Assume that $a \bullet_L b = 0$.

Let $\{v_1, \ldots, v_n\}$ be an orthonormal basis for $\mathbb{R}^n$ such that $b = \beta v_1 \cdots v_q$ for some nonzero scalar β. We take B^* to be the orthogonal complement of B; thus B^* is spanned by $\{v_i\}_{i=q+1}^n$. Either $A \cap B^* = \{0\}$ or $A \cap B^* \neq \{0\}$.

Suppose that $A \cap B^* = \{0\}$. Since a is a p-blade, let us write it in the form $a = a_1 \wedge \cdots \wedge a_p$ where $\{a_i\}_{i=1}^p$ is a basis for A. Since $A \cap B^* = \{0\}$, it is easily shown that the vectors $a_1, \ldots, a_p$ and $v_{q+1}, \ldots, v_n$ are linearly independent. If we set $b^* = \beta v_{q+1} \cdots v_n$, then we have $a \wedge b^* \neq 0$. We then proceed to the following calculation in which r is an integer of uniquely determined parity:

$$\begin{aligned}
a \bullet_L b &= \beta \langle a v_1 \cdots v_q \rangle_{q-p} \\
&= \beta \langle a (v_1 \cdots v_q)(v_1 \cdots v_n) \rangle_{n-q+p} (v_n \cdots v_1) \\
&= (-1)^r \langle a (\beta v_{q+1} \cdots v_n) \rangle_{p+(n-q)} (v_n \cdots v_1) \\
&= (-1)^r (a \wedge b^*)(v_n \cdots v_1).
\end{aligned}$$

Since we can divide out $v_n \cdots v_1$, we must have $a \bullet_L b \neq 0$, but this is a contradiction.

So we must have $A \cap B^* \neq \{0\}$. Choose nonzero $x_0 \in A \cap B^*$. This is a nonzero element of A, and since it belongs to B^*, the orthogonal complement of B, it must satisfy $x_0 \bullet y = 0$ for all $y \in B$. This is our desired conclusion.   □

Proposition 5.25. *Let a and b be blades in $\mathbb{R}^n$ with respective associated vector subspaces A and B. Then the following statements are equivalent:*

1. A and B are perpendicular.

2. $a \bullet_L b = a \bullet_R b = 0$.

3. *The angle between A and B is $\pi/2$.*

Proof. It is easily seen that there is a version of Lemma A.1 that says $a \bullet_R b = 0$ if and only if there exists a nonzero $y_0 \in B$ such that $x \bullet y_0 = 0$ for all $x \in A$. (This is a simple application of reversion and the way it takes the left-hand dot product to the right-hand version and vice-versa.) We see from this that the first and second results of this proposition must be equivalent.

To see the equivalence of the third result to the first two, suppose that we have $\mathrm{grade}(a) \le \mathrm{grade}(b)$. Then

$$\cos(\theta) = \frac{|a \bullet_L b|}{|a|\,|b|}.$$

So $a \bullet_L b$ is zero precisely when $\theta = \pi/2$. $\qquad\square$

A.6 Induced orientation

We shall show that induced orientation of the boundary of a cell depends only on the cell's orientation w, not on our choice of parametrization of the cell.

Note in the proof below that since $\vec{\partial} e$ is an $(m-1)$-vector field on $\partial \mathfrak{I}^m$, when we write $(\vec{\partial} e \circ g)(x)$, we mean $(\vec{\partial} e)(g(x))$.

Recall our assumption in Section 6.3.3 about how cell faces behave under parametrizations:

Cell face assumption. If $\mathcal{M}$ is a $\mathcal{C}^1$ m-cell and $x_1, x_2 : \mathfrak{I}^m \to \mathcal{M}$ are $\mathcal{C}^1$ parametrizations of $\mathcal{M}$, then $x_2^{-1} \circ x_1$ takes k-faces of $\mathfrak{I}^m$ onto k-faces of $\mathfrak{I}^m$ for $k = 0, 1, \ldots, m$.

We begin our proof of independence of parametrization with what amounts to a special case of change of parametrization.

Lemma A.2. *Let $e = e_1 \cdots e_m$, the standard orientation for $\mathfrak{I}^m$, and let $g : \mathfrak{I}^m \to \mathfrak{I}^m$ be a one-to-one onto map such that*

1. *g and g^{-1} are $\mathcal{C}^1$.*

2. *g takes $(m-1)$-faces to $(m-1)$-faces.*

3. *$\wedge^m g'(x)$ is a one-to-one linear transformation for all $x \in \mathfrak{I}^m$.*

4. *For all $x \in \mathfrak{I}^m$, there exists $\alpha > 0$ such that $\wedge^m g'(x) e = \alpha e$.*

Then for every $(m-1)$-face $\mathfrak{I}_{ij}^{m-1}$ of $\mathfrak{I}^m$ and for every $x \in \mathfrak{I}_{ij}^{m-1}$, there exists $\beta > 0$ such that

$$\wedge^{m-1} g'(x)\, \vec{\partial} e(x) = \beta \left(\vec{\partial} e \circ g \right)(x). \tag{A.6}$$

Proof. Choose $x \in \partial \mathcal{I}^m$ and set $y = g(x)$. We know that x belongs to some $(m-1)$-face $\mathcal{I}_{ij}^{m-1}$, and it is sufficient for our purposes to assume that x is an interior point of $\mathcal{I}_{ij}^{m-1}$, that is, that $x = (\chi_1, \ldots, \chi_m)$ where $\chi_i = j$ and $0 < \chi_k < 1$ if $k \neq i$. The image under g of an $(m-1)$-face is again an $(m-1)$-face, so we can assume $g(\mathcal{I}_{ij}^{m-1}) = \mathcal{I}_{i'j'}^{m-1}$. To justify (A.6), we need to show there exists $\beta > 0$ such that

$$
\wedge^{m-1} g'(x) \left((-1)^{i+j} (e_1 \cdots \widehat{e_i} \cdots e_m) \right) = \beta (-1)^{i'+j'} (e_1 \cdots \widehat{e_{i'}} \cdots e_m)
$$
$$
= \beta (\overrightarrow{\partial} e \circ g)(x). \tag{A.7}
$$

We know that $\{g'(x)e_1, \ldots, g'(x)e_m\}$ is a set of linearly independent vectors. Since the vectors $e_1, \ldots, \widehat{e_i}, \ldots, e_m$ are tangent to $\mathcal{I}_{ij}^{m-1}$ and $g(\mathcal{I}_{ij}^{m-1}) = \mathcal{I}_{i'j'}^{m-1}$, we see that the vectors $g'(x)e_1, \ldots, \widehat{g'(x)e_i}, \ldots, g'(x)e_m$ must be linearly independent vectors and tangent to $\mathcal{I}_{i'j'}^{m-1}$ at $g(x)$. The space of $(m-1)$-dimensional tangent vectors to $\mathcal{I}_{i'j'}^{m-1}$ at $g(x)$ consists at every point of the scalar multiples of $e_1 \cdots \widehat{e_{i'}} \cdots e_m$, therefore there must exist β such that (A.7) holds. Now we need only show that $\beta > 0$.

Consider the $(m-1)$-face $\mathcal{I}_{ij}^{m-1}$. We know that if $j = 0$, then $-e_i$ is the outward normal vector to this face, hence e_i must be a normal vector to that $(m-1)$-face that points into the interior of $\mathcal{I}^m$. On the other hand, if $j = 1$, then we easily see that $-e_i$ must be an inward pointing normal vector. More generally, $(-1)^j e_i$ is a normal vector to the $(m-1)$-face $\mathcal{I}_{ij}^{m-1}$ that points into the interior of $\mathcal{I}^m$. Similarly, $(-1)^{j'} e_{i'}$ is an inward pointing vector that is normal to the $(m-1)$-face $\mathcal{I}_{i'j'}^{m-1}$. For τ sufficiently small and positive, $x + \tau((-1)^j e_i)$ is a point in the interior of $\mathcal{I}^m$; and by the way g is defined and its continuity, we may likewise assume that $g(x + \tau((-1)^j e_i))$ is in the interior of $\mathcal{I}^m$. Because of this, the vector $g(x + \tau((-1)^j e_i)) - g(x)$ may be considered as pointing from $g(x)$ in the $(m-1)$-face $\mathcal{I}_{i'j'}^{m-1}$ into the interior of $\mathcal{I}^m$. It follows from these facts that we must have

$$
\left(g(x + \tau((-1)^j e_i)) - g(x) \right) \cdot \left((-1)^{j'} e_{i'} \right) \geq 0.
$$

Clearly we can write $g'(x) e_i = \sum_{k=1}^m \gamma_k e_k$. It follows that

$$
(-1)^{j+j'} \gamma_{i'} = \left(g'(x)(-1)^j e_i \right) \cdot \left((-1)^{j'} e_{i'} \right)
$$
$$
= \left(\lim_{\tau \to 0} \frac{1}{\tau} \left(g(x + \tau((-1)^j e_i)) - g(x) \right) \right) \cdot \left((-1)^{j'} e_{i'} \right) \geq 0. \tag{A.8}
$$

We next calculate

$$
\alpha e = (-1)^{i+j} \wedge^m g'(x) ((-1)^{i+j} e)
$$
$$
= (-1)^{i+j} (-1)^{i-1} \left(g'(x) e_i \right) \wedge \left(\wedge^{m-1} g'(x) \overrightarrow{\partial} e(x) \right)
$$
$$
= (-1)^{j-1} \left(\sum_{k=1}^m \gamma_k e_k \right) \wedge \left(\beta (\overrightarrow{\partial} e \circ g)(x) \right)
$$
$$
= (-1)^{j-1} \gamma_{i'} \beta \, e_{i'} \wedge \left(\beta (\overrightarrow{\partial} e \circ g)(x) \right)
$$

$$= (-1)^{j-1}\gamma_{i'}\beta(-1)^{i'-1}(-1)^{i'+j'}e = \gamma_{i'}\beta(-1)^{j+j'}e.$$

We thus obtain the equation $\alpha = (-1)^{j+j'}\gamma_{i'}\beta$. It follows from this equation and (A.8) and the fact that $\alpha > 0$ that $(-1)^{j+j'}\gamma_{i'} > 0$ and therefore $\beta > 0$ as well. This is the desired result. $\qquad\square$

Recall that if $g : \mathcal{I}^m \to \mathcal{M}$ is a $\mathcal{C}^1$ parametrization of an m-cell that agrees with the orientation w of that cell, then at $x = g(t) \in \mathcal{M}$, we have

$$\overrightarrow{\partial}w(x) = \frac{g'(t)\,\overrightarrow{\partial}e(x)}{|g'(t)\,\overrightarrow{\partial}e(x)|}.$$

However since we have not yet proved that induced orientation is independent of parametrization, instead of writing $\overrightarrow{\partial}w$, we shall write $\overrightarrow{\partial}_g w$ for right-hand induced orientation defined using g.

Proposition A.8. *Suppose that $g_1, g_2 : \mathcal{I}^m \to \mathcal{M}$ are parametrizations of a $\mathcal{C}^1$ m-cell $\mathcal{M}$ and that both induce the orientation w on $\mathcal{M}$ in the sense that*

$$w(y) = \frac{\wedge^m g_i'(x_i)\,(e_1\cdots e_m)}{|\wedge^m g_i'(x_i)\,(e_1\cdots e_m)|}$$

for $i = 1, 2$ whenever $g_i(x_i) = y$. Then $\overrightarrow{\partial}_{g_1} w = \overrightarrow{\partial}_{g_2} w$.

Proof. Let us denote the standard orientation $e_1\cdots e_m$ of $\mathcal{I}^m$ as e. Recall that for this orientation we have a definition of the induced orientation $\overrightarrow{\partial}e$ on $\partial\mathcal{I}^m$, namely,

$$\overrightarrow{\partial}e = (-1)^{i+j}(e_1\cdots\widehat{e}_i\cdots e_m) \quad \text{on } \mathcal{I}^{m-1}_{ij}.$$

Notice that $g_2 = g_1 \circ g$ where $g = g_1^{-1} \circ g_2 : \mathcal{I}^m \to \mathcal{I}^m$. Choose $x \in \mathcal{I}^m$ and set $y = g(x) \in \mathcal{I}^m$ and $z = g_2(x) = g_1(y) \in \mathcal{M}$. We know there exist $\alpha > 0$, $\beta > 0$, and $\gamma \in \mathbb{R}$ such that

$$\wedge^m g_1'(y)\,e = \alpha\,w(z), \quad \wedge^m g_2'(x)\,e = \beta\,w(z), \quad \text{and} \quad \wedge^m g'(x)\,e = \gamma e.$$

(Of course, α, β, γ depend on x.) By the chain rule (see Exercise 8 of Section 5.7), we have

$$\wedge^m g_2'(x) = (\wedge^m g_1'(y))\,(\wedge^m g'(x)).$$

If we apply these linear transformations to e, we obtain $\beta\,w(z) = \alpha\,\gamma w(z)$. Thus $\beta = \alpha\,\gamma$, and we must have $\gamma > 0$.

We can now apply Lemma A.2 to g and see that we have

$$\left(\wedge^{m-1}g'(x)\right)\overrightarrow{\partial}e(x) = \eta\,\overrightarrow{\partial}e(y)$$

for some $\eta > 0$. It follows that

$$\left(\wedge^{m-1}g_2'(x)\right)\overrightarrow{\partial}e(x) = \left(\wedge^{m-1}g_1'(y)\right)\left(\wedge^{m-1}g'(x)\right)\overrightarrow{\partial}e(x)$$

$$= \eta \left(\wedge^{m-1} g_1'(y) \right) \vec{\partial} e(y).$$

But if we consult the definition of induced orientation for $\vec{\partial}_g w$, we see that this amounts to $\vec{\partial}_{g_2} w(z) = \lambda \, \vec{\partial}_{g_1} w(z)$ where λ is a positive scalar. Since we are dealing with unit $(m-1)$-vectors, we must have $\vec{\partial}_{g_2} w = \vec{\partial}_{g_1} w$.   □

We thus see that the induced orientation on $\vec{\partial} w$ on the boundary of an m-cell depends only on w, not on the choice of parametrization. The independence of left-hand induced orientation follows from the fact that for any parametrization g we have $\vec{\partial}_g(w^\dagger) = \left(\overleftarrow{\partial}_g w \right)^\dagger$.

A.7   Rate of change of orientation

We assume that

1. $\mathcal{M}$ is a $\mathcal{C}^2$ p-manifold in $\mathbb{R}^n$ with orientation w,

2. $x \colon U \to \mathbb{R}^n$ is a $\mathcal{C}^2$ chart on $\mathcal{M}$ (where U is an open subset of $\mathbb{R}^p$),

3. x agrees with the orientation w, and

4. x induces coordinates $(\tau_1, \ldots, \tau_p)$ on $\mathcal{M}$.

We introduce the abbreviations

$$u_i = \frac{\partial x}{\partial \tau_i} \quad \text{and} \quad u = u_1 \wedge \cdots \wedge u_p.$$

Then $\{u_i\}_{i=1}^p$ is a local tangent frame for $\mathcal{M}$ and $w = u/|u|$. Of course, $\{d\tau_i\}_{i=1}^p$ is the reciprocal frame for $\{u_i\}_{i=1}^p$. We introduce two more notations: Let

$$x_{ij} \overset{\text{def.}}{=} \frac{\partial^2 x}{\partial \tau_i \, \partial \tau_j}$$

and let v_{ij} be the component of the 1-vector x_{ij} that is orthogonal to $\mathcal{M}$ at the base point of the vector. That is,

$$v_{ij} = x_{ij}\big|_{\perp \mathcal{M}} = \frac{\partial^2 x}{\partial \tau_i \partial \tau_j}\bigg|_{\perp \mathcal{M}} = \left(\frac{\partial^2 x}{\partial \tau_i \partial \tau_j} \wedge u \right) u^{-1}. \tag{A.9}$$

In the last expression, we may replace u by w. (See Proposition 5.21.) Since our parametrization is $\mathcal{C}^2$, we see that $x_{ij} = x_{ji}$ and $v_{ij} = v_{ji}$.

Lemma A.3.
$$\frac{\partial u}{\partial \tau_i} = \sum_{j=1}^p \left(v_{ij}(d\tau_j) u + (x_{ij} \cdot d\tau_j) u \right).$$

Proof. We know that

$$\frac{\partial u_i}{\partial \tau_j} = \frac{\partial^2 x}{\partial \tau_j \partial \tau_i} = x_{ij}.$$

Since $u = u_1 \wedge \cdots \wedge u_p$, it follows via the product rule that

$$\frac{\partial u}{\partial \tau_i} = \sum_{j=1}^{p} (-1)^{j-1} x_{ij} \wedge (u_1 \wedge \cdots \wedge \widehat{u_j} \wedge \cdots \wedge u_p).$$

By Proposition 5.15, we also know that

$$d\tau_j = (-1)^{j-1} (u_1 \wedge \cdots \wedge \widehat{u_j} \wedge \cdots \wedge u_p) u^{-1}$$

so that $(d\tau_j)u = (-1)^{j-1} u_1 \wedge \cdots \wedge \widehat{u_j} \wedge \cdots \wedge u_p$. Thus

$$\frac{\partial u}{\partial \tau_i} = \sum_{j=1}^{p} x_{ij} \wedge \big[(d\tau_j) u \big].$$

Consider the vector $d\tau_j$ and the $(p-1)$-blade $(d\tau_j)u$. Since $d\tau_j$ lies in the span of $\{u_k\}_{k=1}^{p}$ (or, equivalently, is parallel to u), we must be able to write u in the form $u = w_1 \cdots w_p = w_1 \wedge \cdots \wedge w_p$ where $w_1, \ldots, w_p$ are orthogonal and $w_1 = d\tau_j$. Then $(d\tau_j)u = |d\tau_j|^2 w_2 \cdots w_p$ which is clearly orthogonal to $d\tau_j$. Because x_{ij} is a 1-vector, we can decompose it thus:

$$x_{ij} = v_{ij} + \lambda_{ij} d\tau_j + y_{ij} \tag{A.10}$$

where v_{ij} is defined in (A.9) to be the component of the x_{ij} orthogonal to $\mathcal{M}$ (or, equivalently, orthogonal to the p-blade u), λ_{ij} is a scalar, and y_{ij} is the component of x_{ij} parallel to the blade $(d\tau_j)u$. It is clear that this is a decomposition into orthogonal vectors.

If we dot (A.10) with $d\tau_j$, we obtain $x_{ij} \cdot d\tau_j = \lambda_{ij} |d\tau_j|^2$, that is,

$$\lambda_{ij} = \frac{x_{ij} \cdot d\tau_j}{|d\tau_j|^2}.$$

Next notice that since v_{ij} is orthogonal to the blade $(d\tau_j)u$, we must have $v_{ij} \wedge [(d\tau_j)u] = v_{ij}(d\tau_j)u$. Similarly, because $d\tau_j$ is orthogonal to the blade $(d\tau_j)u$, we also have $d\tau_j \wedge [(d\tau_j)u] = (d\tau_j)(d\tau_j)u = |d\tau_j|^2 u$. Finally, since y_{ij} is parallel to the blade $(d\tau_j)u$, we see that $y_{ij} \wedge [(d\tau_j)u] = 0$. Combining these facts with (A.10), we obtain

$$\begin{aligned}
\frac{\partial u}{\partial \tau_i} &= \sum_{j=1}^{p} \Big(x_{ij} \wedge [(d\tau_j)u] \Big) \\
&= \sum_{j=1}^{p} \Big(v_{ij} \wedge [(d\tau_j)u] + |d\tau_j|^2 \frac{(x_{ij} \cdot d\tau_j)}{|d\tau_j|^2} u \Big) \\
&= \sum_{j=1}^{p} \Big(v_{ij}(d\tau_j)u + (x_{ij} \cdot d\tau_j)u \Big). \qquad \square
\end{aligned}$$

Lemma A.4.

$$\frac{\partial |u|}{\partial \tau_i} = \sum_{j=1}^{p} (x_{ij} \cdot d\tau_j) |u|.$$

Proof. We notice that

$$\frac{\partial (|u|^2)}{\partial \tau_i} = 2|u| \frac{\partial |u|}{\partial \tau_i}.$$

Another way to write this is

$$\frac{\partial (|u|^2)}{\partial \tau_i} = \frac{\partial (uu^\dagger)}{\partial \tau_i} = \left(\frac{\partial u}{\partial \tau_i}\right) u^\dagger + u \left(\frac{\partial u}{\partial \tau_i}\right)^\dagger.$$

Therefore, keeping mind that $u = |u|w$, we have

$$\frac{\partial |u|}{\partial \tau_i} = \frac{1}{2|u|} \frac{\partial (|u|^2)}{\partial \tau_i} = \frac{1}{2} \left[\frac{\partial u}{\partial \tau_i} w^\dagger + w \left(\frac{\partial u}{\partial \tau_i}\right)^\dagger\right]. \tag{A.11}$$

By virtue of the fact that v_{ij} is orthogonal to $d\tau_j$ and $u = |u|w$, we see that $v_{ij}(d\tau_j)uw^\dagger$ has grade 2. It follows that

$$\left\langle \frac{\partial u}{\partial \tau_i} w^\dagger \right\rangle_0 = \sum_{j=1}^{p} \left\langle v_{ij}(d\tau_j)uw^\dagger + (x_{ij} \cdot d\tau_j)uw^\dagger \right\rangle_0$$

$$= \sum_{j=1}^{p} (x_{ij} \cdot d\tau_j)|u|.$$

A similar remark applies to $w\left(\partial u/\partial \tau_i\right)^\dagger$. It then follows from (A.11) that

$$\frac{\partial |u|}{\partial \tau_i} = \left\langle \frac{\partial |u|}{\partial \tau_i} \right\rangle_0 = \sum_{j=1}^{p} (x_{ij} \cdot d\tau_j)|u|. \qquad \square$$

Proposition 7.15. *Let $\mathcal{M}$ be a $\mathcal{C}^2$ p-manifold in $\mathbb{R}^n$ with orientation w. Using the notation and coordinates above, we have*

1. $\dfrac{\partial w}{\partial \tau_i} = \displaystyle\sum_{j=1}^{p} \left(\dfrac{\partial^2 x}{\partial \tau_i \partial \tau_j}\Big|_{\perp \mathcal{M}}\right) (d\tau_j)w$ *and*

2. $\vec{\nabla}_{\mathcal{M}} w = (-1)^{p-1} \left(\displaystyle\sum_{i,j=1}^{p} \left(\dfrac{\partial^2 x}{\partial \tau_i \partial \tau_j}\Big|_{\perp \mathcal{M}}\right) (d\tau_i \cdot d\tau_j)\right) w.$

Proof. Since $u = |u|w$, by the product rule and Equation (A.11), we have

$$\frac{\partial u}{\partial \tau_i} = |u| \frac{\partial w}{\partial \tau_i} + \frac{\partial |u|}{\partial \tau_i} w = |u| \frac{\partial w}{\partial \tau_i} + \sum_{j=1}^{p} (x_{ij} \cdot d\tau_j)u.$$

Comparing this with Lemma A.3, we see that

$$\frac{\partial w}{\partial \tau_i} = \sum_{j=1}^{p} v_{ij}(d\tau_j)w$$

which gives us the first of our desired conclusions. Then we compute

$$
\begin{aligned}
\vec{\nabla}_{\mathcal{M}} w &= \sum_{i=1}^{p} \frac{\partial w}{\partial \tau_i} \, d\tau_i \\
&= (-1)^{p-1} \sum_{i,j=1}^{p} v_{ij} (d\tau_i)(d\tau_j) w.
\end{aligned}
\tag{A.12}
$$

We know that

$$
v_{ij}(d\tau_i)(d\tau_j) = v_{ij} \, (d\tau_i \wedge d\tau_j) + v_{ij} \, (d\tau_i \cdot d\tau_j).
$$

Since $v_{ij} = v_{ji}$, it is easily seen that

$$
\sum_{i,j=1}^{p} v_{ij}(d\tau_i \wedge d\tau_j) = 0.
$$

Therefore (A.12) reduces to our desired formula for $\vec{\nabla}_{\mathcal{M}} w$ and we are done. $\qquad \square$

A.8   The normal identity

Proposition 7.31. *Suppose that $\mathcal{M}$ is a $\mathcal{C}^r$ p-manifold (where $r \geq 2$ and $p \geq 1$) in $\mathbb{R}^n$ and $x_0 \in \mathcal{M}$. Then for every $\xi > 0$ there is an open neighborhood V in $\mathbb{R}^n$ of x_0 and a map $I_{\mathcal{M}} : V \to V$ such that the following hold:*

1. *$V \subseteq B(x_0, \xi)$.*

2. *For all $x \in V \cap \mathcal{M}$ and $y \in V$, $I_{\mathcal{M}}(y) = x$ if and only if $y \in V \cap H_x\mathcal{M}$.*

3. *$I_{\mathcal{M}}$ is $\mathcal{C}^{r-1}$.*

Proof. If $p = n$, then $N_x\mathcal{M} = \{0\}$ whenever $x \in \mathcal{M}$, and thus $H_x\mathcal{M} = \{x\}$. If we take $I_{\mathcal{M}}$ to be the identity map on $\mathbb{R}^n$, then the result is trivially true.

Therefore, from this point on, we assume $p < n$.

Choose a $\mathcal{C}^r$ chart $x : W \to \mathbb{R}^n$ on $\mathcal{M}$ with $x_0 = x(t_0)$. We know $\{\partial_{e_i} x(t_0)\}_{i=1}^{p}$ is a basis for $T_{x_0}\mathcal{M}$. Recalling that $\{e_i\}_{i=1}^{n}$ is the standard basis for $\mathbb{R}^n$, we may suppose, without loss of generality, that

$$
\{\partial_{e_i} x(t)\}_{i=1}^{p} \cup \{e_{p+1}, \dots, e_n\}
\tag{A.13}
$$

is a linearly independent set (hence a basis for $\mathbb{R}^n$) at $t = t_0$. More than that, by continuity, (A.13) must be a basis for $\mathbb{R}^n$ for all t sufficiently close to t_0.

Next construct the reciprocal frame $\{m_i(x)\}_{i=1}^{n}$ for (A.13) where $x = x(t)$. By the definition of a reciprocal frame, this is a basis for $\mathbb{R}^n$ such that

$$
\delta_{ij} =
\begin{cases}
(\partial_{e_i} x) \cdot m_j & \text{for } i \leq p, \\
e_i \cdot m_j & \text{for } p < i,
\end{cases}
$$

where δ_{ij} is Kronecker's delta. Of course, each m_j is a function of x and thus of t, $m_j = m_j(x) = m_j(x(t))$. Since

$$(\partial_{e_i} x) \cdot m_j = 0 \quad \text{for } i \leq p \text{ and } j > p, \tag{A.14}$$

we see that

$$\{\partial_{e_i} x(t)\}_{i=1}^{p} \cup \{m_{p+1}(x), \ldots, m_n(x)\} \tag{A.15}$$

must be a basis for $\mathbb{R}^n$ for t sufficiently close to t_0.

On an open neighborhood of the point $(t_0, 0^{n-p})$, we now construct a map Y between subsets of $\mathbb{R}^n$. Let $t = (\tau_1, \ldots, \tau_p) \in \mathbb{R}^p$ and $t' = (\tau_{p+1}, \ldots, \tau_n) \in \mathbb{R}^{n-p}$. Set

$$Y(t, t') = x(t) + \tau_{p+1} m_{p+1}(x) + \cdots \tau_n m_n(x) \tag{A.16}$$

where we understand that $x = x(t)$. Notice that $x_0 = Y(t_0, 0^{n-p})$. We know that the map x is $\mathcal{C}^r$ and each $\partial_{e_i} x$ must be at least $\mathcal{C}^{r-1}$. Since the reciprocal vectors m_i are constructed from the vectors $\partial_{e_i} x$ by an algebraic process, we see that Y must be at least $\mathcal{C}^{r-1}$.

It is straightforward to calculate that

$$\frac{\partial Y}{\partial \tau_i}(t_0, 0) = \begin{cases} \partial_{e_i} x(t_0) & \text{if } i = 1, \ldots, p \\ m_i(x_0) & \text{if } i = p+1, \ldots, n. \end{cases}$$

Since this is a linearly independent set and $\partial_{e_i} x$ and m_j are $\mathcal{C}^{r-1}$ (with $r \geq 2$), we see that $\det(Y')$ must be nonzero for (t, t') sufficiently close to $(t_0, 0)$. Therefore by the inverse function theorem, there exist open sets U_1 and V_1 of $\mathbb{R}^n$ such that $Y: U_1 \to V_1$ is a $\mathcal{C}^{r-1}$ diffeomorphism and $(t_0, 0^{n-p}) \in U_1$.

Case 1. Let us now suppose that x_0 is an interior point of $\mathcal{M}$ and that $\xi > 0$ is a given value.

Because x and Y are continuous maps, there exists $\varepsilon > 0$ such that

$$\begin{aligned} B(t_0, \varepsilon) &\subseteq W = \text{domain of } x, \\ B((t_0, 0^{n-p}), \varepsilon) &\subseteq U_1, \\ Y(B((t_0, 0^{n-p}), \varepsilon)) &\subseteq V_1 \cap B(x_0, \xi). \end{aligned} \tag{A.17}$$

It follows that $x_0 \in x(B(t_0, \varepsilon)) \subseteq \mathcal{M}$. (See Figure A.1 for how these open balls relate to one another and how they map under Y.)

Now let

$$\begin{aligned} U &= B((t_0, 0^{n-p}), \varepsilon), \\ V &= Y(U). \end{aligned}$$

Since $U \subseteq U_1$ and $V \subseteq V_1$, we see that $Y: U \to V$ is still a $\mathcal{C}^{r-1}$ diffeomorphism. Because x_0 is an interior point of $\mathcal{M}$, $x(W) \cap \mathcal{M}$ is an open subset of $\mathcal{M}$, and Y is continuous, we can impose a further restriction on ε and require that

$$\begin{aligned} V \cap x(W) &= Y(B((t_0, 0^{n-p}), \varepsilon)) \cap x(W) \\ &= V \cap \mathcal{M}. \end{aligned} \tag{A.18}$$

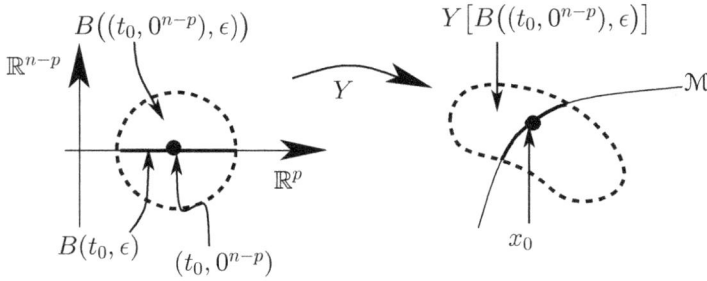

Figure A.1: Y mapping open balls

We now develop the ways in which certain sets are related to one another.

Suppose that $x = x(t)$ where $t \in B(t_0, \varepsilon)$. We know that $\dim T_x \mathcal{M} = p$, so $\dim N_x \mathcal{M} = n - p$. Since $m_{p+1}(x), \ldots, m_n(x)$ are linearly independent and orthogonal to $\mathcal{M}$ at x, it follows that $\{m_i(x)\}_{i=p+1}^n$ must be a basis for $N_x \mathcal{M}$. We know that $H_x \mathcal{M} = x + N_x \mathcal{M}$, and in (A.16), the definition of Y, the scalars $\tau_{p+1}, \ldots, \tau_n$ range over all values of $\mathbb{R}$; it follows that for t held constant, we have

$$Y(t \times \mathbb{R}^{n-p}) = H_x \mathcal{M}.$$

However since we have the constraint $Y : U \to V$, the relation we really want is

$$Y\big(U \cap (t \times \mathbb{R}^{n-p})\big) = V \cap H_x \mathcal{M} \quad \text{where } t \in B(t_0, \varepsilon). \tag{A.19}$$

Next notice that

$$U \cap (\mathbb{R}^p \times 0^{n-p}) = B\big((t_0, 0^{n-p}), \varepsilon\big) \cap (\mathbb{R}^p \times 0^{n-p}) = B(t_0, \varepsilon) \times 0^{n-p} \tag{A.20}$$

and

$$Y\big(B(t_0, \varepsilon) \times 0^{n-p}\big) = x\big(B(t_0, \varepsilon)\big),$$

hence

$$Y(U \cap (\mathbb{R}^p \times 0^{n-p}) = x\big(B(t_0, \varepsilon)\big). \tag{A.21}$$

From (A.20), we have $U \cap (\mathbb{R}^p \times 0^{n-p}) \subseteq W \times 0^{n-p}$, so we must have

$$U \cap (\mathbb{R}^p \times 0^{n-p}) = U \cap (W \times 0^{n-p}).$$

By this fact and (A.18), we see that

$$Y\big(U \cap (\mathbb{R}^p \times 0^{n-p})\big) = V \cap \mathcal{M}. \tag{A.22}$$

Combining (A.21) and (A.22) gives us

$$x\big(B(t_0, \varepsilon)\big) = V \cap \mathcal{M}. \tag{A.23}$$

We are now ready to define the normal identity $I_{\mathcal{M}}: V \to V \cap \mathcal{M}$. Let $\pi: \mathbb{R}^n \to \mathbb{R}^p$ be the map $\pi(t,t') = t$. Notice that in the following diagram, all the maps are onto:

$$V \xleftarrow{Y} U \xrightarrow{\pi} B(t_0, \varepsilon) \xrightarrow{x} V \cap \mathcal{M}.$$

We set

$$I_{\mathcal{M}} \overset{\text{def.}}{=} x \circ \pi \circ Y^{-1}. \tag{A.24}$$

By construction, $I_{\mathcal{M}}$ is $\mathcal{C}^{r-1}$, and we know from (A.17) that $V \subseteq B(x_0, \xi)$, so the only thing we need to check is the relation of $I_{\mathcal{M}}$ to H_x.

Choose $y \in V$ and $x \in V \cap \mathcal{M}$. We know that

$$U \cap (t \times \mathbb{R}^{n-p}) = B\big((t_0, 0^{n-p}), \varepsilon\big) \cap (t \times \mathbb{R}^{n-p}),$$

and this will be nonempty if and only if $(t, 0^{n-p})$ lies in $B\big((t_0, 0^{n-p}), \varepsilon\big)$, that is, if and only if $t \in B(t_0, \varepsilon)$. Thus we can write

$$U = \bigcup_{t \in B(t_0, \varepsilon)} \big(U \cap (t \times \mathbb{R}^{n-p})\big),$$

and by (A.19) this becomes

$$V = \bigcup_{t \in B(t_0, \varepsilon)} \big(V \cap H_{x(t)} \mathcal{M}\big).$$

Since the sets $t \times \mathbb{R}^{n-p}$ are pairwise disjoint, the same must be true for the sets $V \cap H_{x(t)} \mathcal{M}$. Thus for our $y \in V$, there exists a unique $t \in B(t_0, \varepsilon)$ such that $y \in H_{x(t)} \mathcal{M}$. We can write this y in the form

$$y = x(t) + \tau_{p+1} m_{p+1}(x(t)) + \cdots + \tau_n m_n(x(t)) = Y(t, t')$$

where $t' = (\tau_{p+1}, \ldots, \tau_n) \in \mathbb{R}^{n-p}$ is uniquely determined for y. By the definition of $I_{\mathcal{M}}$, we have $I_{\mathcal{M}}(y) = x(t)$. We see that $I_{\mathcal{M}}(y) = x$ if and only if $V \cap H_x = V \cap H_{x(t)}$. Thus $I_{\mathcal{M}}(y) = x$ if and only if $y \in H_x$.

Case 2. Suppose that x_0 is not an interior point of $\mathcal{M}$, that it is a boundary point. We know that we can construct a slightly larger $\mathcal{C}^r$ manifold $\mathcal{M}^+$ containing $\mathcal{M}$ such that x_0 is an interior point of $\mathcal{M}^+$. We now find an open neighborhood V of x_0 in $\mathbb{R}^n$ and construct the normal identity $I_{\mathcal{M}^+}: V \to V$ for $\mathcal{M}^+$. This will be a $\mathcal{C}^{r-1}$ map and will automatically have the properties desired for $I_{\mathcal{M}}$, so we may take $I_{\mathcal{M}^+}$ to be our $I_{\mathcal{M}}$.

We note that this map is not uniquely defined since there must exist an infinite number of ways to expand $\mathcal{M}$ to $\mathcal{M}^+$. □

Appendix B

Some Answers and Hints

Exercises 2.1.

6. The empty set is both open and closed.

Exercises 2.2.

5. We answer the first three problems; the Jacobian matrix of $f'(x)$ is denoted $[f'(x)]$.

(**a**) We see that $f : \mathbb{R} \to \mathbb{R}^2$. Since $\mathbb{R}$ has a basis with a single element, namely the number 1, we denote this basis element as e_1 so that our answer looks like a standard directional derivative. Of course $\partial_{e_1} f$ is in $\mathbb{R}^2$ where there is a basis $\{e_1, e_2\}$.

$$\partial_{e_1} f(\theta) = \frac{\partial f}{\partial \theta} = -\sin(\theta)e_1 + \cos(\theta)e_2,$$

$$[f'(\theta)] = \begin{pmatrix} -\sin(\theta) \\ \cos(\theta) \end{pmatrix}.$$

(**b**)

$$\partial_{e_i} f(\tau) = \frac{\partial f}{\partial \tau} = \phi_1'(\tau)e_1 + \cdots + \phi_n'(\tau)e_n,$$

$$[f'(\tau)] = \begin{pmatrix} \phi_1'(\tau) \\ \vdots \\ \phi_n'(\tau) \end{pmatrix}.$$

(**c**)

$$\partial_{e_i} f(\chi_1, \chi_2) = \frac{\partial f}{\partial \chi_i} = e_i + \frac{\partial \phi}{\partial \chi_i} e_3,$$

$$[f'(\chi_1, \chi_2)] = \begin{pmatrix} 1 & 0 \\ 0 & 1 \\ \frac{\partial \phi}{\partial \chi_1} & \frac{\partial \phi}{\partial \chi_2} \end{pmatrix}.$$

Exercises 2.3.

1. (b) $\dfrac{\partial x}{\partial \tau} = -\sin(\tau)\, e_1 + \cos(\tau)\, e_2 + e_3.$

(c) x_0 corresponds to $\tau = \pi/4$, so a basis for $T_{x_0}\mathcal{M}$ is given by

$$\left.\frac{\partial x}{\partial \tau}\right|_{\tau=\frac{\pi}{4}} = -\frac{1}{\sqrt{2}}\, e_1 + \frac{1}{\sqrt{2}}\, e_2 + e_3.$$

2. (a) One answer: $x(\chi_1, \chi_2) = \chi_1\, e_1 + \chi_2\, e_2 + \left(\chi_1^2 + \chi_2^2\right) e_3.$
(b) One basis for $T_{x_0}\mathcal{M}$ is $\{e_1 + e_3,\, e_2 + e_3\}.$

3. Hint: A chart for $\mathcal{M}$ is

$$x(\chi_1, \ldots, \chi_n) = \left(\sum_{i=1}^{n} \chi_i\, e_i\right) + \phi(\chi_1, \ldots, \chi_n)\, e_{n+1}.$$

4. $\mathcal{M}$ is an ellipse, and we can cover it with four charts. Let $U_1 = (-\alpha_1, \alpha_1)$ and $U_2 = (-\alpha_2, \alpha_2)$, intervals in $\mathbb{R}$, and set

$$\xi_i = \sqrt{1 - \frac{\chi_i^2}{\alpha_i^2}} \quad \text{for } i = 1, 2.$$

We define our charts, $x_1, x_2 : U_1 \to \mathcal{M}$ and $x_3, x_4 : U_2 \to \mathcal{M}$, by

$$\begin{aligned}
x_1(\chi_1) &= \chi_1\, e_1 + \alpha_2 \xi_1\, e_2, \\
x_2(\chi_1) &= \chi_1\, e_1 - \alpha_2 \xi_1\, e_2, \\
x_3(\chi_2) &= \alpha_1 \xi_2\, e_1 + \chi_2\, e_2, \\
x_3(\chi_2) &= -\alpha_1 \xi_2\, e_1 + \chi_2\, e_2.
\end{aligned}$$

Exercises 3.1.

1. Hint: Choose an orthonormal basis $\{u_i\}_{i=1}^{n}$ for $\mathbb{R}^n$. Write the matrix $A = (a_1, \ldots, a_n)$ with respect to the $\{u_i\}_{i=1}^{n}$ basis. Then show that

$$\operatorname{vol}(a_1, \ldots, a_n)^2 = \det(AA^T).$$

2. $\sqrt{2}.$
3. 1.
4. $|\sin(\theta)|.$
5. $4|\beta|.$
7. Hint: First consider the case where your fixed basis of $\mathbb{R}^n$ is orthonormal. In the general case, how does one use matrices to change from an arbitrary basis to an orthonormal one?

Exercises 3.2.

3. No, they have opposite orientations.

4. Hint: Show that $\det(a_i \cdot b_j) = |a_1|^2|a_2|^2 - (a_1 \cdot a_2)^2$. Next carry out some case analysis such as whether or not one of the a_i's is zero, whether or not a_1, a_2 are linearly dependent, etc.

5. Because they both lie in the same m-dimensional subspace and $\det(e_i \cdot a_j) = 1$, we see that $(e_1, \ldots, e_m)$ and $(a_1, \ldots, a_m)$ have the same orientation.

7. No, the orientations are not comparable.

Exercises 3.3.

2. Partial solution:

Any linear combination of a_1 and a_2 is also a linear combination of $a_1 + a_2$ and a_2:

$$\lambda_1 a_1 + \lambda_2 a_2 = (\lambda_1 - \lambda_2)(a_1 + a_2) + 2\lambda_2 a_2.$$

Any linear combination of $a_1 + a_2$ and a_2 is also a linear combination of a_1 and a_2:

$$\lambda_1(a_1 + a_2) + \lambda_2 a_2 = \lambda_1 a_1 + (\lambda_1 + \lambda_2)a_2.$$

Therefore the two ordered pairs (a_1, a_2) and $(a_1 + a_2, a_2)$ span the same vector subspace.

We now consider only the case where a_1 and a_2 are independent.

Using the fact that we can add a scalar times a row (column) vector to another row (column) vector in a determinant, we calculate that

$$\det \begin{pmatrix} (a_1 + a_2) \cdot (a_1 + a_2) & a_2 \cdot (a_1 + a_2) \\ (a_1 + a_2) \cdot a_2 & a_2 \cdot a_2 \end{pmatrix}$$

$$= \det \begin{pmatrix} a_1 \cdot (a_1 + a_2) & a_2 \cdot (a_1 + a_2) \\ a_1 \cdot a_2 & a_2 \cdot a_2 \end{pmatrix}$$

$$= \det \begin{pmatrix} a_1 \cdot a_1 & a_2 \cdot a_1 \\ a_1 \cdot a_2 & a_2 \cdot a_2 \end{pmatrix} = |a_1|^2 |a_2|^2 \sin^2(\theta)$$

where θ is the angle between a_1 and a_2. Equality of volumes follows from this. Equality of orientations follows provided we can show that $\sin(\theta) \neq 0$.

Exercises 3.4.

3. There are six such simple 2-vectors.

$$(e_{i_1} \wedge e_{i_2}) \cdot (e_{j_1} \wedge e_{j_2})^\dagger = \begin{cases} 1 & \text{if } (i_1, i_2) = (j_1, j_2), \\ 0 & \text{if } (i_1, i_2) \neq (j_1, j_2). \end{cases}$$

6. Hint: Look at Problem 3 of Section 3.3.

Exercises 4.1.

3. (a) $2(e_1 \wedge e_2)$.

(b) 0.

(c) Three vectors constructed from e_1 and e_2 cannot be linearly independent. The answer is 0.

(d) $-6(e_1 \wedge e_2) + 9(e_1 \wedge e_3) - 3(e_2 \wedge e_3)$.

Exercises 4.3.

1. Hint: Show that if $\{v_i\}_{i=1}^n$ is a basis for $\mathbb{R}^n$, then it suffices to prove this for $v_I \wedge v_J$ where the multi-indices $I = (i_1, \ldots, i_p)$ and $J = (j_1, \ldots, j_q)$ have no indices in common. Then recall that $v_i \wedge v_j = -v_j \wedge v_i$ when $i \neq j$.

2. Hint: First show that if we have a basis $\{u_i\}_{i=1}^n$ for $\mathbb{R}^n$, then it is sufficient to prove the result for $u_I \wedge u_J$ where I and J are multi-indices with no elements in common. Next note that if we have $i_1 < \cdots < i_p$ and $j_1 < \cdots < j_q$ where the sets $\{i_1, \ldots, i_p\}$ and $\{j_1, \ldots, j_q\}$ have nothing in common, then

$$\left[(u_{i_1} \wedge \cdots \wedge u_{i_p}) \wedge (u_{j_1} \wedge \cdots \wedge u_{j_q})\right]^\dagger = (-1)^m (u_{i_1} \wedge \cdots \wedge u_{i_p}) \wedge (u_{j_1} \wedge \cdots \wedge u_{j_q})$$

where

$$m = \frac{(p+q)(p+q-1)}{2}.$$

Exercises 4.4.

1. The orientations are $\pm a / |a|$.

2. The orientations are $\pm(e_1 \wedge \cdots \wedge \widehat{e_i} \wedge \cdots \wedge e_n)$ where the symbol $\widehat{e_i}$ means that e_i is missing.

3. The orientations are $\pm \frac{1}{\sqrt{30}} \left[5(e_1 \wedge e_2) + (e_1 \wedge e_3) - 2(e_2 \wedge e_3)\right]$.

4. Let w_i be the orientation of P_i. We will have $\cos(\theta_i) = 1/\sqrt{3}$ if we set

$$w_1 = e_2 \wedge e_3,$$
$$w_2 = -e_1 \wedge e_3,$$
$$w_3 = e_1 \wedge e_2.$$

5. The vector space associated with A_n is given by $\sum_{i=1}^n \chi_i = 0$. The vectors $e_1 - e_2, \ldots, e_1 - e_n$ lie in this space and are linearly independent. Hence

$$a = (e_1 - e_2) \wedge \cdots \wedge (e_1 - e_n)$$

is a blade parallel to A_n. If we multiply a out, it is a sum of terms of the form

$$(-1)^{n+i} e_1 \wedge \cdots \wedge \widehat{e_i} \wedge \cdots \wedge e_n.$$

Then

$$a_n = (-1)^n a = \sum_{i=1}^n (-1)^i e_1 \wedge \cdots \wedge \widehat{e_i} \wedge \cdots \wedge e_n$$

is a blade parallel to A_n.

By Proposition 4.7, $|a_n| = \sqrt{n}$. Hence A_n has an orientation

$$w_n = \frac{1}{\sqrt{n}} (e_1 - e_2) \wedge \cdots \wedge (e_1 - e_n)$$

$$= \frac{1}{\sqrt{n}} \sum_{i=1}^{n} (-1)^i e_1 \wedge \cdots \wedge \widehat{e_i} \wedge \cdots \wedge e_n.$$

The angle between A_n and P_i is given by

$$\cos(\theta_{i,n}) = w_n \cdot w_{i,n}^\dagger = \frac{1}{\sqrt{n}}$$

which is independent of i. As $n \to \infty$, we see that $\cos(\theta_{i,n}) \to 0$ so that $\theta_{i,n} \to \pi/2$.

Exercises 4.5.

1. We use the chart to compute a tangent vector to H:

$$\frac{dx}{d\tau} = -\sin(\tau) e_1 + \cos(\tau) e_2 + e_3 = -\chi_2 e_1 + \chi_1 e_2 + e_3.$$

Then

$$w = \frac{\frac{dx}{d\tau}}{\left| \frac{dx}{d\tau} \right|} = \frac{1}{\sqrt{2}} \left(-\chi_2 e_1 + \chi_1 e_2 + e_3 \right).$$

2. $w(\xi_1, \xi_2, \xi_3) = \dfrac{(e_1 \wedge e_2) - 2\xi_2(e_1 \wedge e_3) - 2\xi_1(e_2 \wedge e_3)}{\sqrt{1 + 4\xi_1^2 + 4\xi_2^2}}.$

3. $w = \xi_3 (e_1 \wedge e_2) - \xi_2 (e_1 \wedge e_3) + \xi_1 (e_2 \wedge e_3).$

Exercises 4.6.

1. $2\pi\sqrt{2} (e_1 \wedge e_2).$

2. $2\pi\sqrt{n-1}.$

3. The area of $\mathcal{T}$ is $\int_{\mathcal{T}} 1 = 4\pi^2$.

4. Hint: Show that if $y(\theta) = \rho_1 e_1 + \rho_0 \left(\cos(\theta) e_1 + \sin(\theta) e_3 \right)$, then

$$x(\theta, \xi) = \left(y(\theta) \cdot e_1 \right) \left[\cos(\xi) e_1 + \sin(\xi) e_2 \right] + \left(y(\theta) \cdot e_3 \right) e_3$$

is a parametrization of $\mathcal{T}$.

The area of $\mathcal{T}$ is $\int_{\mathcal{T}} 1 = 4\pi^2 \rho_0 \rho_1$.

5. $\pi (e_1 \wedge e_2).$

6. Partial answer:

$$w(x) = w(\chi_1, \chi_2, \xi_1, \xi_2)$$
$$= \frac{1}{\rho_0 \rho_1} \left[\chi_2 \xi_2 (e_1 \wedge e_3) - \chi_2 \xi_1 (e_1 \wedge e_4) - \chi_1 \xi_2 (e_2 \wedge e_3) + \chi_1 \xi_1 (e_2 \wedge e_4) \right].$$

7. Hint: Suppose that I and J are multi-indices of length p. Show that there are scalar constants λ_{IJ} such that

$$u_I = \sum_{J \in \mathcal{O}_p} \lambda_{IJ} v_J.$$

Next show that we must have

$$\psi_J = \sum_{I \in \mathcal{O}_p} \lambda_{IJ} \phi_I$$

at least when the functions are evaluated at points $x_0 \in \mathcal{M}$.

Exercises 5.1.

1. Basis for $\mathbb{G}^3$:

$$1,$$
$$e_1, \ e_2, \ e_3,$$
$$e_1 \wedge e_2, \quad e_1 \wedge e_3, \quad e_2 \wedge e_3,$$
$$e_1 \wedge e_2 \wedge e_3.$$

Exercises 5.2.

3. (a) If $i \neq j$, then $u_i u_j = u_i \wedge u_j = -u_j \wedge u_i = -u_j u_i$. If $i = j$, then the root of the multi-index (i, i) is $\emptyset$, and $u_\emptyset = 1$.

4. (f)

$$u_1 \cdots u_n + \sum_{j=1}^{i-1} (-1)^{i+j-1} u_1 \cdots \widehat{u}_j \cdots \widehat{u}_i \cdots u_n$$

$$+ \sum_{j=1}^{i+1} (-1)^{i+j} u_1 \cdots \widehat{u}_i \cdots \widehat{u}_j \cdots u_n.$$

Exercises 5.3.

4. (a) Partial answer: Think of $(\lambda + \xi u)^k$ and the binomial theorem. Next set

$$M_r = \{i : i = 0, 1, \ldots, k \text{ and } i \equiv r \,(\text{mod}\, 4)\}.$$

Then

$$\lambda_k = \sum_{i \in M_0} \binom{k}{i} \lambda^{k-i} \xi^i - \sum_{i \in M_2} \binom{k}{i} \lambda^{k-i} \xi^i,$$

$$\xi_k = \sum_{i \in M_1} \binom{k}{i} \lambda^{k-i} \xi^i - \sum_{i \in M_3} \binom{k}{i} \lambda^{k-i} \xi^i.$$

Exercises 5.4.

1. $(e_1 - e_2)^{-1} = (e_1 - e_2)/2$.

2. $x^{-1} = (\alpha - \beta e)/(\alpha^2 + \beta^2)$.

4.

$$w(x) = x(e_1 \cdots e_{m+1}) = \sum_{i=1}^{m+1} (-1)^i \chi_i (e_1 \cdots \widehat{e_i} \cdots e_{m+1}).$$

The other orientation is the negative of this one.

6. Sketch of the answer: There must exist an orthonormal basis $\{u_i\}_{i=1}^n$ for $\mathbb{R}^n$ such that $a = |a| u_1 \cdots u_k$. If we write x in the form $x = \sum_{i=1}^n \chi_i u_i$, then

$$axa^{-1} = (-1)^k \left(-\sum_{i=1}^k \chi_i u_i + \sum_{i=k+1}^n \chi_i u_i \right).$$

Exercises 5.6.

1. $m_1 = e_1 - \cot(\theta) e_2$ and $m_2 = \csc(\theta) e_2$.

2.

$$m_i = \begin{cases} e_i - e_{i+1} & \text{if } 1 \le i \le n-1, \\ e_n & \text{if } i = n. \end{cases}$$

6. We are given the basis $\{v_i\}_{i=1}^p$ for U and $\{n_i\}_{i=1}^p$ the reciprocal basis. Let W be the orthogonal complement to U in V. That is,

$$W = \{x \in V : x \cdot y = 0 \text{ for all } y \in U\}.$$

Construct a basis $\{v_i\}_{i=p+1}^q$ for W and let $\{n_i\}_{i=p+1}^q$ be the reciprocal basis. Then $\{v_i\}_{i=1}^q$ is a basis for $V = U \oplus W$ and $\{n_i\}_{i=1}^q$ is the reciprocal basis.

Exercises 5.7.

5.

$$\underline{f}(x) = \lambda_0 + (\lambda_1 \cos(\theta) - \lambda_2 \sin(\theta))e_1$$
$$+ (\lambda_1 \sin(\theta) + \lambda_2 \cos(\theta))e_2 + \lambda_3 e_1 e_2.$$

6. $x = \lambda_0 + \lambda_2 e_2 + \lambda_3 e_3 + \lambda_{23}\, e_2 e_3.$

7. $\overline{f}(e_1 e_2) = \cos(\theta)\, e_1 e_2.$

Exercises 5.8.

3. (a)

$$P(x) = x - \frac{1}{3}(\chi_1 + \chi_2 + \chi_3)(e_1 + e_2 + e_3),$$
$$Q(x) = \frac{1}{3}(\chi_1 + \chi_2 + \chi_3)(e_1 + e_2 + e_3).$$

(b) $\underline{P}(y) = \frac{1}{3}(\xi_1 + \xi_2 + \xi_3)(e_1 e_2 - e_1 e_3 + e_2 e_3).$

5. $\pi/4.$

Exercises 6.1.

1. We establish only the formula $\partial_u \langle f \rangle_k = \langle \partial_u f \rangle_k$. Suppose we are working in $\mathbb{R}^n$. Recall that f, even if it is presented as a multivector field on a manifold $\mathcal{M}$, has an extension of its domain to an open subset of $\mathbb{R}^n$. This extension may exist only locally, around the point x at which we are calculating the directional derivative, but that is sufficient for our purposes. Because f's domain has this extension, we may write the function (locally) as

$$f = \sum_{i=0}^n \sum_{I \in \mathcal{O}_i} \phi_I e_I$$

in terms of ordered multi-indices and the standard basis $\{e_i\}_{i=1}^n$ for $\mathbb{R}^n$ where the functions ϕ_I are understood to be real-valued. Then

$$\langle f \rangle_k = \sum_{I \in \mathcal{O}_k} \phi_I e_I.$$

It follows that

$$\partial_u \left(\langle f \rangle_k \right) = \sum_{I \in \mathcal{O}_k} (\partial_u \phi_I) e_I.$$

(Remember that the e_I multi-vectors are just constants.) If we return to the expansion of f and differentiate, we see that

$$\partial_u f = \sum_{i=0}^n \sum_{I \in \mathcal{O}_i} (\partial_u \phi_I) e_I,$$

and hence

$$\langle \partial_u f \rangle_k = \sum_{I \in \mathcal{O}_k} (\partial_u \phi_I) e_I.$$

This establishes the desired result.

9. $\vec{\nabla} f = k n x^{k-1}.$

10. $\vec{\nabla} \phi(x) = 2x.$

11. (a) Hint: $\vec{\nabla}_{\mathcal{M}} (\chi_1^2 + \chi_2^2 + \chi_3^2) = \vec{\nabla}_{\mathcal{M}}(1).$
(c) $f(x) = -\frac{1}{2} \sin(2\phi) \, d\theta.$

12. (a) $-\chi_1 \cos(\chi_2).$
(b) $-\chi_1 \sin(\chi_2).$
(c) $(e^{\chi_2} - e^{\chi_1}) \, d\chi_1 \, d\chi_2.$
(d) $(r - 2) |x|^r d\chi_1 \, d\chi_2.$
(e) $(r + 2) |x|^r.$

14. In $\mathbb{R}^2$, let $f(x) = x = \chi_1 e_1 + \chi_2 e_2$ and $g(x) = e_1 e_2.$

Exercises 6.2.

1. We can set $r = (m-1)(m+2)/2.$

Exercises 6.3.

4. We consider only the first result in connection with $\vec{\partial}$.
Notice that the outward normal unit vector n to $\partial \mathcal{J}^p$ depends only on the set $\mathcal{J}^p$ and does not depend on the choice of orientation of $\mathcal{J}^p$. The defining equation for the induced orientation is $\vec{\partial} w = nw$. If we switch the sign of the orientation, we have

$$\vec{\partial}(-w) = n(-w) = -nw = -\vec{\partial} w,$$

which is the desired result.

6. We suppose that $\mathcal{M}$ is a $\mathcal{C}^1$ m-cell with orientation w and that we are given the equation $\vec{\partial} w = nw$. We know that both w and $\vec{\partial} w$ are unit blades, and by Proposition 5.12, we have $|nw| = |n| \, |w| = |n|$. So $|\vec{\partial} w| = |nw|$ yields $|n| = 1$.

Since n is a unit vector, we know that $n^{-1} = n^\dagger = n$, so if we multiply $\vec{\partial} w = nw$ by n and invoke Proposition 5.14, we see that we have

$$n(\vec{\partial} w) = w = n \cdot_L \vec{\partial} w + n \wedge \vec{\partial} w.$$

Since $n(\vec{\partial} w) = w$, we see that $n(\vec{\partial} w)$ has grade m. But $n \cdot_L \vec{\partial} w$ has grade $m - 2$ and $n \wedge \vec{\partial} w$ has grade m. Thus $n(\vec{\partial} w) = n \wedge \vec{\partial} w$. By Proposition 5.24, this means that n and $\vec{\partial} w$ are orthogonal.

9. (a) Hint: This is a standard result about gradients in an introductory calculus course.

(b) A unit vector that is orthogonal to $\mathcal{M}$ is

$$n(x) = \frac{\vec{\nabla}\phi(x)}{|\vec{\nabla}\phi(x)|} = \frac{1}{\sqrt{1+2\chi_3^2}}\,(\chi_1 e_1 + \chi_2 e_2 - \chi_3 e_3).$$

where $\phi(x) = \chi_1^2 + \chi_2^2 - \chi_3^2 - 1$. Since $n e_1 e_2 e_3$ must be a unit 2-blade that is orthogonal to n, we can set

$$w(x) = n(x)e_1 e_2 e_3 = \frac{1}{\sqrt{1+2\chi_3^2}}\,(\chi_1\,(e_2 e_3) - \chi_2\,(e_1 e_3) - \chi_3\,(e_1 e_2)).$$

(Keep in mind in these equations that we always assume x satisfies $\phi(\chi_1, \chi_2, \chi_3) = 0$.)

Exercises 6.4.

5. The change of ϕ over $\mathcal{M}$ is

$$\begin{aligned}
&\mathrm{vol}(\mathcal{M})\,e_2 e_3 \quad \text{if } i = 1, \\
-&\mathrm{vol}(\mathcal{M})\,e_1 e_3 \quad \text{if } i = 2, \\
&\mathrm{vol}(\mathcal{M})\,e_1 e_2 \quad \text{if } i = 3.
\end{aligned}$$

The change of f over $\mathcal{M}$ is $\mathrm{vol}(\mathcal{M})\,e_1 e_2 e_3$ for all i.

6. (b)
$$d\rho \wedge d\theta = \frac{1}{\rho}\,d\chi_1 \wedge d\chi_2 = \frac{e_1 e_2}{\sqrt{\chi_1^2 + \chi_2^2}}.$$

(c) $\int_{\mathcal{M}} f\,w^{\dagger} = \int_{-\pi/2}^{\pi/2}\int_1^2 \rho^2\,d\rho\,d\theta = \frac{7}{3}\pi.$

Exercises 6.5.

2. The change of f over $\mathcal{M}$ is

$$\left(\sum_{i=1}^{n}(-1)^{i-1}v_i e_1 \cdots \widehat{e_i} \cdots e_n\right)\mathrm{vol}(\mathcal{M}).$$

3. (a) θ_0.
(b) $\phi(\chi_1, \chi_2) = \arctan(\chi_2/\chi_1)$.

4. (a) 0.
(b) 0.

293

(c) The change of f over $\mathcal{A}$ is

$$\int_{\partial\mathcal{A}} f \, \vec{\partial} w = \int_{\mathcal{A}} (\vec{\nabla} f) \, w = -4 \int_{\mathcal{A}} \frac{1}{|x|^2} \, d\chi_1 \, d\chi_2 \, w.$$

Change the integral from cartesian to polar coordinates:

$$\chi_1 = \rho \cos(\theta),$$
$$\chi_2 = \rho \sin(\theta).$$

Notice that $|x| = \rho$, $w^\dagger = -w$, and

$$d\chi_1 \, d\chi_2 = d\chi_1 \wedge d\chi_2 = \rho \, (d\rho \wedge d\theta).$$

Invoking Proposition 6.17, we see that

$$\int_{\mathcal{A}} (\vec{\nabla} f) \, w = 4 \int_{\mathcal{A}} \frac{1}{\rho} \, (d\rho \wedge d\theta) \, w^\dagger$$
$$= \int_0^{2\pi} \int_{\rho_0}^{\rho_1} \frac{1}{\rho} \, d\rho d\theta = 8\pi \ln\left(\frac{\rho_1}{\rho_0}\right).$$

Exercises 7.1.

1. Suppose $\mathrm{sub}(a) \subseteq \mathrm{sub}(b)$. We can find an orthonormal basis $u_1, \dots, u_q$ for $\mathrm{sub}(b)$ and scalars λ, η such that $b = \eta \, u_1 \cdots u_q$ and $a = \lambda u_p \cdots u_1$. Then

$$a \cdot_L b = \lambda \eta \, u_{p+1} \cdots u_q = ab.$$

The equality $b \cdot_R a = ba$ follows similarly.

3. We see from the displayed system that

$$a_1 = 3e_1 + e_2,$$
$$a_2 = 2e_1 - 4e_2,$$
$$b = e_1 + 2e_2.$$

We calculate from this that

$$a_1' = \frac{2}{7} e_1 + \frac{1}{7} e_2,$$
$$a_2' = \frac{1}{14} e_1 - \frac{3}{14} e_2,$$
$$\chi_1 = \frac{4}{7}, \quad \chi_2 = -\frac{5}{14}.$$

4. Hints: From Proposition 5.25 we know that

$$(A_0 \cdots \widehat{A_i} \cdots A_n) \cdot (A_0 \cdots \widehat{A_j} \cdots A_n) = 0 \quad \text{for } 1 \le i, j \text{ when } i \neq j.$$

By Proposition 7.3 we have

$$\sum_{i=1}^{n}(-1)^{n}A_{0}\cdots\widehat{A_{i}}\cdots A_{n} = 0.$$

If we dot this with $A_{0}\cdots\widehat{A_{j}}\cdots A_{n}$ where $j \neq 0$, we can show that

$$(A_{1}\cdots A_{n})\cdot(A_{0}\cdots\widehat{A_{j}}\cdots A_{n}) = (-1)^{j+1}|A_{0}\cdots\widehat{A_{j}}\cdots A_{n}|^{2}.$$

Now square the expression $A_{1}\cdots A_{n}+\sum_{i=1}^{n}(-1)^{i}A_{0}\cdots\widehat{A_{i}}\cdots A_{n}$ using the dot product.

Exercises 7.2.

4. Trace of $f = (-1)^{k}(n-2k)$.

5.

$$\frac{\partial \chi_{1}}{\partial \theta} = \cos(\theta)\cos(\phi) \qquad\qquad \frac{\partial \chi_{1}}{\partial \phi} = -\sin(\theta)\sin(\phi)$$

$$\frac{\partial \chi_{2}}{\partial \theta} = \cos(\theta)\sin(\phi) \qquad\qquad \frac{\partial \chi_{2}}{\partial \phi} = \sin(\theta)\cos(\phi)$$

$$\frac{\partial \chi_{3}}{\partial \theta} = -\sin(\theta) \qquad\qquad \frac{\partial \chi_{3}}{\partial \phi} = 0.$$

$$
\begin{aligned}
P_{\mathcal{M}}(d\chi_{1}) = {} & \Big(\cos^{2}(\theta)\cos^{2}(\phi)+\sin^{2}(\theta)\Big)e_{1} \\
& + \Big(\cos^{2}(\theta)\sin(\phi)\cos(\phi)-\frac{1}{\sin(\phi)}\sin^{2}(\theta)\cos(\phi)\Big)e_{2} \\
& -\sin(\theta)\cos(\theta)\cos(\phi)e_{3}, \\
P_{\mathcal{M}}(d\chi_{2}) = {} & \Big(\cos^{2}(\theta)\sin(\phi)\cos(\phi)-\frac{1}{\sin(\phi)}\sin^{2}(\theta)\cos(\phi)\Big)e_{1} \\
& + \Big(\cos^{2}(\theta)\sin^{2}(\phi)+\frac{1}{\sin^{2}(\phi)}\sin^{2}(\theta)\cos^{2}(\phi)\Big)e_{2} \\
& -\sin(\theta)\cos(\theta)\sin(\phi)e_{3}, \\
P_{\mathcal{M}}(d\chi_{3}) = {} & -\sin(\theta)\cos(\theta)\cos(\phi)e_{1} \\
& -\sin(\theta)\cos(\theta)\sin(\phi)e_{2} \\
& +\sin^{2}(\theta)e_{3}.
\end{aligned}
$$

7. $\Big(P_{\mathcal{M}}(e_{1})\Big)(\chi_{1},\chi_{2},\chi_{3}) = (\chi_{2}^{2}+\chi_{3}^{2})e_{1}-\chi_{1}\chi_{2}e_{2}-\chi_{1}\chi_{3}e_{3}.$

9. $N_{\mathcal{M}} = -\dfrac{1}{2}\operatorname{sech}(\chi_{1})\left(e^{\chi_{1}}e_{1}-e_{3}\right)$ where e_{1},e_{2},e_{3} is the standard basis for $\mathbb{R}^{3}$.

11. This is essentially Lemma A.4 of Appendix A.

Exercises 7.3.

1. Hint: Show that $\overrightarrow{\nabla}_{\mathcal{M}}(\chi_i \chi_j) = \chi_j \, d\chi_i + \chi_i \, d\chi_j$.

5. (b) Sketch of the proof: We take $e_1, \ldots, e_{p+1}$ to be the standard basis for $\mathbb{R}^{p+1}$, and it is its own reciprocal basis. Using this basis and the coordinates $(\chi_1, \ldots, \chi_{p+1})$, we see that

$$\overrightarrow{\nabla}\phi = \sum_{i=1}^{p+1} \frac{\partial \phi}{\partial \chi_i} e_i.$$

Let $(\xi_1, \ldots, \xi_p)$ be the coordinates on $\mathcal{M}$. If x is the coordinatization that assigns these coordinates to $(\chi_1, \ldots, \chi_{p+1}) \in \mathcal{M}$, we have

$$x(\xi_1, \ldots, \xi_p) = \sum_{i=1}^{p+1} \chi_i \, e_i.$$

Treating $\phi = \phi(\chi_1, \ldots, \chi_{p+1})$ as a function of $(\xi_1, \ldots, \xi_p)$, since $\phi(\chi_1, \ldots, \chi_{p+1}) = 0$ on $\mathcal{M}$, we find that

$$\frac{\partial \phi}{\partial \xi_i} = \sum_{j=1}^{p+1} \frac{\partial \phi}{\partial \chi_j} \frac{\partial \chi_j}{\partial \xi_i} = (\overrightarrow{\nabla}\phi) \cdot \frac{\partial x}{\partial \xi_i} = 0.$$

This tells us that $\overrightarrow{\nabla}\phi$ will be orthogonal to the tangent space of $\mathcal{M}$ at all points of $\mathcal{M}$. Set $e = e_1 \cdots e_{p+1}$. Then $e(\overrightarrow{\nabla}\phi)$ will be a p-blade that is tangent to $\mathcal{M}$. We set $w = e(\overrightarrow{\nabla}\phi)/|\overrightarrow{\nabla}\phi|$ and take this to be the orientation of $\mathcal{M}$. Then using the formula $N_{\mathcal{M}} = -w^\dagger(\overrightarrow{\nabla}_{\mathcal{M}} w)$ for the spur, it is straightforward to compute that

$$N_{\mathcal{M}} = -\frac{1}{|\overrightarrow{\nabla}\phi|^2} \left(\overrightarrow{\nabla}\phi \right) \left(\overrightarrow{\nabla}_{\mathcal{M}}(\overrightarrow{\nabla}\phi) \right).$$

(c) Hint: At the point $x_0 = (0, \ldots, 0, \rho_0)$, we can take the basis of $T_{x_0}\mathcal{M}$ to be $e_1, \ldots, e_p$ when calculating $\overrightarrow{\nabla}_{\mathcal{M}}\phi$. The answer is $N_{\mathcal{M}} = (p/\rho_0) e_{p+1}$.

Exercises 7.4.

3. Hint: Use the product rule for the differential operators $\partial/\partial\chi_i$ as applied to z^k and construct a proof by induction.

Exercises 7.5.

1. (a) $\overleftarrow{d}_{\mathcal{M}} f = \dfrac{4\chi_1 \chi_2}{(\chi_1^2 + \chi_2^2)^2}$.

(b) $\overleftarrow{d}_{\mathcal{M}} f = 0$.

(c) $\overleftarrow{d_{\mathcal{M}}} f = \left(\sum_{i=1}^{p} \frac{\partial \psi_i}{\partial \chi_i} \right) d\chi_1 \wedge \cdots \wedge d\chi_p.$

Exercises 7.6.

1. $I_{\mathcal{M}}(\chi_1, \chi_2, \chi_3) = \left(\dfrac{\chi_1}{\sqrt{\chi_1^2 + \chi_2^2}}, \dfrac{\chi_2}{\sqrt{\chi_1^2 + \chi_2^2}}, 0 \right).$

4. We can take the normal identity to be

$$I_{\mathcal{M}}(\chi_1, \chi_2) = \left(\frac{\chi_1}{\rho}, \frac{\chi_2}{\rho} \right), \quad \text{where } \rho = \sqrt{\chi_1^2 + \chi_2^2},$$

on $\mathbb{R}^2 - 0$. To see if f is normally extended, we need to see if the equation $f(\chi_1, \chi_2) = f(\chi_1/\rho, \chi_2/\rho)$ holds. It follows that ϕ is normally extended while ψ is not.

Bibliography

1. Rafał Abłamowicz and Garret Sobczyk, editors. *Lectures on Clifford (Geometric) Algebras and Applications*. Birkhäuser, 2004.

2. F. Brackx, R. Delanghe, and F. Sommen. *Clifford Analysis*. Pitman Advanced Publishing Program, 1982.

3. Alan Bromborsky. An Introduction to Geometric Algebra and Calculus. https://www.mathschoolinternational.com/Math-Books/Geometric-Algebra/Books/An-Introduction-to-Geometric-Algebra-and-Calculus-by-Alan-Bromborsky.pdf, November 2014.

4. Henri Cartan. *Differential Forms*. Dover, 1970.

5. W. K. Clifford. Applications of Grassman's Extensive Algebra. *Amer. Journal of Math.*, I:350–8, 1878.

6. C. A. Deavours. The Quaternion Calculus. *American Mathematical Monthly*, 80(9):995–1008, November 1973.

7. Dietmar Hildenbrand. *Foundations of Geometric Algebra Computing*, volume XXVIII of *Geometry and Computing*. Springer, 2013.

8. Chris Doran and Anthony Lasenby. *Geometric Algebra for Physicists*. Cambridge University Press, 2003.

9. Leo Dorst, Daniel Fontijne, and Stephen Mann. *Geometric Algebra for Computer Science (Revised Edition): An Object-Oriented Approach to Geometry*. The Morgan Kaufmann Series in Computer Graphics. Morgan Kaufmann, 2007.

10. Eckhard Hitzer, Jacques Helmstetter, and Rafał Abłamowicz. Square Roots of -1 in Real Clifford Algebras. In Eckhard Hitzer and Stephen J. Sangwine, editors, *Quaternion and Clifford Fourier Transforms and Wavelets*, volume 27 of *Trends in Mathematics*, pages 123–153. Springer Basel, 2013.

11. Eckhard Hitzer and Stephen J. Sangwine, editors. *Quaternion and Clifford Fourier Transforms and Wavelets*, volume 27 of *Trends in Mathematics*. Springer Basel, 2013.

12. Eckhard Hitzer and Stephen Sangwine. Multivector and Multivector Matrix Inverses in Real Clifford Algebras. *Applied Mathematics and Computation*, 311(C):375–389, October 2017.

13. Harley Flanders. *Differential Forms with Applications to the Physical Sciences*. Dover, 1989.

14. Matthew R. Francis and Arthur Kosowsky. The Construction of Spinors in Geometric Algebra. arXiv:math-ph/0403040v2, October 2004.

15. Garret Sobczyk. *Matrix Gateway to Geometric Algebra, Spacetime and Spinors*. Independently published, 2019.

16. H. Grassmann. *Extension Theory*. American Mathematical Society, 2000. Translated by Lloyd C. Kannenberg.

17. David Hestenes. Multivector calculus. *Journal of Mathematical Analysis and Applications*, 24(2):313–325, 1968.

18. David Hestenes. A unified language for mathematics and physics. In J. S. R. Chisholm and A. K. Commons, editors, *Clifford Algebras and their Applications in Mathematical Physics*, pages 1–23, Boston, 1986. Dordrecht.

19. David Hestenes. Universal Geometric Algebra. *Simon Stevin*, 62(3-4), September-December 1988.

20. David Hestenes. Mathematical Viruses. In A. Micali, R. Boudet, and J. Helmstetter, editors, *Clifford Algebras and their Applications in Mathematical Physics*, volume 47 of *Fundamental Theories of Physics*, pages 3–16. Springer, 1992.

21. David Hestenes. Differential Forms in Geometric Calculus. In F. Brackx, R. Delanghe, and H. Serras, editors, *Clifford Algebras and their Applications in Mathematical Physics*, pages 269–285. Kluwer, 1993.

22. David Hestenes and Garret Sobczyk. *Clifford Algebra to Geometric Calculus*. D. Reidel, 1984.

23. David Hestenes and Renatus Ziegler. Projective Geometry and Clifford Algebra. *Acta Applicandae Mathematicae*, 23:25–63, 1991.

24. Kenneth Hoffman. *Analysis in Euclidean Spaces*. Dover, 2007.

25. John H. Hubbard and Barbara Burke Hubbard. *Vector Calculus, Linear Algebra, and Differential Forms, a Unified Approach*. Matrix Editions, 4 edition, 2009.

26. Jr. Jayme Vaz and Jr. Roldã o da Rocha. *An Introduction to Clifford Algebras and Spinors*. Oxford University Press, 2019. Paperback, ISBN 9780198836285.

27. Mehrdad Khosravi and Michael D. Taylor. The wedge product and analytic geometry. *The American Mathematical Monthly*, 115(7):623–644, 2008.

28. Klaus Gürlebeck, Klaus Habetha, and Wolfgang Sprößig. *Application of Holomorphic Functions in Two and Higher Dimensions*. Birkhäuser, 2016.

29. John M. Lee. *Riemannian Manifolds An Introduction to Curvature*, volume 176 of *Graduate Texts in Mathematics*. Springer, 1997.

30. Alan Macdonald. An elementary construction of the geometric algebra. *Advances in Applied Clifford Algebras*, 12(1):1–6, 2002. `http://faculty.luther.edu/\%7Emacdonal/`.

31. Alan Macdonald. *Linear and Geometric Algebra*. Alan Macdonald, 2010.

32. Alan Macdonald. *Vector and Geometric Calculus*. Alan Macdonald, 2012.

33. Macdonald, Alan. A Survey of Geometric Algebra and Geometric Calculus. *Advances in Applied Clifford Algebras*, 27:853–891, March 2017.

34. Saunders MacLane and Garrett Birkhoff. *Algebra*. Macmillan, New York, 1967.

35. Piotr Mikusiński and Michael D. Taylor. *An Introduction to Multivariable Analysis from Vector to Manifold*. Birkhäuser, Boston, 2002.

36. Miroslav Josipović. *Geometric Multiplication of Vectors: An Introduction to Geometric Algebra in Physics*. Birkhäuser, 2019.

37. Pertti Lounesto. *Clifford Algebras and Spinors*. Cambridge University Press, 2001.

38. John Ryan. Clifford analysis. In Rafał Abłamowicz and Garret Sobczyk, editors, *Lectures on Clifford (Geometric) Algebras and Applications*, pages 53–89. Birkhäuser, 2004.

39. Anton R. Schep. A Simple Complex Analysis and an Advanced Calculus Proof of the Fundamental Theorem of Algebra. *The American Mathematical Monthly*, 116(1):67–68, 2009.

40. Georgi E. Shilov. *Linear Algebra*. Dover Publications, Inc., 1977.

41. John Snygg. *A New Approach to Differential Geometry Using Clifford's Geometric Algebra*. Birkhäuser, 2012.

42. Garret Sobczyk. *New Foundations in Mathematics, The Geometric Concept of Number*. Birkhäuser, 2013.

43. Michael Spivak. *Calculus on Manifolds*. Westview Press, 1971.

44. Shlomo Sternberg. *Curvature in Mathematics and Physics*. Dover, 2012.

45. Swanhild Bernstein, editor. *Topics in Clifford Analysis: Special Volume in Honor of Wolfgang Sprößig*. Trends in Mathematics. Birkhäuser, 2019.

46. M. D. Taylor. A Crash Course in Geometric Algebra/Calculus. `https://www.mdeetaylor.com/wp-content/uploads/2013/01/GA_quick_course.pdf`, 2014.

Index

www.ingramcontent.com/pod-product-compliance
Lightning Source LLC
Chambersburg PA
CBHW081804200326
41597CB00023B/4138